The SCIENTIFIC LITERATURE

THE
SCIENTIFIC
LITERATURE

—•• *A Guided Tour* ••—

edited with commentaries by

Joseph E. Harmon *and* Alan G. Gross

THE UNIVERSITY OF CHICAGO PRESS

CHICAGO AND LONDON

The scientific literature

JOSEPH E. HARMON is a senior writer and editor at Argonne National Laboratory. He is the coauthor of *Communicating Science: The Scientific Article from the 17th Century to the Present*, with Alan Gross and Michael Reidy.

ALAN G. GROSS is a professor in the department of rhetoric at the University of Minnesota. He has published widely on rhetoric of science and on rhetorical theory. He is the author or coauthor of several books, including *The Rhetoric of Science*.

The University of Chicago Press, Chicago 60637
The University of Chicago Press, Ltd., London
©2007 by The University of Chicago
All rights reserved. Published 2007
Printed in the United States of America

16 15 14 13 12 11 10 09 08 07   1 2 3 4 5

ISBN-13: 978-0-226-31655-0 (cloth)
ISBN-13: 978-0-226-31656-7 (paper)
ISBN-10: 0-226-31655-6 (cloth)
ISBN-10: 0-226-31656-4 (paper)

Library of Congress Cataloging-in-Publication Data

Harmon, Joseph E.
   The scientific literature : a guided tour / Joseph E. Harmon and Alan G. Gross.
      p.   cm.
   Includes bibliographical references and index.
   ISBN-13: 978-0-226-31655-0 (cloth : alk. paper)
   ISBN-13: 978-0-226-31656-7 (pbk. : alk. paper)
   ISBN-10: 0-226-31655-6 (cloth : alk. paper)
   ISBN-10: 0-226-31656-4 (pbk. : alk. paper)
   1. Scientific literature—History.   2. Communication in science.
3. Science—History—17th century.   4. Science—History—18th century.
5. Science—History—19th century.   6. Science—History—20th century.
7. Science—History—21st century.   I. Gross, Alan G.   II. Title
   Q225.5.H37 2007
   500—dc22
                                                        2006016547

⊗ The paper used in this publication meets the minimum requirements of the American National Standard for Information Sciences—Permanence of Paper for Printed Library Materials, ANSI Z39.48-1992.

I would like to dedicate The Scientific Literature *to my new grandson, Adam Brent Griffith*

ALAN GROSS

Contents

9 SELECT MODERN CLASSICS

Discovering Crucial Facts

Providing Theoretical Explanations

Performing Thought Experiments

Turning to Technology

BIBLIOGRAPHY

Illustrations

Introduction

Every student of nature has delighted at the rainbow formed by sunlight passing through a glass prism. Few ever have the occasion, however, to read Isaac Newton's original text explaining how prisms work and why rainbows form in the sky—published in the eightieth issue (1672) of *Philosophical Transactions*, the main outlet for articles by members of the Royal Society of London for Improving Natural Knowledge:

> Why the Colours of the *Rainbow* appear in falling drops of Rain, is also from hence evident. For, those drops, which refract the Rays, disposed to appear purple, in greatest quantity to the Spectators eye, refract the Rays of other sorts so much less, as to make them pass beside it; and such are the drops in the inside of the *Primary* Bow, and on the outside of the *Second* or Exteriour one. So those drops, which refract in greatest plenty the Rays, apt to appear red, toward the Spectators eye, refract those of other sorts so much more, as to make them pass beside it; and such are the drops on the exterior part of the *Primary*, and interior part of the *Secondary* Bow.

Newton's 1672 article is one of a select company of texts chosen for this book to exemplify the adaptability and range of this remarkably robust medium for communicating new facts and explanations about the natural world.

Newton's article is a historically important example of a literary genre still in its formative years. It was a mere seven years earlier, in January 1665, that a Parisian aristocrat, Denys de Sallo, launched the *Journal des Sçavans* (Journal of the Learned), one of whose stated purposes was "to make known experiments that might serve to explain natural phenomena." Two months later in England, the secretary of the Royal Society of London, Henry Oldenburg, inaugurated the regular publication of technical and scientific letters in *Philosophical Transactions*—"Giving some accompt of the present undertakings, studies, and labours of the ingenious in many considerable parts of the world."

These gentlemanly communications caught on quickly because they expedited transmittal of discoveries about the natural world from individual scientists to an international community of like-minded readers. Before that time, scientists communicated their discoveries in written form either by personal letters that were passed around and sometimes read at scientific meetings or, most prestigious of all, through books. Both of these genres had shortcomings: the letter because of its limited distribution, the book because of its long gestation period.

A long gestation period also characterizes the present volume. It began life as an exhibition held in the Special Collections Department of the Joseph Regenstein Library at the University of Chicago in the year 2000. We were pleased with the reception of the exhibition and thought it worthwhile to convert its text into book form. To do so, we have revised and greatly expanded the original. The exhibition itself was a byproduct of our collaborative research on a book with Michael Reidy, *Communicating Science: The Scientific Article from the 17th Century to the Present* (Oxford University Press, 2002), begun in the early 1990s. Indeed, this new work may be viewed as, in part, a transformation of the exhibition into a book and, in part, a follow-up effort to *Communicating Science*.

The aim of our latest enterprise is broadly pedagogical: to convey to the general reader and the student of science the written and visual expression of science over time in all its variety. Our approach was to extract brief selections from the scientific literature over the last four centuries and provide each selection with explanatory commentary. We chose more than one hundred items to reflect points of interest concerning the history of science, the rhetorical strategies scientists employ to convey their discoveries, and the ways they choose to argue for claims of new knowledge. Obviously, ours is not the only way to represent the sweep of science as it develops over time; but we think it is a legitimate approach, one way of learning the ways that scientists communicate and communicated. Along the way we also believe readers will gain some insight into the history of science as a discipline as well as a social institution, and will maybe even learn some science.

In assembling this anthology-like book, we kept several guidelines in mind. First, we focused on the equations, tables, and visuals as well as words of science, because we view them all as important in scientific communication and its history. Very early on, authors of scientific texts recognized that only a fusion of equations, tables, visuals, and words can overcome the limitations of each in disclosing and explaining relevant aspects of the material universe.

Second, we do not simply reproduce interesting or historically significant texts and images drawn from the scientific literature. Each selected passage or image comes with a commentary. These provide not only context-setting historical and scientific information, but also brief analyses of the texts and images as literary and rhetorical creations. For each selection, we comment upon one or more of the following: the writing style, visual representation, presentation of text and visuals, and ways of arguing.

Third, we wanted our compendium to represent a sampling from "many considerable parts of the world," not just English-speaking lands. With a few exceptions, our selections were originally written in English, French, Latin, or German. The first three were the dominant languages of the scientific literature from the fifteenth through the eighteenth centuries. During this period, important research articles by natural philosophers in countries like Italy, Spain, Denmark, and Sweden—though sometimes published in the vernacular—were usually communicated to an international audience in one of the three major languages of early science. Moreover, the two centers of scientific activity during this early period, England and France, greatly preferred the vernacular over Latin. The early *Philosophical Transactions,* for example, printed some articles in Latin, but the vast majority appeared in English. The early scientific journals from the German states did favor Latin well into the eighteenth century. But coincident with their emergence as a center of nineteenth-century research, particularly in the disciplines of chemistry and physics, German joined the ranks of major scientific languages. During the twentieth century, as science has become a global phenomenon, scientific articles in Chinese, Japanese, and Russian also have become prominent. Nonetheless, what linguists refer to as "scientific English" is now the universal language of international scientific communication. For instance, the *Chinese Journal of Physics*, when launched in 1933, had its main texts in English and only the abstracts of its articles in Chinese.

Fourth, we set out to tell a story focused on the origin of the scientific journal in the seventeenth century and the development of the research article to the present time. In so doing, we sought to represent the full range of science and its communication, not just the high points. That approach permitted us to include passages from articles by illustrious scientists like Christiaan Huygens, Isaac Newton, Antoine Lavoisier, Dimitri Mendeleev, and Albert Einstein along with much lesser-known ones not found in any other anthologies or collected writings—including a few highly eccentric scientist-authors, such as one chemist who wrote his entire article in the form of a song, and another who wrote an erratum ruthlessly castigating his earlier published article in the same journal.

Given that the scientific article is one of the most robust literary genres around today, and that this literature carries enormous prestige within our society and benefits from a long and distinguished history, it seems unfortunate that relatively few outside the scientific community have an inkling of what this literature is really like. Our book is meant to provide the curious general reader with an entree to this privileged form of scholarly discourse.

All books have their limitations, and ours is certainly no exception. Because our selections are many, they are also relatively brief. This is the result of the page limitations of any book that aims at broad coverage, but it is also our desire to avoid the technicalities that would defeat the comprehension of the general reader, the non-specialist, and even specialists reading outside their field of expertise. Fortunately, the complete articles we use are readily available in any moderately sized university library; indeed, a great many are now available to one and all on the World Wide Web (see sites listed at end of book).

Our survey has another limitation. Our choices represent the development of Western science throughout its last four centuries of spectacular growth. That time frame—plus, to be honest, our own limited skills in translation— excludes the great flowering of Greek, Chinese, and Islamic science, which deserve anthologies of their own. And although we represent the scientific work of both men and women, articles by women scientists appearing before the twentieth century are few and far between. We can only imagine the increase of scientific productivity that would have resulted if the bias against women in science had not been in place. Marie Curie, Barbara McClintock, and Christiane Nüsslein-Volhard are three Nobel Prize winners we excerpt, three scientists who indicate the measure of our loss in earlier centuries.

Ours is thus not a complete history, but more of a Michelin Guide; if it succeeds, it is meant to inform, amuse, and tantalize rather than fully to satisfy the inquiring mind. Less modestly, we also hope that the excerpts we present and our commentary on them do more than merely whet the appetite of our readers; we hope they provide an incentive to look further into the meaning and significance of the scientists' achievements and the verbal and visual means by which they convey those achievements. For those who wish to explore further, we have appended a list of fifty books we regard as essential reading in the relatively new academic field of science studies, favoring those with a literary or rhetorical slant. We follow that general list with chapter bibliographies of the key secondary sources we consulted in preparing this book.

We divide our book into nine chapters of extracts with commentaries. These chapters fall into two parts that are nearly equal in length, but not in time frame. The first part (chapters 1 through 4) covers the "early" scientific literature, from narratives describing the natural world observed with the aid of a microscope or telescope in the seventeenth century up to the radioactivity papers that launched the nuclear age at the end of the nineteenth century. The second part (chapters 5 through 9) spans the last hundred years, starting with Einstein's relativity papers and ending with Web-based publications in the early twenty-first century.

In the first part of our book, the initial two chapters center on the two learned societies that assisted at the birth of the scientific periodical in the mid-seventeenth century, the Royal Society of London and the Royal Academy of Sciences in Paris. The Royal Academy was the chief scientific society on the Continent. Its members first published their articles in the *Journal de Sçavans*, a private enterprise, and later in the *Mémoires de l'Académie Royale des Sciences*, sponsored by the Academy itself. The Royal Society was the English counterpart of the Royal Academy. Its members published in the *Philosophical Transactions*, a private enterprise closely associated with the Society. Although we sample work from all three sources, we do not want to convey the idea that science was exclusively an English or a French enterprise. While it is clear that in the seventeenth and through most of the eighteenth century England and France dominated the scientific scene, science was beginning elsewhere as an organized enterprise, particularly in the German states and in the United States. Consequently, in the third chapter, we extract selections from articles that represent the beginnings of German and American science.

Up to the end of the eighteenth century, those interested in the serious pursuit of science (the English word "scientist" had not yet been coined) saw themselves as generalists. Of course, individuals focused on particular topics according to their talents and interests; and certainly, there were people who identified themselves as astronomers, an ancient vocation. But what was Gottfried Leibniz? Was he a philosopher, a mathematician, or a physicist? What was Isaac Newton? Was he a physicist, a mathematician, a theologian, a chemist or an alchemist? Amateurs also abounded. Benjamin Franklin and Thomas Jefferson were politicians, but they were also men of science (not a likely combination today). By the end of the eighteenth century, however, science began to specialize, a transformation signaled by the appearance of journals devoted solely to individual disciplines such as botany or chemistry. We sample this early specialized literature to end the third chapter.

The fourth chapter presents classic articles written from the end of the eighteenth century through the nineteenth. Earlier, we were careful not to prefer the extraordinary to the exclusion of the everyday; we included, for example, a run-of-the-mill botanical article by Martin Lister alongside the first scientific article by Isaac Newton, and the famous engraving of Robert Boyle's air pump alongside a little-known engraving of twin strangled fetuses by Jean Méry. But we devote this fourth chapter to great science produced by great scientists, beginning with James Hutton's theory of the earth, published in 1788, and ending one hundred and ten years later with a publication in which the Curies announce the discovery of radium. Our selection here represents personal favorites; readers may have their own.

In the second part of our book, we turn to the "modern" era, which we arbitrarily define as from the year of Einstein's first relativity articles (1905) to the present. Here, we focus less on the history of communication than on the nature of modern communicative practices. To begin, the fifth chapter deals with the components of scientific articles other than words: equations, tables, and pictures. There is a compelling reason for this. The humanists among us are used to texts that are just words. But in modern science, words are seldom enough; in most cases, science cannot be communicated effectively unless the authors also resort to equations, tables, and pictures. Equations strip nature to its basic regularities; tables display the data that brings us closer to the laboratory and to nature; line and bar graphs display relationships among that data; photographs and realistic drawings bring us in touch with the sensuous experience that is the foundation of all empirical knowledge; and schematics convey the essence of new facts and theories in a visual language.

If it is a mistake to ignore equations, tables, and pictures, it is also a mistake to think that we can understand how scientists communicate if we look at only one sort of article. In fact, the rules for communicating science vary with the goal of the writing scientist. Is she conveying a new theory or research method? Or an experimental result? Or the results of observation? Or perhaps she is writing a review, not conveying original science, but an evaluation of the original literature on a particular subject? Each type of communication brings with it different reader expectations on the content and organization.

Despite this variety, there is an organizational backbone common to the communication of all original research, one that permits scientist-readers to discover whether they need to read an article and, if they do, what they need to focus on. Reading the modern scientific article is not like novel reading; its goal is not delight but exploitation. The division into discrete parts facilitates this exploitation. The title and abstract tell scientists whether an article is worth their time. If it is, the organizational backbone allows them easily to skip to Results, to Discussion, or to a particularly salient figure or equation. This backbone is not rigid but flexible; it is not a backbone, really, so much as a guide to composition and reading. We devote the sixth chapter to this basic structure and its use in different types of articles.

Our seventh chapter tackles modern scientific writing style. By and large, in writing for journal publication, scientists do not seek the literary effects of the poet or novelist. What typifies this style is impersonality and careful attention to technical details. These qualities are clearly out of place in communications meant for general readers. True, individual scientists like Roald Hoffmann, Richard Feynman, Stephen Jay Gould, and Lewis Thomas have written with flair in their popular science works. When they have written up their research for journal publication, however, they have seldom deviated from the standard style of science.

In this style, words and phrases like "renormalization," "Bose-Einstein condensate," "MS/GC analysis," and "DNA sequence for the genome of bacteriophage ΦX_{174} of approximately 5,375 nucleotides" are the rule, not the exception. There are seldom exaggeration, humor, emotional outbursts, or circumstantial details like exactly who did what, when, and where. The flesh-and-blood people doing the research are largely invisible outside their names and institutional affiliations after the title. The emphasis is on detailed methodological descriptions, measurement, quantitative-based assertions, and precise discrimination. Yet, despite the cognitive complexity, the sentences themselves tend to be relatively short and simple in structure, as least compared with those in centuries before the twentieth. The selections that begin the sixth chapter illustrate our meaning better than these generalities.

The same chapter continues with three sorts of aberration from the standard style: being playful, being belligerent, and writing with style. Our examples of playfulness stretch from a pun on three authors' names to the visual whimsy hidden in a schematic from an article on photosynthesis. But these minor detours from the norm pale before our examples from the *Journal of Organic Chemistry*, which published an entire article in the form of a poem, and the *Journal of Histochemistry and Cytochemistry*, which published an article in the form of a musical composition. Our examples of belligerence include an astronomer's and geologist's anger and exasperation directed toward colleagues whom they believed had constructed fanciful theories out of flimsy empirical evidence, as well as the neurotic self-laceration in a published erratum by a physical chemist. Among our examples of literary gracefulness is that of the minor entomologist and major twentieth-century novelist Vladimir Nabokov, who records his discovery of a new butterfly in an article and later commemorates becoming "godfather to an insect" in a poem, and the paleontologist and virtuoso essayist Stephen Jay Gould, who makes an evolutionary argument by an extended analogy based on an architectural structure and a literary work by a famous French savant from the eighteenth century. We include these aberrant selections to provide a glimpse into the more human side of modern science, a side normally hidden from public view.

Our survey of twentieth-century science continues in the penultimate chapter with the argumentative give-and-take, the verbal sparring, involved in two controversies: the quarrel between R. A. Fisher and Sewall Wright over the interaction of selection and population size in evolution, and the clash between Freudian dream theory and one of its more recent rivals, the "activation-synthesis hypothesis." Controversy is part of science and has been so at least since the time that Copernicus, Galileo, and Newton put pen to parchment or paper. We cannot avoid representing it if our book is to be true to its subject. Moreover, it is true that science is not and cannot be the mere accumulation of facts and theory that Francis Bacon imagined in his *New Atlantis*; like most knowledge-seekers, scientists often disagree with one another. As we shall see in the chapter on controversy, such disagreements can be a spur to further discovery and invention.

But we do not want to leave the reader with the impression that scientists are particularly irascible. While they often disagree, they are seldom overtly disagreeable in print. Scan the pages from any scientific journal from any century and you will find nary a discouraging word. To restore the reader's sense of balance, the ninth and final chapter turns to excerpts of classic papers from the twentieth century, including Albert Einstein's famous first article on special relativity, Enrico Fermi's announcement of a nuclear chain reaction at the University of Chicago, and Kroto and his colleagues' report on a new carbon structure they called "buckminsterfullerene" as a tribute to one of the truly innovator thinkers of the twentieth century. We close this final chapter with an article that, while only a few years old, surely deserves inclusion as a classic— the mapping of the human genome. This article is also representative of a new trend in scientific communication: it first went public on the World Wide Web. With the increasing exploitation of this vast electronic network, the twenty-first century, what Stevan Harnard calls the "post-Gutenberg era," may very well witness the extinction of the original scientific "paper" appearing in print.

Finally, success of an anthology of poetry or fiction hinges on the pleasures gained in reading the selected texts. But the pleasures of the literary text are not those science commonly affords. It is those different pleasures that we emphasize in the book as a whole: that is to say, the scientists' pleasure at discovery, and our subsequent pleasure derived from vicariously looking over their shoulders as, pen in hand or computer at hand, they rummage among the communicative resources of their profession in an attempt to choose the most effective way of conveying what they have just learned and convincing a highly critical audience that some new claim is credible. Our goal is that readers will also experience the deep satisfaction we have had from being exposed to the original scientific communications of the past and near present.

1

FIRST ENGLISH
PERIODICAL

And new Philosophy calls all in doubt,
The Element of fire is quite put out,
The Sun is lost, and th'earth, and no man's wit
Can well direct him where to look for it.
And freely men confess that this world's spent,
When in the Planets and the Firmament
They seek so many new; they see that this
Is crumbled out again to 'his Atomies.
'Tis all in pieces, all coherence gone,
All just supply, and all relation . . .

From "An Anatomy of the World" (1611) by John Donne

WITHOUT QUESTION, printed books and not scientific articles
communicated the "new Philosophy" and revelations about the
natural world that spawned the scientific revolution and called "all
in doubt." Among those making the honor roll, one would have to include
Copernicus's *De Revolutionibus Orbium Coelestium* (On the Revolutions of Celestial Spheres), Bacon's *Novum Organum* (New Tool of Reasoning), Galileo's

Sidereus Nuncius (Message of the Stars), Harvey's *Motion of the Heart and Blood in Animals*, Descartes's *Discours de la Méthode* (Discourse on the Method), and Newton's *Philosophiae Naturalis Principia Mathematica* (Mathematical Principles of Natural Philosophy). In fact, there is little reason to believe that prospective authors viewed the early journal literature as anything more than an ancillary mode of communication, much as e-mail is considered today.

Besides books, the other major predecessor and rival for communicating new science in the seventeenth century was the "learned letter," most famously illustrated by Galileo's letters on sunspots and the orbits of the planets. As the ideas of the scientific revolution spread in England and on the Continent, the accelerated pace of scientific activity compelled natural philosophers to communicate their recent findings through personal correspondence within and between countries. But these are not "letters" in the traditional sense of the word; authors wrote these epistles on some scientific or technical topic with the understanding that they would be passed on to others. Thus the actual intended audience was interested members of the scientific community at large, though short passages within them may personally address the primary recipient.

To disseminate the information in these learned letters more efficiently, industrious scholars became centers for spreading the latest technical news at home and abroad. Their job was to receive letters, make copies, and pass them on to other interested scholars. After the emergence of scientific societies, the job of "trafficker in intelligence" became more formalized in that the societies themselves appointed a secretary to handle correspondence and circulate newsworthy learned letters among society members and friends.

It was in March 1665 that Henry Oldenburg, the first secretary of the Royal Society of London, launched the first English scientific periodical, *Philosophical Transactions*. Its introductory sentence established the importance of published communication to the advancement of science and still rings true today: "there is nothing more necessary for promoting the improvement of Philosophical Matters, than the communicating to such, as apply their Studies and Endeavours that way."

The early *Philosophical Transactions* reflect what was communicated during meetings of the Royal Society of London. This society was a fairly large, loose-knit group of men living in and around London. Some were extraordinarily talented. Some simply had an above-average curiosity about the natural world and the social status that provided the leisure to purchase and read scientific texts and attend meetings. The Royal Society members met regularly in London to learn about and discuss the latest scientific news from home and abroad. Attendance at meetings averaged a mere twenty to thirty members, although considerably more might show up on special occasions. The early scientific

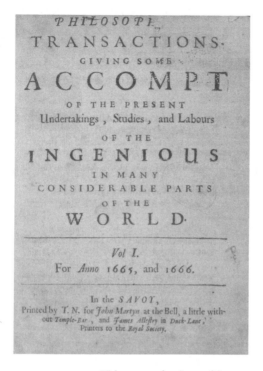

FIGURE 1-1 · Title page to first issue of first
volume, *Philosophical Transactions*.
Courtesy of Special Collections Research Center,
University of Chicago Library.

societies like the Royal Society—part elite social club, part research institute,
part publishing house—constitute the major institutional factor in shaping the
scientific article in the first century after its origin.

The initial *Philosophical Transactions*, issued monthly, were only about twenty
pages long and printed in runs of 1,200 copies. The first issue has a definite international and socially diverse flavor, with contributors from different nations
and walks of life. In it we find reports (what Oldenburg called "relations,"
"accounts," and "narratives") attributed to Robert Hooke, official curator of
experiments for the Royal Society; the astronomer Adrien Auzout, "a *French
Gentleman* of no ordinary Merit and Learning"; the Dutch mathematician
and inventor Christiaan Huygens; the English explorer Robert Holmes; an
unnamed "inquisitive Physician" from Germany; and an "understanding and
hardy Sea-man."

European savants immediately took note of this journal's introduction. A mere two weeks after its first issue (March 30, 1665) the French *Journal des Sçavans* reported that the English, taking their cue from the French of course, had founded "a journal . . . under the title *Philosophical Transactions* . . . to make known to all the world that which is discovered of novelty in natural philosophy," and that this journal contained articles by Royal Society members, "who produce each day an infinity of beautiful works."

Under Oldenburg's stewardship (1665–77) and beyond, *Philosophical Transactions* remained true to its subtitle, opening its pages to contributors "from many considerable parts of the world." The author index reads like a who's who of seventeenth-century scientists in England and Europe: Isaac Newton, Edmond Halley, Robert Boyle, Robert Hooke, John Wallis, Antoni van Leeuwenhoek, Christiaan Huygens, Johannes Hevelius, and Gottfried Leibniz. In so doing, editor Oldenburg shepherded into print articles on important seventeenth-century scientific topics like the mechanisms behind the ocean tides, the properties of respiration and combustion, the properties of light, and methods for determining longitude at sea.

With the death of Oldenburg, the *Transactions* lost its founding father and guiding light, and it temporarily ceased publication in 1679. It was replaced by the *Philosophical Collections,* edited by Robert Hooke, second secretary of the Royal Society and a past critic of Oldenburg and his journal. Only seven issues of this journal appeared over its four years, and it never achieved the stability and reputation of its predecessor. In 1683, the Society resuscitated the *Transactions* first with Robert Plot, then Edmond Halley, as editor, and the periodical eventually returned to some semblance of its former glory.

How did early English scientists write and argue within the early *Philosophical Transactions*? For one thing, among the founding members of the Royal Society, there was a programmatic effort to avoid what Royal Society historian Thomas Sprat called "fine Speaking." But practice did not always follow theory, as our selection from Martin Lister exemplifies, with its elaborate sentence structure and liberal use of metaphor and analogy.

In general, the earliest scientific articles ask the reader to trust the author rather than the details of the science. They draw upon qualitative experience more than experiment and measurement in support of theory. Hence, a more personal voice is evident, and ideas of relevance are interpreted liberally. In the midst of describing his experiments with a prism in the seventeenth-century *Philosophical Transactions*, for instance, Isaac Newton confides that "Amidst these thoughts I was forced from *Cambridge* by the Intervening Plague, and it was more then two years, before I proceeded further." Today, such circumstantial detail would never appear.

As to organization, no attempt is made to codify and modularize in the modern manner (see chapter 6). This absence is not meant to imply, however, that the parts of the modern experimental article—its introduction, materials and methods, and results and discussion—are not generally present in one form or another. This is readily apparent in our selections from Robert Boyle, Benjamin Franklin, Henry Cavendish, and Caroline Herschel.

Nor had the natural philosophers of England settled on a preferred way of arguing. Among our selected passages, for example, Newton argues against speculative hypotheses—those which no crucial experiment could settle—while John Arbuthnot argues by statistical analysis for divine providence, a claim beyond confirmation with a crucial experiment. And while a skeptical Mr. Hill expresses doubts on a report of a merman sighted in a Virginia river, offering a more plausible explanation, a gullible Mr. Toyard from France argues for the achievement of man-powered flight, apparently based on hearsay alone.

We open this chapter with a small sampling of books and letters written by Royal Society authors, then turn to *Philosophical Transactions* articles spanning the period 1665 to 1800. It is worth bearing in mind that the boundaries among books, letters, and articles of this time were not as sharply defined as they are today. Some journal articles are printed as letters beginning "Dear Sir" and ending "Your humble servant," but otherwise differ little, if at all, from articles in the same publication that do not follow such epistolary conventions. Moreover, some letters and articles are as long as typical books, while some books are nothing more than collections of loosely connected articles or letters with a scientific slant.

EARLY BOOKS AND LETTERS

Boyle's World-Wide Letter

Robert Boyle, 1660. *New Experiments Physico-mechanicall, Touching the Spring of the Air, and Its Effects, Made, for the Most Part, in a New Pneumatical Engine.* Oxford: H. Hall (printer to Oxford University).

Sir Robert Boyle was a founding member of the Royal Society of London, inaugurated in 1660. Along with other early members like Robert Hooke, Isaac Newton, and Edmond Halley, they made this fledgling scientific society into one of the most extraordinary groups of scientific minds ever assembled. Boyle was also instrumental in the founding of Oldenburg's *Philosophical Transactions* in 1665. Before that time, society members had only letters and books as avenues for spreading the written word about their work. The above-cited selection qualifies on both counts. On the one hand, it reports a diverse collection of forty-three experiments, mostly performed with the aid of an air pump, and is printed in the form of a fairly long book. On the other hand, it is a "letter" addressed to the author's nephew, Charles Lord Viscount of Dungarvan, eldest son to the Earl of Corke.

In the preface, Boyle concedes his treatise is "far more prolix then becomes a letter" and addresses "why I publish to the World a Letter, which by its Stile and diverse passages, appears to have been written For, as To a particular person." His answer to that question has two components, both related to the importance of publication to scientific progress:

> The one, That the Experiments therein related, having been many of them try'd in the presence of Ingenious Men; and by that means having made some noise among the *Virtuosi* [fellow science enthusiasts] ... I could not quite without trying more then one *Amanuensis*, give out half as many Copies of them as were so earnestly desired, that I could not civilly refuse them. The other, That intelligent Persons in matters of this kinde perswaded me, that the publication of what I had observ'd touching the Nature of the Air, would not be useless to the World; and that in an Age so taken with novelties as is ours, these new Experiments would be grateful to the Lovers of free and real Learning.

Whatever the kind of communication—book, unpublished letter, or published journal article—visual representations routinely complement the seventeenth-century written texts. At the very front of Boyle's book-length letter to his nephew appears a picture of his regal air pump. Constructed with the aid of Robert Hooke, this air pump consisted of two main components: a large glass bulb (about thirty "Wine Quarts" in volume) on top of a pumping apparatus. For purposes of experiment, Boyle or his assistants inserted test materials (burning candles, different species of animals, Torricelli apparatus) through a port on the top. A hollow brass cylinder directly below the globe contained a piston, which was worked by the crank at the bottom and sucked air out of the globe, something like a reverse bicycle pump. This device was not without problems: the globe cracked or even imploded, the cylinder bent because of the negative pressure, the device leaked and would not hold a vacuum for long, and the crank was extremely difficult to turn as one approached the evacuated state. Also, the air pump was expensive, with only a handful in existence two decades after its invention (about 1650, by the German physicist Otto von Guericke). A reasonable estimate for the cost of the first pump is twenty-five pounds, more than the annual salary of Robert Hooke as Curator

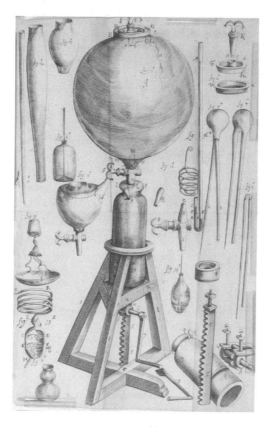

FIGURE 1-2 · Air pump.
Courtesy of Special Collections Research Center,
University of Chicago Library.

of Experiments for the Royal Society. From his many experiments with this
state-of-the-art device, Boyle concluded that air is essential to both respiration
and combustion, that sound does not occur in a vacuum, and that a key property
of air is its elasticity or "spring."

The air pump is a typical scientific instrument—the remote ancestor of the
centrifuge and the particle accelerator. Experiments with such instruments, by
creating a condition not available in nature, permit scientists to argue into place
some natural fact or law. In the history of science, revolutionary new theories
(the heliocentric universe, uniformitarianism, natural selection, relativity, etc.)
tend to grab most of the headlines. But the advance of science is propelled at

least as much, if not more, by the creation and continued refinement of new
research tools.

Hooke Looks a Flea in the Eye

....

Robert Hooke, 1665. *Micrographia; or, Some Physiological Descriptions of Minute Bodies
Made by Magnifying Glasses, with Observations and Inquiries Thereupon*. London:
J. Martyn and J. Allestry (printers to the Royal Society), pp. 210–11.

In addition to the air pump, the premier scientific instruments in the sixteenth
and seventeenth centuries were the telescope and microscope. The two master
technicians of the microscope were the Dutchman Antoni van Leeuwenhoek (of
whom more later) and the Englishman Robert Hooke. At one time a research
assistant to Robert Boyle, Hooke himself was a brilliant experimentalist, inventor, and theorist. He also possessed a contentious personality, especially when
it came to defending a knowledge claim he believed his intellectual property.
According to an annotated version of *Philosophical Transactions* issued in 1809,
"It is said he was rather deformed in his person, of a penurious disposition, and
extremely jealous of his reputation as an original discoverer."

Published under the sponsorship of the Royal Society of London and a masterpiece of seventeenth-century bookmaking, *Micrographia* contains Hooke's
multifarious microscopical observations of such objects as the point of a needle,
a printed period, pores in cork, eels in vinegar, the eye of a fly, the hair of a
cat, and the body of a flea and louse. Hooke complements many of his verbal
observations with realistic engravings. One of the most spectacular is the often-
reproduced magnificent engraving of the lowly flea. The original figure is on a
foldout sheet, measuring nearly a foot and a half in length. Our reduced copy
is but a pale imitation.

Though the mouth is distorted, Hooke's achievement is remarkable, given
that the primitive resolution of his compound microscope necessitated that he
construct the whole image by carefully examining isolated regions of his specimen. As he says himself in the preface: "The Glasses I used were of our English
make, but though very good of the kind, yet far short of what might be expected . . . for though Microscopes and Telescopes, as they now are, will magnify
an Object about a thousand times bigger then it appears to the naked eye; yet
the Apertures of the Object-glasses are so very small, that very few Rays are
admitted, and even of those few there are so many false, that the Object appears
dark and indistinct."

Hooke saw his project of accurate and minute description as a partial ful-
fillment of the plan upon which the Royal Society of London was founded:

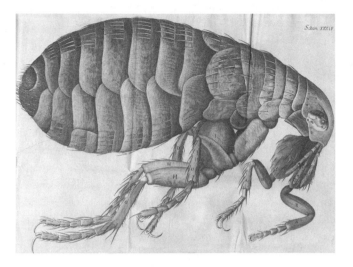

FIGURE 1-3 · Flea from Hooke's *Micrographia*. Courtesy of Special
Collections Research Center, University of Chicago Library.

"For the members of the Assembly having before their eyes so many fatal Instances of the errors and falshoods, in which the greatest part of mankind has so
long wandred, because they rely'd upon the strength of human Reason alone,
have begun anew to correct all Hypotheses by sense, as Seamen do their dead
Reckonings by Coelestial Observations."

To aid the verbal description of his drawing, Hooke added letters to map
the principal body parts—a labeling method used in geometric diagrams for
centuries before. Hooke's description accompanying this figure is almost as
good as the drawing itself:

The strength and beauty of this small creature, had it no other relation at all
to man, would deserve a description.

For its strength, the *Microscope* is able to make no greater discoveries of
it then the naked eye, but only the curious contrivance of its leggs and
joints, for the exerting that strength, is very plainly manifested, such as no
other creature, I have yet observ'd, has any thing like it; for the joints of it
are so adapted, that he can, as 'twere, fold them short one within another,
and suddenly stretch, or spring them out to their whole length, that is, of
the fore-leggs, the part A, of the 34 *Scheme*, lies within B, and B within C,
parallel to, or side by side each other; but the parts of the two next, lie quite

contrary, that is, D without E, and E without F, but parallel also; but the parts of the hinder leggs, G, H and I, bend one within another, like the parts of a double jointed Ruler, or like the foot, legg and thigh of a man; these six leggs he clitches up altogether, and when he leaps, springs them all out, and thereby exerts his whole strength at once.

But, as for the beauty of it, the *Microscope* manifests it to be all over adorn'd with a curiously polish'd suit of sable Armour, neatly jointed, and beset with multitudes of sharp pinns, shap'd almost like Porcupine's Quills, or bright conical Steel-bodkins; the head is on either side beautify'd with a quick and round black eye K, behind each of which also appears a small cavity, L, in which he seems to move to and fro a certain thin film beset with many small transparent hairs, which probably may be his ears; in the forepart of his head, between the two fore-leggs, he has two small long jointed feelers, or rather smellers, MM, which have four joints, and are hairy, like those of several other creatures; between these, it has a small *proboscis*, or *probe*, NNO, that seems to consist of a tube NN, and a tongue or sucker O, which I have perceiv'd him to slip in and out. Besides these, it has also two chaps or biters PP, which are somewhat like those of an Ant, but I could not perceive them tooth'd; these were shap'd very like the blades of a pair of round top'd Scizers, and were opened and shut just after the same manner; with these Instruments does this little busie Creature bite and pierce the skin, and suck out the blood of an Animal, leaving the skin inflamed with a small round red spot. These parts are very difficult to be discovered, because, for the most part, they lye covered between the fore-legs. There are many other particulars, which, being more obvious, and affording no great matter of information, I shall pass by, and refer the reader to the Figure.

Scientific visuals like Hooke's and Boyle's are surely a joy to behold, a delight to the eye. But it is important to bear in mind an important distinction between the visual creations of art and science. Scientific visuals normally support some new claim about the physical universe and come accompanied by explanatory text like Hooke's. Pictures by themselves are seldom enough, and their aesthetic appeal is usually more by accident than design.

A Flea's Anatomy, Uncensored

Antoni van Leeuwenhoek, 1693. "Concerning the colors of parrot feathers and wool, and the anatomy and copulation of fleas." From J. Heniger, ed., *The Collected Letters of Antoni van Leeuwenhoek*. Amsterdam: Swets and Zeitlinger (1976), Vol. 9, Letter 126, pp. 241 and 247.

Detailed observations of nature, sometimes as part of experiments, many times not, dominate the literature of early modern science. One of the great observers with a microscope was Leeuwenhoek, a tradesman from Delft whose passion for building microscopes and studying the microscopic world made him famous in England and Europe. Through his microscope he observed and identified bacteria, sperm cells, blood cells, and much more.

Leeuwenhoek was a prolific writer of scientific letters. Written in Dutch, most of his letters were sent to the Royal Society and translated into English or Latin for publication in *Philosophical Transactions*. Typical is the following passage in which Leeuwenhoek describes his observations of the reproductive organ of the male flea:

Fig. 12. I assumed LMN to be the Male Organ of the Flea, where at M will be seen the knotty part of the Male Organ, and I also assumed that without such a part the Flea could not remain joined to the Female; nay, what is more, that this knot begins to swell and grow bigger during copulation, just as happens when Dogs are copulating. On the said Organ, between L and N, there are two long parts covered with little hairs . . . . P and Q in the said Fig. 12 indicate different parts, which are not seen in the Flea when it is alive, for just as the latter, when it does not want to use the supposed Male Organ, draws it into its body, so likewise it puts the parts P and Q against L, and as a result they cannot be seen . . .

I selected among several Males one whose abdomen was the most transparent, in order that the Draughtsman might depict some of the manifold Arteries that were to be seen therein. This Flea, as it was fixed before the magnifying glass, was still alive, and I saw that deep inside the Body the Male Organ was also equipped with a knotty part, from which I concluded that the Flea could thrust its Male Organ out of its Body until the said internal knot reached the terminal part of its Body.

This Flea also brought a short portion of its Male Organ several times out of and into its Body, and this Organ, in coming out, passed through or between the two parts which are indicated in Fig. 12. by P and Q . . . . And during this passage the knotty part alters its form, for I then saw in two different places, on the underside of the knotty part of the Male Organ, two small long parts shooting out, which were about four times as long as they were thick and whose length was about the same as the thickness of the Male Organ below the knot. And when the Flea began to draw the Male Organ back again into the Body, the said Parts, which had been thrust out obliquely downwards, were drawn in again, while in addition and at the same time another third small part or Member was thrust out of the Male Organ as it

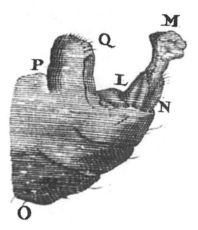

FIGURE I-4 · Flea anatomy from
Leeuwenhoek letter. Reprinted with
permission from *The Collected Letters of
Antoni van Leeuwenhoek* (Amsterdam:
Swets and Zeitlinger, J. Heniger, ed.).

was carried through the said parts. This thrust-out Member corresponded
very closely in length and thickness to the aforesaid Parts, and this latter
Member was thrust out very nearly straight upwards, and was also brought
inwards again as the Male Organ was drawn in. On seeing this, I wondered
whether the Male seed might not be conveyed through this latter thrust-out
member.

Although the instrument Leeuwenhoek used had high power and excellent
resolution, it was less a microscope than a superior magnifying glass, a single lens
ground by Leeuwenhoek himself, mounted between two metal plates about two
inches long and an inch wide. His observations of the male flea's reproductive
system are a testament to his versatility, his patience, and his excellent eyesight.
He was sixty-one at the time.

The advance in detail and accuracy on Robert Hooke's admirable pioneering work is palpable. One of Leeuwenhoek's more remarkable observations and
deductions concerns what he calls the "third small part or Member," and identifies as the flea's penis. The advance over Hooke is remarkable in another way.
More than intellectual curiosity about the world of the very small motivated
Leeuwenhoek's observations of the flea. He was interested in a larger theoretical

question, the doctrine of spontaneous generation, the widely held belief that lower forms of life could be generated from inanimate materials, such as animal feces. If fleas had reproductive equipment analogous to that of human beings, was it not obvious that the origin of fleas was not inanimate matter, but other fleas?

PHILOSOPHICAL TRANSACTIONS

A Seventeenth-Century Solution to the Problem of Longitude

Henry Oldenburg and Christiaan Huygens, 1665. "A narrative concerning the success of pendulum-watches at sea for the longitudes." *Philosophical Transactions*, Vol. 1, pp. 13–15.

We find two distinct voices in the early issues of the *Transactions*: one is that of the editor Henry Oldenburg reporting to his readers what he had heard through word of mouth or been informed about through a letter; the other is that of the actual authors of the many letters sent to the editor. In later scientific articles, both of these voices disappear, to be superseded by another voice, an impersonal one in which Nature speaks through the scientist.

This particular article in the first issue contains both voices. It begins with editor Oldenburg relating a test of a pendulum clock conducted at sea on a ship skippered by Major Robert Holmes in 1664. The clock had been invented by Christiaan Huygens; several of them had been fitted to go to sea by the Earl of Kincardin. Both gentlemen were fellows of the Royal Society. Oldenburg informs his readers:

> The said *Major* . . . being come to the Isle of St. *Thomas* under the *Line*, accompanied with four Vessels, having there adjusted his Watches, put to Sea, and sailed Westward, seven or eight hundred Leagues, without changing his course; after which, finding the Wind Favorable, he steered towards the Coast of *Africk*, North-North-East. But having sailed upon that *Line* a matter of two or three hundred Leagues, the Masters of the other Ships, under his Conduct, apprehending that they would want Water, before they could reach that Coast, did propose to him to steer their Course to the *Barbadoes*, to supply themselves with Water there. Whereupon . . . it was found, that those Pilots did differ in their reckonings from that of the Major, one of them eighty Leagues, another about an hundred, and the third, more; but the Major judging by his *Pendulum-Watches*, that they were only some thirty Leagues distant from the Isle of *Fuego*, which is one of the Isles of *Cape Verde*, and that they may reach it next day, and having great confidence

in the said Watches, resolved to steer their Course thither, and having given
the order so to do, they got the very next day about Noon a sight of the said
Isle of *Fuego*, finding themselves to sail directly upon it, and so arrived at it
that Afternoon, as he had said.

Upon learning of the successful expedition, an excited Huygens wrote to
Oldenburg updating his recent activities with his new invention. Oldenburg
published this first-person letter immediately after his account of Holmes's field
test:

> Major *Holmes* at his return, hath made a relation concerning the usefulness
> of *Pendulums,* which surpasseth my expectation: I did not imagine that the
> Watches of this first Structure would succeed so well, and I had reserved my
> main hopes for the New ones. But feeling that those have already served so
> successfully, and that the other are yet more just and exact, I have the more
> reason to believe, that the invention of *Longitudes* will come to its perfec-
> tion.

The remainder of Huygens's letter mainly concerns his seeking a patent
for his clock, improving it, and publishing details of its operation. Holmes's
successful test had obviously thrilled Huygens. Later testing after Holmes,
however, revealed that his pendulum clocks only worked when sailing under
mild weather conditions: rough seas disrupted the pendulum timing. Alas, a
device to determine longitude had not yet "come to perfection."

How to Improve a Telescope

Adrien Auzout, 1665. "Opinion respecting the apertures of object glasses, and their
relative proportions, with the several lengths of telescopes." *Philosophical Transactions*,
Vol. 1, p. 55.

While Cartesian plots of data, such as the line graphs so common in current
experimental articles, were not invented until the late eighteenth century and
not applied with any regularity until the twentieth, tables with systematically
arranged rows and columns of data existed from well before the earliest scientific
articles. Copernicus made frequent use of them in his revolutionary 1543 book
on the heliocentric universe, for example. Like Cartesian plots, such tables
allow the reader to quickly grasp relationships among several variables. Tables
facilitate comparisons in a way beyond the powers of a succinct prose. They

guide the eye by employing alignment, vertical and horizontal lines of data, and white space.

Reproduced on the next page is the first table appearing in *Philosophical Transactions* (fourth issue), reporting Auzout's values for obtaining a clear image with a telescope, one with different grades of resolution. In telescopes with lengths from four inches to four hundred feet, Auzout computed the sizes needed for excellent, good, and ordinary object lenses (also called apertures, as distinct from eye-pieces). A key conclusion from this table is that the ratio of the focal length of the telescope to aperture size is approximately the ratio of its square roots.

What is the point of this project? Auzout wants what all astronomers of the time wanted: a clear, very highly magnified image with a refracting telescope, one that employs lenses rather than mirrors. This quest led to the creation of longer and longer instruments. But increased length was a problematic solution to the astronomer's problems; can you imagine a telescope four hundred feet long? That is longer than a football field.

Experiments with Humans—Ethics of Blood Transfusions

Henry Oldenburg, 1667. "Of more tryals of transfusion, accompanied with some considerations thereon, chiefly in reference to its circumspect practice on man; together with a farther vindication of this invention from usurpers." *Philosophical Transactions*, Vol. 2, pp. 517-24.

Experimenters in the seventeenth century faced an ethical dilemma that bedevils the scientific/medical community to this day: when is it proper to transfer medical treatments from animal subjects to human patients? Henry Oldenburg tackled that issue in his 1667 article reviewing the state of experiments on blood transfusion.

The motivation behind this article is twofold: to call for caution in applying this technology to humans and to address the dispute over whether the English or French deserve credit for the invention of blood transfusion. The message is clear; not too surprisingly, Oldenburg sides with the English on the priority dispute. Oldenburg does give the French some backhanded credit for having been the first to perform transfusion with humans, even though a few European patients died as a consequence:

> Before we dismiss this *subject*, something is to be said of . . . why the Curious in *England* make a demurr in practicing this Experiment upon *Men*. The

(56)

the *Reader*, that he hath found, that the *Apertures*, which *Optick-Glasses* can bear with distinctness, are in about a *subduplicate proportion* to their *Lengths*; whereof he tells us he intends to give the reason and demonstration in his *Diopticks*, which he is now writing; and intends to finish, as soon as his Health will permit. In the mean time, he presents the *Reader* with a *Table* of such *Apertures*; which is here exhibited to the Consideration of the Ingenious, there being of this *French* Book but one Copy, that is known, in *England*.

A *TABLE* of the *Apertures of Object-Glasses*.

The *Points* put to some of these *Numbers* denote *Fractions*.

Lengths of Glasses Feet, Inches	For excellent ones. Inch. Lines.	For good ones. Inch. Lines.	For ordinary ones. Inch. Lanes.	Lengths of Glasses Feet, Inches.	For excellent ones. Inch. Lines.	For good ones. Inch. Lines.	For ordinary ones. Inch. Lines.
4	4	4	3	25	3 4 2	10 2	4.
6	5	5	4	30	3 8 3	2 2	7
9	7	6	5	35	4 0 3	4. 2	10
1 0	8.	7	6	40	4 3 3	7 3	.
1 6	9	8.	7	45	4 6 3	10	2.
2 0	11	10	8	50	4 9 4	0 3	4.
2 6 1	0	11	9	55	5 0 4	3 3	6.
3 0 1	1 1	0	10	60	5 2 4	6 3	8.
3 6 1	2. 1	1	11	65	5 4 4	8 3	10
4 0 1	4 1	2 1	0	70	5 7 4	10 4	.
4 6 1	5 1	3 1	.	75	5 9 5	0 4	2.
5 0 1	6 1	4 1	1.	80	5 11 5	2 4	5
6 1	7 1	5 1	2	90	6 4 5	6 4	7.
7 1	9 1	6 1	3	100	6 8 5	9 4	10
8 1	10 1	8 1	4	120	7 5 6	5 5	3
9 1	11. 1	9 1	5	150	8 0 7	0 5	11
10	2 1 1	10 1	6	200	9 6 8	6	9
12	2 4 2	0 1	8	250	10 6 9	2 7	8.
14	2 6 2	2	9.	300	11 6 10	0 8	5
16	2 8 2	4 1	11.	350	12 6. 10	9 9	0
18	2 10 2	6 2	1	400	13 4 11	6 9	8
20	3 0 2	7 2	2,				

FIGURE 1-5 · First table appearing in *Philosophical Transactions*. Courtesy of Special Collections Research Center, University of Chicago Library.

above-mentioned ingenious Monsieur *Denys* has acquainted the World, how this degree was ventured upon at *Paris*, and what good success it there met with: And the *Journal des Scavans* glorieth, that the *French* have advanced this Invention so far, as to trie it upon *Men*, before any *English* did it, and that with good success.

We readily grant, *they* were the first, we know of, that actually thus *improved* the Experiment; but then they must give us leave to inform them of this Truth, that the Philosophers in *England* had practiced it long ago upon *Man*, if they had not been so tender in hazarding the Life of Man (which they take so much pain, to preserve and relieve) nor so scrupulous to incurre the Penalties of the Law, which in *England*, is more strict and nice in cases of this concernment, than those of many other Nations are.

Throughout most of this article, Oldenburg maintains the polite tone characteristic of scientific literature in general, even in cases of open conflict. In his title, however, Oldenburg calls the French "usurpers," and at the article's close he moves from accusation to ironic sneer, followed by a conciliatory final paragraph—all delivered in a richly metaphoric language:

It seems strange, that so surprising an Invention should have been conceived in *France*, as they will have it, ten years ago, and lain there so long in the Womb, till the way of Midwiving it into the world was sent hither from *London*: To say nothing of the disagreement, there seems to be about the French *Parent* of this *foetus* . . .

But whoever the Parent be, that is not so material, as that all lay claim to this Child, should joyn together their endeavors and cares to breed it up for the service and relief of human life, if it be capable of it; and this is the main thing aimed at and solicited in this Discourse; not written to offend or injure any, but to give every one his due, as near as can be discerned by the *Publisher*.

Insult, irony, metaphor, derogation—even today priority disputes incite scientists to burst the bounds of polite scientific discourse.

Animal Experimentation, Eighteenth-Century Style

Robert Boyle, 1670. "New pneumatical experiments about respiration." *Philosophical Transactions*, Vol. 4, pp. 2011–56.

In consecutive issues of the 1670 *Philosophical Transactions*, Boyle published two long articles reporting a series of experiments in which he inserted a small zoo of animals (including birds, snakes, frogs, and kittens) under his air-pump receiver evacuated of air. The selection below (first two paragraphs of Boyle's first article) begins with a question a child might ask: do ducks have a special respiratory system allowing them to hold their breath under water longer than other birds? Here is how Boyle posed and answered his own question:

Nature having, as *Zoologists* teach us, furnished *Ducks* and other water-Fowl with a peculiar structure of some vessels about the heart, to enable them, when they have occasion to Dive, to forbear for a pretty while respiring under water without prejudice: I thought it worth the tryal, whether such Birds would much better than other Animals endure the absence of the Air in our exhausted Receiver. The accounts of which tryals were, when they were made, registered as follows.

Experiment the I.

We [namely, Boyle and his unnamed assistants] put a full grown *Duck* (being not then able to procure a fitter) into a Receiver, whereof she fill'd, by our guess, a third part or somewhat more, but was not able to stand in any easy posture in it; then pumping out the Air, though she seemed at first (which yet I am not too confident of, upon a single tryal,) to continue well somewhat longer than a Hen in her condition would have done; yet within the short space of one minute she appeared much discomposed, and between that and the second minute, her struggling and convulsive motions increased so much, that, her head also hanging carelessly down, she seemed to be just at the point of death; from which we presently reduced her by letting in the Air upon her: So that, this duck being reduced in our Receiver to a gasping condition, within less than two minutes, it did not appear, that, notwithstanding the peculiar contrivance of nature to enable these water-Birds to continue without respiration for some time under water, this Duck was able to hold out considerably longer than a Hen, or other Bird not-Aquatick, might have done: and to manifest that it was not closeness and narrowness of the vessel, in reference to so bulky an Animal, that produced in the subject of our tryal the great and suddain change above-recited, we soon after included the same Bird in the same Receiver, and having by a special way cemented it on very close, we suffered her to stay thus shut up with the Air for five times as long as formerly (by our guess, helped by a watch;) without perceiving her to be discomposed; and she would probably have continued longer in the same condition, if my patience and leisure would have held out so long, as she could have done in that prison.

To answer his opening question, Boyle tells us a relatively easy-to-follow story. He places a duck in his air pump (pictured at the beginning of this chapter) and evacuates the air. The bird shows signs of distress almost as quickly as other animals. End of story? Not quite. The smallness of the air pump's "prison" cell could have caused the duck to have problems breathing even when some air is available. Sir Robert thus runs a control experiment with air present in the receiver. This time the duck has no apparent breathing problems. Final conclusion: the duck is not much better equipped than a hen to breathe under water. Boyle follows this story with about fifty more like it, in terms of style and content. He seems relatively indifferent to animal suffering, an attitude apparently common in his time. William Harvey's vivisections to better understand the circulation of the blood, we must remember, were carried out without anesthetic.

Boyle's basic narrative style is typical for the seventeenth and eighteenth centuries: detail is piled on detail so that the reader may re-experience the event. The author presents a series of experiments, each described separately, some giving the dates they were performed. These strings of experiments tell a tale of discovery guided by improvisation: one experiment suggests another, which suggests another, and so on. No larger, theoretical concerns seem to loom in the background, the goal being only *"to bring some more light on the doctrine of RESPIRATION."*

A Revolutionary Paper in Optics

....

Isaac Newton, 1672. "New theory about light and colors." *Philosophical Transactions*, Vol. 6, pp. 3075–87.

Isaac Newton's reputation as a scientist rests solidly on two books: *Philosophiae Naturalis Principia Mathematica* (1687) and *Opticks; or, A Treatise of the Reflections, Refractions, Inflections and Colours of Light* (1704). But his first published article (1672) is also deservedly famous; it is widely considered by historians as the first major scientific article ever published. Henry Oldenburg, the *Transactions* editor, recognized the importance of this twelve-page article by the then-young and relatively obscure mathematics professor, because it is the sole research article in the eightieth issue and was printed less than two weeks after its reception in handwritten manuscript form. The journal's contents page gives an excellent abstract of the article's major claims. (Two years—1671 and 1672—are given for the date of publication because, until 1752, the English and Continental calendars did not match.)

Newton's main contention was that white light, far from being simple, as previously believed, was a compound of all the colors of the spectrum, a

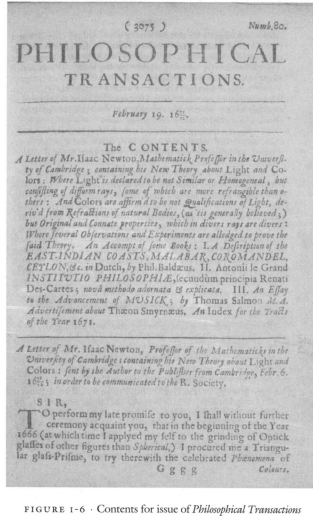

FIGURE 1-6 · Contents for issue of *Philosophical Transactions* carrying Newton's first scientific article. Courtesy of Special Collections Research Center, University of Chicago Library.

compound that could be decomposed by passing white light through a prism and recomposed through reversing that passage in a second prism. The claim had two basic components: first, Newton's experimental results with prisms and, second, his theoretical explanation for them. The latter also has two components: light appears to be a body composed of particles, and rays of differently colored light refract at different angles.

Newton's optics article is written in the style of a letter aimed at the members of the Royal Society of London and anyone else who might be interested in such technical matters, but sent care of Henry Oldenburg. In January of the same year, Oldenburg had asked Newton to send details on the reflecting telescope he had designed. In return, Oldenburg got that and considerably more: a "new theory of light."

In this letter's opening paragraphs, Newton escorts his readers right into the darkened chamber where he undertook his experiments with a triangular prism:

in the beginning of the Year 1666 (at which time I applyed my self to the grinding of Optick glasses of other figures than *Spherical,*) I procured me a Triangular glass–Prisme, to try therewith the celebrated *Phenomena* of *Colours.* And in order thereto having darkened my chamber, and made a small hole in my window–shuts, to let in a convenient quantity of the Suns light, I placed my Prisme at his entrance, that it might be thereby refracted to the opposite wall. It was at first a very pleasing divertisement, to view the vivid and intense colours produced thereby; but after a while applying my self to consider them more circumspectly, I became surprised to see them in an *oblong* form; which, according to the received laws of Refraction, I expected should have been *circular.*

They were terminated at the sides with streight lines, but at the ends, the decay of light was so gradual, that it was difficult to determine justly, what was their figure, yet they seemed *semicircular.*

Comparing the length of this coloured *Spectrum* with its breadth, I found it about five times greater; a disproportion so extravagant, that it excited me to a more then ordinary curiosity of examining, from whence it might proceed. I could scarce think, that the various *Thickness* of the glass, or the termination with shadow or darkness, could have any Influence on light to produce such an effect; yet I thought it not amiss, first to examine those circumstances, and so tryed, what would happen by transmitting light through parts of the glass of divers thicknesses, or through holes in the window of divers bignesses, or by setting the Prisme without so, that the light might pass through it, and be refracted before it was terminated by the hole: But I found none of those circumstances material. The fashion of the colours was in all these cases the same.

Like a good reporter, Newton establishes right at the start who did what, why, when, and where. He also records his emotional response: "It was at first a very pleasing divertisement," "I became surprised," "it excited me to a more then ordinary curiosity." This touch lends an intimacy and individuality to his communication generally lacking in the dispassionate scientific prose of today. (In the modern scientific paper, who did what, where, and when normally appear in the byline along with the list of authors after the title. Such circumstantial details seldom find their way into the main text; see chapter 7.)

After those introductory paragraphs, Newton briefly describes, in what one would assume is chronological order, the diverse experiments he performed in order to determine what "could have any influence on light to produce such an effect." Interestingly, thanks to first-rate detective work by historians of science, we now know the deceptiveness of Newton's Baconian tale of discovery leading from his serendipitous finding of an experimental discrepancy to landmark knowledge claims. Newton's private notebooks and manuscripts for his Cambridge University lectures reveal that he probably formed a theory about the properties of light long before his crucial experiment. It would appear that his experiments were carried out to find support for his initial flash of insight.

The initial reception of Newton's article by the Royal Society fellows was largely favorable. His article had an early impact on thinking about light and color in England and on the Continent. Much to Newton's eventual dismay and irritation, however, several prominent virtuosi would challenge his findings: most notably, two respected Jesuit scientists, Ignace Pardies and Francis Linus, and two scientists of the first rank, Robert Hooke and Christiaan Huygens. They questioned Newton on several fronts: the experimental design, the results it purported to generate, and most extensively, his theoretical explanation. Pardies and Hooke offered a plausible alternative to Newton's explanation, while Huygens was troubled by the lack of a postulated mechanism for the effect Newton produced. This controversy marks the first important public differences of opinion aired in the scientific journal literature. We next briefly discuss the argumentative exchanges between Pardies and Newton.

A Frenchman Criticizes the Young Newton

Isaac Newton, 1672. "Mr. Newton's answer to the foregoing letter [from Ignace Pardies, May 21, 1672]." *Philosophical Transactions*, Vol. 7, pp. 5014–18. Translated from the Latin.

Newton's "theory of light" begins with a research problem: light passed through a prism has an oblong shape, at odds with the oval predicted by the received laws of seventeenth-century optics. This discrepancy called for some new explanation. His *experimentum crucis*, or crucial experiment, showed that white light is a compound of all the radiant colors, each with a different index of refraction. This experiment explained the discrepancy identified in his research problem.

The first serious published criticism to Newton's optics article came from a French Jesuit and professor of mathematics in the Parisian college of Clermont, Ignace Pardies, who wrote two letters on the subject. The second was particularly telling, not because of the cogency of its criticisms but because it provoked Newton to define more carefully both the crucial experiment in his original paper (including, finally, an explanatory diagram) and to clarify his way of doing science.

In his second letter, published in the pages immediately preceding Newton's rebuttal, Pardies objects that the oblong shape found by Newton's prism has an alternative explanation. For that explanation, he draws upon a "hypothesis" of Newton's great rival, Robert Hooke: namely, that light behaves as waves or pulses, not particles. Pardies argues that the effect Newton attributes to *refraction* may be caused by *diffraction*, a phenomenon only recently discovered by the Italian scientist Francisco Grimaldi and defined as a sideways bending at the wave front. As Pardies notes,

> the greater length of the image may be otherwise accounted for, than by the different refrangibility of the rays. For according to that hypothesis, which is explained at large by Grimaldi, and in which it is supposed that light is a certain substance very rapidly moved, there may take place some diffusion of the rays of light after their passage and decussation [crossing] in the hole. Also on that other hypothesis, explained by Mr. Hooke, colours may be explained by a certain diffusion and expansion of the undulations, made on the sides of the rays beyond the hole by the influence and continuation of the subtle matter.

As it turned out, Pardies was mistaken in his counterargument, even if wave theory was admitted to be correct—a concession Newton was unwilling to make. But Newton had a more important point to make concerning hypotheses. They are useful as ways of devising experiments, but one must be cautious about using them to explain the results of these experiments:

> [It] is to be observed that the doctrine which I explained concerning refraction and colours, consists only in certain properties of light, without regarding any hypotheses, by which those properties might be explained. For the best and safest method of philosophizing [doing science] seems to be first, to inquire diligently into the properties of things, and establishing those properties by experiments and then to proceed more slowly to

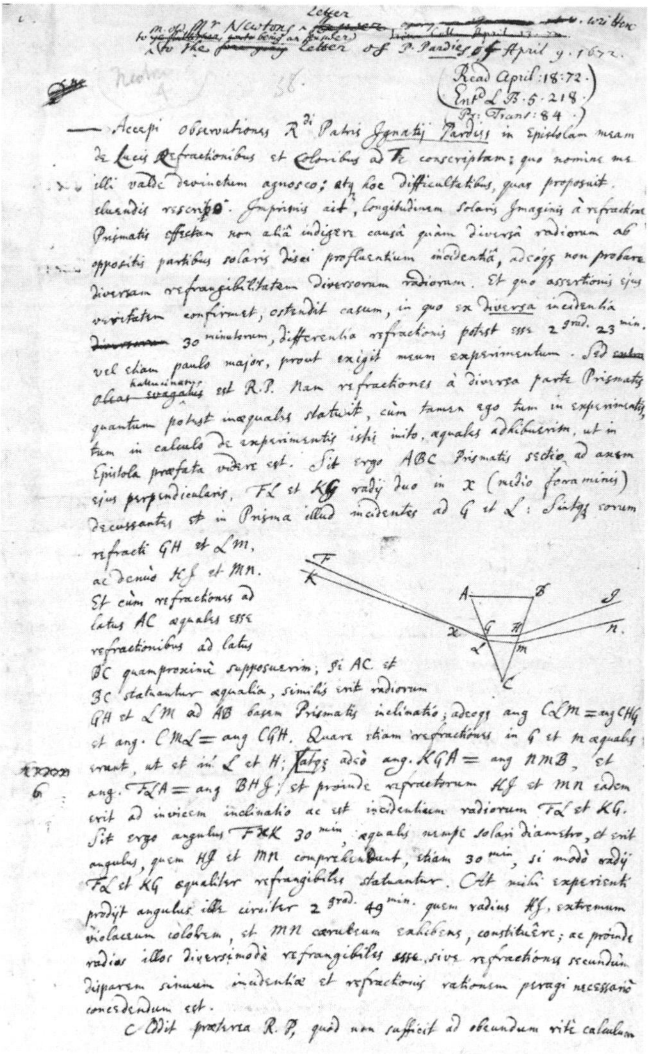

Mr Newtons Letter to the Letter of P. Pardies of April 9. 1672.

Read April: 18: 72.
Ent. L B. 5. 218.
Ph. Trans: 84.

Accepi observationes R^di Patris Ignatij Pardies in Epistolam meam de Lucis Refractionibus et Coloribus ad Te conscriptam; quo nomine me illi valde devinctum agnosco; atq hoc difficultatibus, quas proposuit, eluendis rescribo. Imprimis ait, Longitudinem Solaris Imaginis à refractione Prismatis effectam non aliâ indigere causâ quàm diversâ radiorum ab oppositis partibus solaris disci profluentium incidentiâ, adeoq non probare diversam refrangibilitatem diversorum radiorum. Et quo assertionis ejus veritatem confirmet, ostendit casum, in quo ex diversa incidentia 30 minutorum, differentia refractionis potest esse 2 grad. 23 min. vel etiam paulo major, prout exigit meum experimentum. Sed hallucinatus est R.P. Nam refractiones à diversa parte Prismatis quantum potest inæquales statuit, cùm tamen ego tum in experimentis tum in calculo de experimentis isti inito, æquales adhibuerim, ut in Epistola præfata videre est. Sit ergo ABC Prismatis sectio ad axem ejus perpendicularis, FL et KG radij duo in x (medio foraminis) decussantes et in Prisma illud incidentes ad G et L: Sintq eorum refracti GH et LM, ac denuò HI et MN. Et cùm refractiones ad latus AC æquales esse refractionibus ad latus BC quamproximè supposuerim; si AC et BC statuantur æqualia, similis erit radiorum GH et LM ad AB basem Prismatis inclinatio; adeoq ang CLM = ang CHG et ang. CML = ang CGH. Quare etiam refractiones in G et M æquales erunt, ut et in L et H; atq adeò ang. KGA = ang NMB, et ang. FLA = ang BHI, et proinde refractorum HI et MN eadem erit ad invicem inclinatio ac est incidentium radiorum FL et KG. Sit ergo angulus FXK 30 min, æqualis nempe solari diametro, et erit angulus, quem HI et MN comprehendunt, etiam 30 min, si modò radij FL et KG equaliter refrangibiles statuantur. At mihi experienti prodijt angulus ille circiter 2 grad. 49 min. quem radius HI, extremum violaceum colorem, et MN cæruleum exhibens, constituere; ac proinde radios illos diversimodè refrangibiles esse, sive refractiones secundùm disparem sinuum incidentiæ et refractionis rationem peragi necessariò concedendum est.

Addit præterea R. P. quòd non sufficit ad obeundum ritè calculum

FIGURE 1-7 · First page of letter on optics from Isaac Newton to Ignace Pardies, 1672. Reprinted with permission of the Royal Society.

hypotheses for the explanation of them. For hypotheses should be subservient only in explaining the properties of things, but not assumed in determining them; unless so far as they may furnish experiments. For if the possibility of hypotheses is to be the test of the truth and reality of things, I see not how certainty can be obtained in any science; since numerous hypotheses may be devised, which shall seem to overcome difficulties. Hence it has been here thought necessary to lay aside all hypotheses, as foreign to the purpose, that the force of the objection should be abstractedly considered, and receive a more full and general answer.

At this point, Newton re-explains his crucial experiment, showing clearly that it supports his claim of white light's "frangibility," its tendency to split into its component parts, each a different color. Having shown this, Newton indicates what he thinks is the proper and improper use of hypotheses in explaining the phenomenon he has discovered. His crucial experiment supports a ray theory in which light consists of the translationary radiation of tiny particles, but of course, any existing hypothesis may be suitably modified to fit the experiments. As far as Newton is concerned, it is this flexibility that shows what is wrong with giving hypotheses pride of place in forming explanations:

After the properties of light shall, by these and such like experiments, have been sufficiently explored, by considering its rays as collateral or successive parts of it, of which we have found by their independence that they are distinct from one another; hypotheses are thence to be judged of, and those to be rejected which cannot be reconciled with the phaenomena. But it is an easy matter to accommodate hypotheses to this doctrine. For if anyone wish to defend the Cartesian hypothesis, he need only say that the globules are unequal, or that the pressures of some of the globules are stronger than others, and that hence they become differently frangible, and proper to excite the sensation of different colours. And thus according to Hooke's hypothesis, it may be said that some undulations of the aether are larger and denser than others. And so of the rest.

In these passages, the philosopher of science in Newton is doing boundary work; he is declaring an important class of arguments outside the bounds of good scientific practice.

A French Test Pilot

Monsieur Toyard, 1681. "An account of the Sieur Bernier's way of flying."
Philosophical Collections, Vol. 1, pp. 14–18.

Robert Hooke claimed to have invented thirty ways of flying, so it comes as no surprise that he would include an article on flying in the first issue of *Philosophical Collections* (the temporary replacement for *Philosophical Transactions* after Oldenburg's death), for which he served as editor in its brief lifetime.

The illustration on the next page is taken from a letter sent by a Monsieur Toyard, describing some Frenchman's design of a flying machine "consisting of 4 wings, to be moved by the strength of the arms and legs of the man that flies." Hooke's correspondent reports that this method was practiced in France between 1670 and 1680. The writer goes into great detail on how this machine works, and the concluding paragraph teeters on the edge of reasonableness, but then takes off into a dream-like fantasy, crossing the borderline between science and science fiction:

> But he [the inventor Bernier] pretends not nevertheless to be able to raise himself from the Earth by this his Machine, nor to sustain himself any long time in the Air, by reason of the want of strength and quickness in his Arms and Legs, which is necessary to move these kinds of Wings frequently and efficaciously enough; but yet he is confident that from a place pretty elevated into the Air, he shall be able to pass over a River of considerable breadth, having already done as much from several heights, and at several distances. — He began his trials first by springing out himself from a stool, then from the top of a Table, then from a pretty high Window, then from a Window in a second Story, and at last from a Garret, from whence he flew over the Houses of the Neighbours; practicing thus with it by little and little, till he had brought it to the perfection it now hath.

A Plant Described in a Flowery Style

Martin Lister, 1697. "An account of the nature and differences of the juices, more particularly, of our English vegetables." *Philosophical Transactions*, Vol. 19, pp. 365–83.

In 1667, Thomas Sprat, a full-time cleric and part-time historian of science, expressed vehement disgust with "this trick of Metaphor, this volubility of Tongue, which makes so great a noise in the World" (quoted more fully later in this chapter). Nonetheless, some early science writers did manage to make effective use of metaphor and simile in the service of scientific communication. Take the following metaphorically rich botanical description by Martin Lister, a physician and frequent early contributor to *Philosophical Transactions* with a keen interest in a wide variety of sciences, not only botany, but also entomology, paleontology, geology, chemistry, and medicine:

> In the middle of *July,* I drew and gathered of the Milk of *Lactuca syl. costa spinosa*, *C. B.* and of all our *English* Plants, that I have met with, this most freely and plentifully affords it [juices that coagulate]. It springs out of the

FIGURE 1-8 · Apparatus for flying. University of Minnesota Library.

Wound thick as Cream and Ropes, and is White, and yet the Milk which came out of the Wounds, made towards the top of the Plant, was plainly streaked or mixt with a purple Juice, as though one had dashed or sprinkled Cream with a few drops of Claret. And indeed, the Skin of the Plant thereabouts was purplish also, perhaps with Veins. Again, in the Shell I drew it, it turned still yellower and thicker, and by and by curdled, that is, the white and thick caseous part did separate from a thin purple Whey. So the Blood also of Animals, whilst warm remains liquid and alike, but so soon as cold, it cakes and has a *Serum* or Whey separated from it; the Cake is made of glutinous Fibers, and therefore if the hot or new drawn Blood be well stirred or beaten, it will not break. *Qu.* If the same stirring the Milk for *ex.* of *Lac. syl.* in drawing it into the Shell, will hinder its coagulating or parting with a Serum or Whey? also, the caseous part of the Milk of Animals is glutinous and stringy. Further, this *Serum* came freely from the other, by squeezing betwixt my Fingers: and the Curds I washed in Spring water, which became immediately like Rags and tough (Draw this Milk immediately, or let in fall off the Plant, into a Shell of fair Water, or other Menstruum, as Vinegar, *S. V.* Spirit of Vitriol or Sulphur, &c.) and remained still white and dry. As for the purple Whey after a Days insolation, it stifined and became hard, and was easily formed into Cakes; which Cakes were yet very brittle, and would easily crumble into Powder. About *December* following, I broke one of the Cakes, made of the caseous part of the Milk of this Plant; it then proved very brittle, and shined upon breaking like Rosin; it was then of a dark brown Colour; moreover, it burned with a lasting Flame, like Rosin or Wax, and that being melted by Heat, it would draw out into long tough strings, like Bird-lime. On the contrary, the purplish Powder, which was the Whey, if put into the Flame of a Candle, would scarce burn with a Flame at all, but soon be turned into a Coal. Lastly, the purple Powder did taste very bitter; whereas the caseous part was as insipid as Wax.

While this image-rich text is not typical of early scientific prose, the sentence structure is: the "plain style" at the time most definitely did not mean the relatively short, syntactically simple sentences modern readers associate with newspapers.

Mathematics Proves That God Exists

John Arbuthnot, 1710. "An argument for divine providence, taken from the constant regularity observ'd in the births of both sexes." *Philosophical Transactions*, Vol. 27, pp. 186–90.

Scientific arguments for the existence of a deity long predate the modern argument that neither natural selection nor chance can possibly explain all the complex designs found in nature. Physician to Queen Anne and literary collaborator with the poet and satirist Alexander Pope, John Arbuthnot wrote a *Philosophical Transactions* article that rests its case for the existence of a divine being on mathematical probability and population statistics. He begins with the assumption that a divine being purposely designed the equal distribution of males and females in the world to ensure their continued existence:

Among innumerable Footsteps of Divine Providence to be found in the Works of Nature, there is a very remarkable one to be observed in the exact Ballance that is maintained, between the Numbers of Men and Women; for by this means it is provided, that the Species may never fail, nor perish, since every Male may have its Female, and of a proportionable Age. This Equality of Males and Females is not the Effect of Chance but Divine Providence, working for a good End, which I thus demonstrate.

From this starting point a mathematical argument unfolds, proving that on the basis of probability theory alone, "in the vast Number of Mortals, there would be a small part of all possible Chances, for its happening at any assignable time, that an equal Number of Males and Females should be born." His argument here rests on probability calculations with the binomial theorem,

$$(x + a)^n = \sum_{k=0}^{n} \frac{n!}{k!(n-k)!} x^k a^{n-k}$$

That finding is hardly strong enough to justify a larger claim of divine intervention as the explanation. For that, Arbuthnot needs a miracle. So he adds another wrinkle to his argument: he notes that "the external Accidents to which Males are subject (who must seek their food with danger) do make a great havoc of them, and that this loss exceeds that of the other Sex." To repair that discrepancy, he asserts, the divine being tilted the balance in favor of male births, as evidenced by the fact that male had slightly exceeded female births in London every year for an eighty-two year period (1629–1710). He calculated the probability against this annual male predominance in the birth rate to be astronomically high (1 in about 50×10^{24}!). The only apparent explanation for that surprising result was the purposeful design of a divine being. (In a 1781 article, using birth data from Paris showing a similar propensity of male over female births, Pierre Simon Laplace would prove that this deviation could be explained by a universal "error curve" derived by Abraham de Moivre in 1773.)

After having presented his proof, Dr. Arbuthnot adds a scholium, a note illustrating a consequence of his proof:

Scholium. From hence it follows, that Polygamy is contrary to the Law of Nature and Justice, and to the Propagation of Human Race; for where Males and Females are in equal number, if one Man takes Twenty Wives, Nineteen Men must live in Celibacy, which is repugnant to the Design of Nature; nor is it probable than Twenty Women will be so well impregnated by one Man as by Twenty.

From the vantage of present time, it would be easy to poke fun at Arbuthnot's mathematical rationalization for religious beliefs and practices. But his is an important early attempt at argument by statistical analysis, even if ultimately wrong-headed.

An Electrical Cure for Paralysis

Benjamin Franklin, 1758. "An account of the effects of electricity in paralytic cases." *Philosophical Transactions*, Vol. 50, pp. 481–83.

Electricity was an important topic in the late eighteenth-century *Philosophical Transactions*, and a prominent contributor on that topic was Benjamin Franklin, a fellow of the Royal Society and the first American with an international reputation in science. Since Franklin had a strong interest in medical care, it was only natural that he would investigate newspaper reports on cures for paralysis by electric shock. This he did with his usual diligence and caution:

> Some years since, when the news-papers made mention of great cures performed in Italy and Germany, by means of electricity, a number of paralytics were brought to me from different parts of Pennsylvania, and the neighbouring provinces, to be electrified; which I did for them at their request. My method was, to place the patient first in a chair, on an electric stool, and drawn a number of large strong sparks from all parts of the affected limb or side. Then I fully charged two six-gallon glass jars, each of which had about three square feet of surface coated; and I sent the united shock of these thro' the affected limb or limbs; repeating the stroke commonly three times each day. The first thing observed was an immediate greater sensible warmth in the lame limbs, that had received the stroke, then in the others: and the next morning the patients usually related, that they had in the night felt a prickling sensation in the flesh of the paralytic limbs; and would sometimes shew a number of small red spots, which they supposed were occasioned by those prickings. The limbs too were found more capable of voluntary motion, and seemed to receive strength. A man, for instance, who could not the first day lift the lame hand from off his knee, would the next day raise it four or five inches, the third day higher; and on the fifth day was able, but with a feeble languid motion, to take off his hat. These appearances gave great spirits to the patients, and made them hope a perfect cure; but I do not remember, that I ever saw any amendment after the fifth day: which the patients perceiving, and finding the shocks pretty severe, they became discouraged, went home, and in a short time relapsed; so that I never knew any advantage from electricity in palsies, that was permanent. And how far the apparent temporary advantage might arise from the exercise in the patients journey, and coming daily to my house, or from the spirits given by the hope of success, enabling them to exert more strength in moving their limbs, I will not pretend to say.
>
> Perhaps some permanent advantage might have been obtained, if the electric shocks had been accompanied with proper medicine and regimen, under the direction of a skilful physician. It may be, too, that a few great strokes, as given in my method, may not be so proper as many small ones; since, by the account from Scotland of a case, in which two hundred shocks from a phial were given daily, it seems, that a perfect cure was made.

Franklin's narrative follows the typical modern sequence: statement of research problem (does electric shock cure paralysis?), method for solving it, results, discussion, and conclusion. Franklin concluded that electricity may have some temporary benefits, but no permanent ones in the patients he tested. Yet, from his small sample he could not generalize with great conviction. In particular, he admits that he could not rule out the possibility that another treatment method than his might actually work. At the very least, in an ideal world, Franklin's article would have put the burden of proof on those making claims of miraculous cures. As we know, the real world does not work that way. Electrotherapy machines offering cures unsubstantiated by Franklin-like research prospered in the eighteenth and nineteenth centuries.

A Rearguard Action in Favor of Phlogiston

Henry Cavendish, 1784. "Experiments on air." *Philosophical Transactions*, Vol. 74, pp. 119–53.

Throughout the eighteenth century, for most chemists, the release of "phlogiston" explained reactions involving combustion. In the later part of this same century, Antoine Lavoisier conducted experiments indicating that he was satisfied that combustion could be explained without having to rely on a substance that had never been actually detected (see chapter 2). In this passage, Henry Cavendish acknowledges the explanatory power of Lavoisier's new chemistry, and sees that the criterion for deciding between it and the existing theory will be an *experimentum crucis*, an experiment that permanently turns the tide in favor of one theory or another. But no such experiment is currently available because, Cavendish asserts, in combustion the result would be the same whether we say that phlogiston is lost, as he does, or that oxygen is added, as does Lavoisier. Given this experimental equivalence, Cavendish is inclined to allow the presumption to rest with existing theory. Still, he feels the need to justify his position (a key phrase is italicized):

It seems, therefore, from what has been said, as if the phaenomena of nature might be explained very well on this principle [of Monsieur Lavoisier], without the help of phlogiston; and, indeed, as adding dephlogisticated air [oxygen] to a body comes to the same thing as depriving it of its phlogiston and adding water to it, and as there are, perhaps, no bodies entirely destitute of water, and as I know no way by which phlogiston can be transferred from one body to another, without leaving it uncertain whether water is not at the same time transferred, it will be very difficult to determine by experiment which of these opinions is the truest; but as the commonly received principle of phlogiston explains all phaenomena, at least as well as Mr. Lavoisier's, I have adhered to that. There is one circumstance also, which though it may appear to many not to have much force, I own has *some weight* with me; it is, that as plants seem to draw their nourishment almost intirely from water and fixed and phlogisticated air [nitrogen and carbon dioxide], and are restored back to those substances by burning, it seems reasonable to conclude, that notwithstanding their infinite variety they consist almost intirely of various combinations of water and fixed and phlogisticated air [carbon dioxide and nitrogen], united according to one of these opinions to phlogiston, and deprived according to the other of dephlogisticated air [oxygen]; so that, according to the latter opinion, the substance of a plant is less compounded than a mixture of those bodies into which it is resolved by burning; and it is more reasonable to look for great variety in the more compound than in the more simple substance.

His argument in favor of phlogiston is in the form of an implicit syllogism: substances that are apparently more complex should consist of more components; because plants are apparently more complex than their ashes, plants should consist of more components than these ashes; therefore, it is more likely that plants contain phlogiston. His argument is weak, as he himself acknowledges: it is little more than a string of loose analogies. But, in an interesting turn, Cavendish uses this very weakness to his argumentative advantage. By suggesting the metaphor of the balance in the expression "some weight," he implicates his scrupulousness in evaluating the evidence for and against the rival theories. This metaphor strengthens the argument in another way: even the slightest weight will tip a balance whose pans are exactly on a level, making it more reasonable to prefer the more slightly plausible alternative, an explanation involving phlogiston.

A Woman Scientist Discovers a Comet

Caroline Herschel, 1787. "An account of a new comet." *Philosophical Transactions*, Vol. 77, pp. 1–3.

In the seventeenth century, the Royal Society of London was open to just about any gentleman who could pay its dues: not only full-time natural philosophers like Hooke, Boyle, and Newton, but also those with an above-average curiosity about things scientific and mathematical, like John Dryden, poet laureate of England. In sharp contrast, women were excluded from membership in the early scientific societies. Indeed, the Royal Society did not elect a woman "fellow" until 1945, and women did not begin publishing in the scientific journal literature with any regularity until well into the twentieth century.

A few women, however, did make it into the pages of the Royal Society's *Transactions* before modern times. The first one of real scientific prominence was Caroline Herschel. Sister and assistant to the famous astronomer William Herschel, she discovered eight new comets, three of which she reported in brief *Philosophical Transactions* articles (1787, 1794, 1796). She also prepared a comprehensive *Catalogue of the Stars* under the auspices of the Royal Society of London.

Reproduced is the first half of her 1787 *Philosophical Transactions* article. Her "imperfect account" reflects the low status in which women were held at that time and place, as she virtually apologizes for having made an important discovery in her brother's absence. At the same time, she demonstrates her mastery of eighteenth-century astronomical observation and reporting:

Sir,

In consequence of the friendship which I know to exist between you and my Brother [William Herschel], I venture to trouble you in his absence with the following imperfect account of a comet.

The employment of writing down the observations, when my Brother uses the 20-feet reflector, does not allow me time to look at the heavens; but as he is now on a visit to Germany, I have taken the opportunity of his absence to *sweep* in the neighbourhood of the sun, in search of comets; and last night, the 1st of August, about 10 o'clock, I found an object resembling in colour and brightness the 27th nebula of the *Connaissance des Temps* [the premier astronomical journal in France, best translated as *Meteorological Observations*], with the difference however of being round. I suspected it to be a comet; but a haziness coming on, it was not possible intirely to satisfy myself or to its motion till this evening. I made several drawings of the stars in the field of view with it, and have enclosed a copy of them, with my observations annexed, that you may compare them together.

August 1, 1786, 9 h. 50′, the object in the center is like a star out of focus, while the rest are perfectly distinct, and I suspect it to be a comet. Tab. I. fig. 1.

Apart from social deference to her famous brother, there is nothing "feminine" about this communication any more than there is anything feminine about George Eliot's novels. Caroline Herschel is simply a good observational

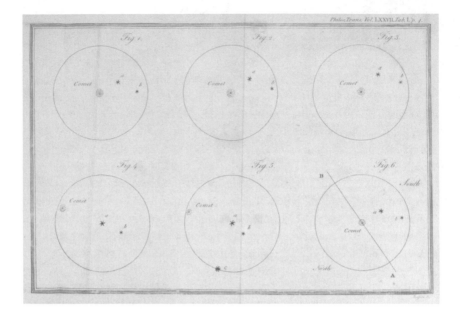

FIGURE 1-9 · Position over time of newly sighted comet.
Courtesy of Special Collections Research Center, University of Chicago Library.

astronomer. The epistolary opening of this communication, which may seem
quaint to us, was well within the range of English practice at the time.

ON EARLY ENGLISH SCIENTIFIC WRITING

Two Pleas for a Plain Style of Writing

Thomas Sprat, 1667. *The History of the Royal Society of London, for the Improving of
Natural Knowledge*. London: J. Martyn, pp. 112–13.

    Robert Boyle, 1661. "A proemial essay, wherein, with some considerations touching
experimental essays in general, is interwoven such an introduction to all those written
by the author, as is necessary to be perus'd for the better understanding of them." From
Certain Physiological Essays Written at Distant Times, and on Several Occasions. London:
Henry Herringman, pp. 11–12.

Thomas Sprat's *History of the Royal Society of London* is more a manifesto for the
seven-year-old society than a history. In it, one will find an extended discussion
of the English scientific style of writing in the seventeenth century. Sprat begins

with a diatribe against contemporary scholarly writing, then offers the Society's antidote:

> Who can behold, without indignation, how many mists and uncertainties, these specious *Tropes* and *Figures* have brought to our Knowledg? How many rewards, which are due to more profitable, and difficult *Arts*, have been still snatch'd away by the easie vanity of *fine Speaking*? For now I am warm'd with this just Anger, I cannot with-hold my Self, from betraying the shallowness of all these seeming Mysteries; upon which, we *Writers*, and *Speakers*, look so bigg. And, in a few words, I dare say, that of all the Studies of men, nothing may be sooner obtain'd, than this vicious abundance of *Phrase*, this trick of *Metaphor*, this volubility of *Tongue*, which makes so great a noise in the World. But I spend words in vain; for the evil is now so inveterate, that it is hard to know whom to blame, or where to begin to reform . . . It will suffice my present purpose, to point out, what has been done by the Royal Society, towards the correcting of its excesses in Natural Philosophy; to which it is, of all others, a most profest enemy.
>
> They have therefore been most rigorous in putting in execution, the only Remedy, that can be found for this extravagance: and that has been the constant Resolution, to reject all the amplifications, digressions, and swellings of style: to return back to the primitive purity, and shortness, when deliver'd so many *things*, almost in equal number of *words*. They have exacted from all their members a close, naked natural way of speaking; positive expressions, clear senses; a native easiness: bringing all things as near the Mathematical plainness, as they can: and preferring the languages of Artizans, Countrymen, and Merchants before that of Wits, or Scholars.

In this passage, Sprat argues that the communication issuing from the Royal Society, particularly its *Transactions,* is not mere rhetoric or "fine Speaking," but consists of exact descriptions of the world as it is (or at least a reasonable facsimile) in plain English. In particular, Sprat frowns upon tropes and figures, that is to say, comparison-based turns of phrase like metaphor and simile. Instead of, say, *a glass sphere the size and shape of an orange*, presumably he would prefer the "Mathematical plainness" of *a glass sphere with three-inch diameter*. Ironically, Sprat's own prose puts rhetorical tropes and figures of speech to good use in making his argument as strongly as possible. But then he is not writing science.

While Sprat's is the most famous early dictum on scientific style, it is not the best balanced. That honor belongs to Robert Boyle:

And . . . as for the style of our Experimental Essays, I suppose you will readily find that I have endeavour'd to write rather in a Philosophical than a Rhetorical strain, as desiring, that my Expressions should be rather clear and significant, than curiously adorn'd . . . And certainly in these Discourses, where our Designe is only to inform Readers, not to delight or perswade them, Perspicuity ought to be esteem'd at least one of the best Qualifications of a style, and to affect needlesse Rhetorical Ornaments in setting down an Experiment, or explicating something Abstruse in Nature, were little lesse improper than it were (for him that designes not to look directly into the Sun it self) to paint the Eye-glasses of a Telescope, whose clearness is their Commendation, in which ev'n the most delightfull Colours cannot so much please the Eye as they would hinder the sight . . . But I must not suffer my self to slip unawares into the Common place of the unfitness of too spruce a style for serious and weighty maters; and yet I approve not that dull and insipid way of writing, which is practic'd by many Chymists, even when they digress from Physiological Subjects: for though a Philosopher [scientist] need not be sollicitous, that his style should delight its Reader with his Floridnesse, yet I think he may very well be allow'd to take a Care, that it disgust not his Reader by its Flatness, especially when he does not so much deliver Experiments or explicate them, as make Reflections or Discourses on them; for on such Occasions he may be allow'd the liberty of recreating his Reader and himself, and manifesting that he declin'd the Ornaments of Language, not out of Necessity, but Discretion, which forbids them to be us'd where they may darken as well as adorn the Subject they are appli'd to. Thus . . . though it were foolish to colour or enamel upon the glasses of Telescopes, yet to gild or otherwise embellish the Tubes of them, may render them more acceptable to the Users, without at all lessening the Clearness of the Object to be look' d at through them.

Spoofing the Royal Society

John Hill, 1751. *A Review of the Works of the Royal Society of London: Containing Animadversions on Such of the Papers as Deserve Particular Observation*. London: R. Griffiths, pp. 62–63.

Unlike its more professional counterpart in Paris, the Royal Society and its *Transactions* were sometimes the target of public ridicule at the hands of wits and scholars. Given the publication of such articles as Toyard on flying-machines and Arbuthnot on the workings of divine providence, this attitude is hardly surprising. As early as 1676, Thomas Shadwell wrote a popular restoration play called "The Virtuoso: A Comedy." The central character, Sir Nicholas Gimcrack, is a caricature of the growing population of science enthusiasts, also called "virtuosi" and "savants." In pursuing his avocation, Sir Gimcrack makes all sorts of absurd observations. For example, through his telescope, he observes on the moon's surface not only "the larger sort of animals, as elephants and camels; but public buildings and ships." So consumed is Sir Gimcrack with the pursuit of new knowledge that he is blind to his wife's seduction by another man.

In a slightly more serious vein, John Hill wrote an entire book wittily castigating the Royal Society for publishing "trivial and downright foolish articles" over the preceding decades, including reports of a miraculous plant that heals fresh wounds ("but to touch it, is to be healed") as well as "incontestable proofs of a strange and surprising Fact, namely, that Fish will live in Water." Here is a particularly sarcastic passage concerning an account of a merman observed in Virginia, 1676:

It has for some odd reason, not easily comprehended by the gentle Reader, pleased the Writers of Miracles and Monsters in the Animal World, to make them almost all Females: The Creature described, as being of the Human Species, and living under Water, has been almost always made a Female by these Gentlemen, and distinguished in our Language by the Name of Mer-Maid: We have an Account in the Philosophical Transactions, however, of one of these Creatures, which is, contrary to this general Custom, declared to be Male. It does not appear indeed from any Part of the Account, that the Author of it saw any Thing about the Creature that might determine its Sex, but however as he has all along spoke of it in the masculine Gender, we think it is as fairly and fully proved that it was a Male, as that it was a Fish. The Author of this marvelous History is Mr. *Thomas Glover*; it stands in the hundred and twenty-sixth Number of the Philosophical Transactions.

Mr. *Glover* tells us, that the Creature appeared to him in *Rapahannock* River in *Virginia*, and that he had so many and so favorable Opportunities of examining its Figure, that he is qualified to give a very good Description of it. It was *larger*, he tells us, *than a Man*, otherwise like, *its skin* tawny, like that of the inhabitants of that Country; its head pyramidical and without Hair; its Eyes large and black, its Eye-brows broad; its Mouth wide, with a black streak at the upper Lip, and turned up at the End like Mustachios; its Neck, Shoulders, Breast and Waist, like a Man; but its tail *like that of a Fish*. Mr. *Glover* informs us that it played upon the Water near him, and looked him in the Face: that its Aspect was *very grim and terrible*, and that it dived and rode up again, and sometime swam just under the Surface of the Water, at which Time he could observe it throw out its Arms and Draw them in again *just as a Man does*... It appeared soon after the publication of this Transactions, that the Creature was no other than an *Indian* of the Country diverting himself with swimming, and with a high Cap upon his Head made of split Wood, in the manner of our Basket Work to keep up his Hair.

To add insult to injury, Hill also wrote a book-length parody of the Royal Society's publications called *Lucina sine Concubitu* (1750), arguing that a woman may conceive a child without "any Commerce with a Man" by exposure to seminal animalcules borne on the West Wind. As part of this elaborate satire, Hill describes a machine that will capture these dangerous particles. It is "a wonderful cylindrical, caloptrical, rotundo-concavo-convex Machine... being hermetically sealed at one End, & electrified according to the nicest Laws of Electricity."

While a thorn in the side of the Royal Society of London, John Hill did have a more constructive side as well: he helped introduce the Linnaean system of botanical classification to England and performed what is probably the first scientific study of the health effects of tobacco use.

Not surprisingly, despite his achievements, the Royal Society never made Hill a fellow. Yet Hill must have gotten some satisfaction from the Society's response to his verbal assault, for in 1752, two years after *Lucina sine Concubitu* and one year after his *Review*, the Society established a committee of five members to read and select articles for publication on the basis of "the importance and singularity of the subjects, or the advantageous manner of treating them." The Society, however, made it clear that this committee and the Society itself was not answerable "for the certainty of the facts, or the propriety of the reasonings"; that remained the author's responsibility. In 1831, *Philosophical Transactions* began sending articles to referees who were not on the committee.

2

FIRST FRENCH
PERIODICALS

JOURNAL DES SÇAVANS (Journal of the Learned) debuted in France, January 1665. Its founder and first editor was a Parisian government official, Denis de Sallo. Before starting his weekly journal, de Sallo had had copyists transcribe passages from books and letters so that he could familiarize himself with a potpourri of topics. Recognizing that others might also profit from his scholarly intelligence gathering, he decided to publish weekly book reviews and news in science, as well as in theology, law, and anything else that might pique the curiosity of the learned in France. In short, all the scholarly news that's fit to print.

Because science was only one subject of many that the *Journal des Sçavans* covered, scholars deservedly bestow the honor of first "science" journal on *Philosophical Transactions*, even though de Sallo founded his journal two months earlier.

The inaugural issue of the *Journal* appeared as a quarto pamphlet of fourteen pages, with the compiler given as "Sieur de Hédouville," a pseudonym for de Sallo. The eight short articles that follow Sallo's prefatory remarks begin with titles in Latin (except for one in French), followed by the main text in French. Only two articles summarize scientific-type reports: one on astronomical observations of Jupiter and Saturn made by Giuseppe Campani with his new telescope lenses, another on René Descartes's distinctions between the functions of the human body and the soul, as well as his observations on how the human fetus forms. The first issue closes with a vivid description of a Siamese twin, extracted from a letter sent by a correspondent in Oxford. Only one article in the second issue is of a scientific nature, a review of a medical text by Thomas Willis, *Cerebri Anatome*, on the anatomy of the brain and nervous system. Indeed, according to one modern scholar's count, the *Journal* reviewed or summarized some 80 publications in its first three months, but carried only four original scientific articles—a sharp contrast to the contents of *Philosophical Transactions*, which was essentially all scientific or technical from the start.

De Sallo's reign as publisher and editor was short-lived, probably because he did not shy away from controversial articles on religion and the law. Notwithstanding a caveat in the preface to the first issue of *Journal des Sçavans* ("No one ought find it strange to see herein opinions different from their own, touching the sciences, since we profess to report the impressions of others without guaranteeing them"), the government censor suppressed publication after a mere three months. Nine months later, in January 1666, publication resumed with the same title but a replacement editor, Jean Gallois, who gave the journal a more scientific slant. While never on a par with *Philosophical Transactions*, this journal did report several important discoveries in the world of seventeenth-century European science, including Christiaan Huygens's invention of the pendulum clock, Ole Roemer's calculation of the speed of light, and Giovanni Cassini's planetary observations.

Up to the early eighteenth century, the *Journal des Sçavans* was the only periodical that routinely published the research findings of the Académie Royale des Sciences (which we hereafter refer to as simply the "Royal Academy") in Paris. During the late seventeenth century, the Royal Academy did bankroll two sumptuous volumes on plants and animals, as well as a large collection of articles on mathematics and physical science, all written by "gentlemen of the Royal Academy." But the Academy had no outlet for regular publication like the monthly *Philosophical Transactions*.

That changed in 1702, when the Academy started the annual publication of its research in what it called *histoire* and *mémoires*. The *histoire* contained summaries or extended abstracts of the annual *mémoires*, which were full-length articles (some running several hundred pages) arranged in roughly chronological order. The Academy apparently never rushed anything into print, with the annual *histoire* and *mémoires* often appearing several years after the time period covered. Indeed, it was not until the 1730s that the Académie published the full fruits of its labors during the seventeenth century, in the eleven volumes of *Histoire et Mémoires de l'Académie Royale des Sciences depuis 1666 jusqu'à 1699*.

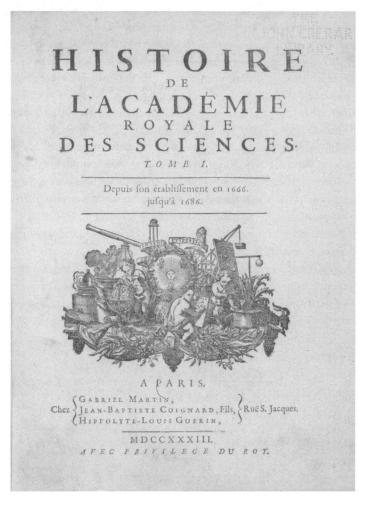

FIGURE 2-1 · Title page from first volume of *Histoire*,
Royal Academy of Sciences in Paris. Courtesy of Special Collections
Research Center, University of Chicago Library.

With good reason, historians of science view the Academy's periodical as the preeminent scientific journal in the eighteenth century. Its many contributors over that century include the upper crust of European science: the naturalist Comte de Buffon, the mineralogist Réné Antoine Ferchault de Réaumur, the geologist Nicholas Desmarest, the mathematician Jean le Rond d'Alembert, the mathematical astronomer Pierre Simon Laplace, and—the jewel in the crown—chemist Antoine Laurent Lavoisier. Its prominence ended during the French revolution, some time after 1793 when the new government suspended Academy activities with the chilling edict: "All academies and literary societies patented or endowed by the nation are eliminated." In the early nineteenth century, *Comptes Rendus* became the main outlet for the revived Academy's publications.

The French Revolution also marks a temporary end to the *Journal des Sçavans*, owing to financial problems. (It was resuscitated briefly in 1797, then permanently in 1816 as *Journal des Savants*.) During the eighteenth century, this journal remained a general scholarly publication, devoting less than half its content to scientific matters. There is really no modern equivalent to this type of learned journal, one in which the writings of scientists rubbed pages with those of scholars having an interest in history, law, theology, and philosophy.

How did the early French scientists write and argue? Differences in composition between the Royal Society of London and the Royal Academy in Paris are reflected in differences in style and substance. The Royal Society was a society of gentlemen; the Royal Academy was a society solely of natural philosophers, the best of the Continent. Neither incredible claims without sound evidence nor literary flourishes were tolerated by the Royal Academy, and the standards of argument were uniformly high, as we shall soon see.

CONCERNING THE ROYAL ACADEMY IN PARIS

The Sun King Shines on His Royal Academy

Jean-Baptiste Du Hamel and Bernard de Fontenelle, 1733. *Histoire de l'Académie Royale des Sciences depuis 1666 jusqu'à 1699*, Vol. 1, pp. 2 and 5–7.

Unlike the English gentleman's club that was the Royal Society, the Royal Academy was a small, tight-knit community of men hand-picked for their scientific talent, living and working together in Paris under the patronage of the sun king, Louis XIV. Science was their principal occupation, and they had world-class research facilities at their disposal. They were a seventeenth-century "think tank," if you will. Indeed, the Academy was referred to as "la Compagnie" (the Company) in its publications. And the consistent quality of the resulting publications reflects the exclusivity of this French scientific institution. In marked contrast to the *Philosophical Transactions*, the *histoire* and *mémoires* of the Academy rarely printed work by anyone but its own members.

The Academy began as the dream of Charles Perrault, better known as the author of *Cinderella*. He proposed a major cultural center drawing upon the learned in different areas of French culture—the sciences being only one such area. The royal minister Jean-Baptiste Colbert embraced the idea of a general academy for the glory of the French Crown. But this dream never materialized. Instead, a more limited academy emerged with a charter to study nature, in part, for its own sake and, in part, to aid society in better controlling the environment.

The account below describes the founding of this extraordinary startup company from the seventeenth century. It comes from a two-volume work combining historical information with short written accounts of the research conducted there from 1666 to 1699. Du Hamel and de Fontanelle begin their history with the burst of activity that signaled the beginning of organized science in France:

It is almost exclusively in this century that we see a renaissance in the study of mathematics and the physical sciences. Descartes and other important personages have achieved such great success that the course of these studies altered dramatically. On account of their efforts, we have left behind these sterile sciences, stagnant for several centuries. Today the tyranny of words is over. The things themselves are the preferred objects of study. Principles are established and applied, and as a consequence of this process, the sciences advance. Reason has taken the place of authority. What has been accepted without question is now scrutinized and often rejected. We dare to look into the things themselves, Nature herself, rather than the ancients. In return, she reveals herself more easily. Often enough, when we are bombarded by new experiences that trigger their exploration, she reveals another of her secrets.

Having relayed the standard version of the "new philosophy," the authors then describe the formation of the Royal Academy on the initiative of the royal minister Colbert:

This royal minister . . . formed the initial project of an Academy composed of all those knowledgeable in every sort of learned and creative endeavor. Historians, grammarians, mathematicians, philosophers, poets, orators were to be equally represented in this august body, gathering together and combining their very different talents. The king's library was designated the common meeting place. Historians would assemble there on Mondays and Thursdays; creative writers, Tuesdays and Fridays; mathematicians and those interested in the physical sciences, Wednesdays and Saturdays. Thus no day during the week would be idle: and since there was some common ground among the different groups, they resolved to hold a general assembly on the first Thursdays of the month, where the Secretaries would report the judgments and decisions of their particular groups, and where each would be able to ask for a clearing up of any difficulties . . .

This project was never realized. At first, the Academy excluded the members who were connected with history. They were not able to refrain from asking questions that were too politically sensitive. Because of a long tradition of separation, those in literature did not think of themselves as part of any universal academy. Since they were nearly all members of the French Academy established by Cardinal Richelieu, they explained to Monsieur Colbert that he need not create two different companies that had the same purpose, the same professional concerns, and nearly all the same members . . .

There remained from the debris of this projected great academy, a group of six or seven mathematicians: Monsieurs Carcavi, Huygens, Roberval, Frenicle, Auzout, Picard and Bluot. They assembled outside Colbert's library, and initiated scientific investigations in the month of June in the year 1666.

It seemed that the heavens wished to favor this Company born of mathematicians by the arrival of two eclipses within 15 days of each other, a very short time span to have two such events, and we know how precious eclipses are to Astronomers from all the usages that they put them to.

JOURNAL OF THE LEARNED

A Monstrous Birth in Oxford
....

Anonymous, 1665. "Extrait d'une letter escrite d'Oxfort, le 12 Novembre 1664" (Extract of a letter written from Oxford, 12 November 1664). *Journal des Sçavans*, pp. 11–12.

Perusing the tables of contents from the early scientific literature, whether from France or England, one notices a minor thematic strain: the reporting of "monsters" in the plant and animal world, and strange events and objects in general. As evidence, one need look no further than the first issues of *Philosophical Transactions* (featuring a monstrous calf) and *Journal des Sçavans* (a Siamese twin). Part of the fascination with the monstrous and bizarre stemmed no doubt from a *National Inquirer* mentality and desire to titillate the audience. But another part derived from genuine curiosity about all things in nature: the normal and abnormal, the everyday and extraordinary. That motivation caused Robert Boyle to analyze his urine, Guillaume Homberg to analyze fecal matter, and Joseph-Guichard du Verney to study a petrified cow brain—all reported on in the scientific literature with studied indifference and the utmost seriousness.

Waxing philosophical about "une rose monstrueuse" in the 1707 volume of *Mémoires de l'Académie Royale des Sciences*, Nicolas Marchant muses on why such monsters are important to science at the time: "we have paid little attention there [to deviations in the plant world]: but a scientist ought neglect nothing, especially when he is able to find in ordinary things reasons behind some surprising effects that different combinations produce in nature. That is what motivated me to report on the structure of a rose that appeared to be unique, and dignified some reflections for those who study nature." At the Royal Academy and Royal Society, the monstrous and strange became clues that might lead to secrets about the natural world.

The example below, from a letter sent to the editor reporting a monstrous birth, closes the first issue of *Journal des Sçavans:*

It was about three weeks ago, near the town of Salisbury, that the wife of a stableman, having given birth to a girl, gave to the world, an hour later, an infant with two heads diametrically opposed, four arms, the same number of hands, one abdomen and two feet. For a long time they were in a quandary how to baptize this creature. But at last, as they decided that she was really two people, they gave her at baptism the names of Marthe and Marie. She took nourishment from the two heads, and excreted in the ordinary way. Of her two visages, one was much more animated and good-humored than the other. This monster only lived for around two days. Marthe, who appeared always more full of life than Marie, died first, and Marie, a quarter of an hour later. Both were autopsied by a physician, who found two perfect heads and two chests. But the bowels were defective, joined to the *ductus Communis*, and having only one *intestinum caecum*, one bladder and one uterus. But there were two livers, two spleens, and two stomachs. They have embalmed this monster, and preserved it carefully. Our doctors have remarked that Jacques Rueff, in his book *de Conceptu & generatione hominis*, printed in Zurich in 1554, speaks of a monster nearly the same as this one. It was born in 1552 near Oxford. The description that this doctor gave of it closely approaches the one they have sent us from Salisbury. Rueff reports that, of the two parts of the monster that he describes, one was much more lively and more good-humored than the other. But what is really unusual is that he maintains that one survived the other by 15 days.

The presentation is matter of fact; the aim is not to titillate the public, nor to solve some limited research problem of interest to Royal Academy members, but by describing the findings from dissection and careful inspection, to better understand human anatomy and the cruelties of nature.

A Chameleon Properly Described

Anonymous, 1669. Review of *Description Anatomique d'un Caméléon, d'un Dromadaire, d'un Castor, d'un Ours, & d'une Gazelle* (Anatomical Description of a Chameleon, a Donkey, a Beaver, a Bear, and a Gazelle) by Claude Perrault. *Journal des Sçavans*, pp. 37–42.

This selection comments upon and summarizes the first book published by the Royal Academy, offering detailed descriptions with illustrations of the five animals listed in the title. After having died in King Louis's menagerie at Versailles, these animals had been dissected and examined at the Academy. In this review appearing in the 1669 *Journal des Sçavans*, the anonymous author (G. P.) noted that "the animals described in this book have been examined with very great exactitude; the writers remark here on a number of things that naturalists have not observed before, or in which they have been wrong." As pointed out in the review, a key motivation behind this publication was correcting errors in the past, a rhetorical move commonplace at the time:

> In the first description, we see that which most authors have said about the chameleon is not found to be true. If we listened to what they said, we would believe that this animal is terrifying. Pliny made it the size of a crocodile. Panarolus armed its back with sharp barbs to defend against enemies: and Solin, so as to render it more frightful, said that it always has the mouth open, whereas the chameleon dissected at the Royal Library was not more than a foot long from tail to snout; moreover, barbs were completely absent from its back; in fact, the spiny dorsal protuberances of these vertebrates are square. And far from having its mouth always open, it always had it so completely closed while alive that it was difficult to see the separation between its lips. But perhaps neither Pliny nor Solin ever saw a chameleon.

The anatomical menagerie grew from the initial five to over twenty animals, published as a book in 1671, then fifty animals in the third tome of the *Mémoires de l'Académie Royale des Sciences depuis 1666 jusqu'à 1699*, released to the public in the 1730s. In each version of this natural history, the chameleon appears first and has the longest verbal description by far, occupying nearly half the 1669 book under review. As author Claude Perrault notes, "There is hardly an animal more famous than the chameleon. Its changing color and peculiar manner of eating have given to students of nature in past centuries much to admire and study."

FIGURE 2-2 · Chameleon before and after dissection. Courtesy of Special Collections Research Center, University of Chicago Library.

The two-panel illustration above, an engraving made for the *Description Anatomique* by the master draftsman Sébastien Leclerc, depicts the chameleon posed in a "natural" setting beside a view of its skeleton and individual organs, along with a cut-away view after dissection. This pair of illustrations, and ones just like it for the other animals, accompany extensive anatomical descriptions. As in Boyle's famous engraving of his air pump given earlier, the drawing shows not only the object as a whole, but its various parts, implicitly differentiating between what *anyone* can see and what a *natural philosopher* sees.

Roemer's Bold Prediction

Ole Roemer, 1676. "Demonstration touchant de la mouvement de la lumière" (Demonstration of the speed of light). *Journal des Sçavans,* pp. 233–36.

In 1705 Edmund Halley predicted that the comet that had streaked across the sky in 1607 and 1682 was one and the same and would return again in 1758. And it did as predicted, about 16 years after Halley's death.

An equally stunning but less heralded prediction from the same era came from the Danish-born Ole Roemer. Before his fellow members of the Royal Academy on September 1676, he predicted that the eclipse of one of the moons of Jupiter would occur on November 9, 1676, at 5 hours, 35 minutes, and 45 seconds; that is to say, exactly ten minutes after the time astronomers had computed on the basis of an eclipse in August. Roemer used the regular pattern in the observed temporal discrepancies for the eclipses of Io to argue that these occurred because light has a definite speed, and is not infinite as previously believed. As Io and the earth grow farther apart, then closer together, the time for light to reach the earth changes accordingly. His prediction was confirmed by the Royal Academy Observatory in Paris. According to Roemer's calculations, the speed of light is approximately 215,000 kilometers per second, reasonably close to the current value of roughly 300,000.

Roemer's work on the speed of light was reported on December 7, 1676, in the *Journal des Sçavans*. Roemer's is an elegant argument, relayed in the third person, as was the journal's style at the time. First, a hitherto unsolved problem of general interest is broached, and then the solution summarized:

> For some time now, scientists have been laboring to decide by observation or experiment whether the action of light is instantaneous no matter what the distance, or whether its transit takes time. Mr. Roemer of the Royal Academy of Sciences has found a way to resolve this problem, drawn from his observations of Jupiter's first moon, Io. By this means he shows that to travel a distance of about 7,500 miles, nearly the diameter of the earth, light takes far less than a second.

But this answer alone is not enough; inquisitive readers want to see the proof. This Roemer provides by imagining Io's circuit around Jupiter as a gigantic celestial clock. But if we tell time by Io, he discovers, we find that our clock is not accurate; as the earth moves away from Jupiter, we lose time; as it moves toward Jupiter, we gain that time back. This would not happen if the speed of light were infinite:

> Let A be the sun, B Jupiter, C the first moon, which disappears in Jupiter's shadow-cone only to reappear at D [consult accompanying figure]. Let EFGHKL be the earth, which we find at various distances from Jupiter during its orbit around the sun.
>
> Let us also suppose that the earth is at L moving toward a second point, K. At this point, Io is visible during its reappearance at D. Subsequently, after about forty-two and a half hours, that is, after a revolution of this moon, the earth now finding itself at K, Io is visible again at D. It is clear that if light is not instantaneous, between its appearance at L and at K, Io will be seen returning at D later than if the earth had been at rest at L. The movement from L to K will be such that Io's observed reappearances will be slowed by as much time as it takes for the light to traverse the distance from L to K.
>
> By contrast, in the side opposite to LK, FG, where the earth is actually moving toward Jupiter, that is, toward the source of light, the cycles of disappearances of the moon will be shortened as, in LK, those of the reappearances have been lengthened.

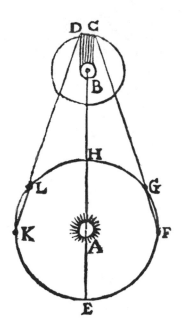

FIGURE 2-3 · Diagram used to
calculate speed of light. University
of Minnesota Library.

In the nearly forty-two and a half hours that the moon takes for each
of its revolutions, the distance between the earth and Jupiter, as the earth
travels in its orbit, changes at least 210 earth diameters. Because of this, it
follows that if for the value of each earth diameter, a second is needed, light
will take three and one half minutes to traverse each of the intervals *GF,* and
KL, a period that will create a difference of seven minutes between the two
revolutions of Io, the one observed in *FG,* the other in *KL,* although no
difference at all may be perceived at each individual sighting.

This lack of perception is a consequence of limitations in the accuracy of
seventeenth-century instrumentation. But Roemer compensates by cumulating
his observations:

It does not follow, however, that light has an infinite velocity: for after
having examined the matter more closely, Mr. Roemer found that what was
not perceptible in two revolutions, became very considerable when many

were taken together. For example, forty revolutions observed from the side
F were perceived as shorter than that of forty others observed from the
other side in any place of the zodiac in which Jupiter is encountered. Indeed,
when Jupiter and the earth are in opposition—a difference of double the
distance from the earth to the sun—the expected eclipse of Io is consistently
twenty-two minutes late.

Roemer's calculations then led to a prediction with a very satisfying result:

> The need for this new calculation of the speed of light is established by
> all the observations that have been made at the Royal Academy and at the
> Observatory for the last eight years, and lately has been confirmed by the
> eclipse of Io observed at Paris on last November 9, at 5 h. 35' 45" p.m., ten
> minutes later than had been expected, deducing this new time from those
> times that had been observed in August, when the earth was much nearer to
> Jupiter. Mr. Roemer had predicted this at the Academy at the beginning of
> September.

In the final stages of his argument, Roemer deals with and dismisses other
possible explanations:

> But in order to remove any lingering doubts concerning this inequality
> caused by the finite velocity of light, Mr. Roemer showed that it could not
> come from any orbital eccentricity or other cause usually brought forward
> to explain the irregularities of our moon and of the other celestial bodies.
> Yes, Io has an eccentric orbit; yes, its revolutions are advanced or retarded
> to the extent that Jupiter approaches or distances itself from the sun; yes,
> the revolutions of the Primum Mobile are unequal. Nevertheless these three
> causes of inequality do not prevent the first from clearly manifesting it-
> self.

The defects of his instruments led Roemer to an under-calculation of the velocity
of light. But while his instruments were crude by modern standards, the written
argument by which he achieved his result was, and remains, exemplary.

Newton Makes a Change

Anonymous, 1688. Review of *Philosophiae Naturalis Principia Mathematica*
by Isaac Newton. *Journal des Sçavans,* pp. 153–54.

Isaac Newton's *Philosophiae Naturalis Principia Mathematica* (Mathematical Principles of Natural Philosophy) stands as an intellectual feat unsurpassed by an individual investigator in the history of science. In the first two books of this awe-inspiring work, Newton established the laws of motion, applicable to any body, and in Book III used these laws to explain such otherwise inexplicable phenomena as the planetary orbits and the tides. But this eminence does not mean that the *Principia* was not a work in progress, that its first edition represented Newton's final thoughts. Some time between the first and second editions of his great work, Newton emended the crucial beginning of Book III. He crossed out the heading "Hypotheses" and substituted "Rules of Reasoning in Philosophy." He wanted no misunderstanding as to the nature of his laws of motion; they were not hypotheses, he says in a letter to his editor, Roger Cotes, but principles "deduced from phenomena and made general by induction, which is the highest evidence that a proposition can have in [natural] philosophy."

Newton might have been compelled to make that change in response to an early French review of the first edition:

Mr. Newton's is the most perfect work of mechanics that can be imagined; it is not possible to make demonstrations more precise or more exact than those that he gives in the two first books concerning gravity, levity, elastic force, the resistance of fluid bodies, and the attractive and impulsive forces that are the principal foundation of any natural science. But we must admit that these demonstrations belong to mechanics proper and not to natural science itself: the author himself recognizes at the end of page 4, and beginning of 5, that he has not considered his principles as a natural scientist, but merely as a mathematician.

He admits the same thing at the beginning of the third book in which he endeavors, nevertheless, to explain the system of the world. But this system is only derived from hypotheses that are, for the most part, arbitrary, and that as a consequence are able to serve only as a foundation to a treatise on pure mechanics. That appears most evident when we look at the example of the ebb and flow of the seas. Mr. Newton bases the explication of the inequality of the tides on the principle *that all celestial bodies gravitate towards one another*; from that he draws the consequence that, when the moon and sun are conjoined or opposed, their gravitational forces on the earth combine and being combined, produce a very large flux and reflux of the earth's seas; on the other hand, when the sun and the moon are in quadrature [that is, at an angle of ninety degree to each other], these forces are separated: the sun raises the earth's waters while the moon lowers them. As a consequence,

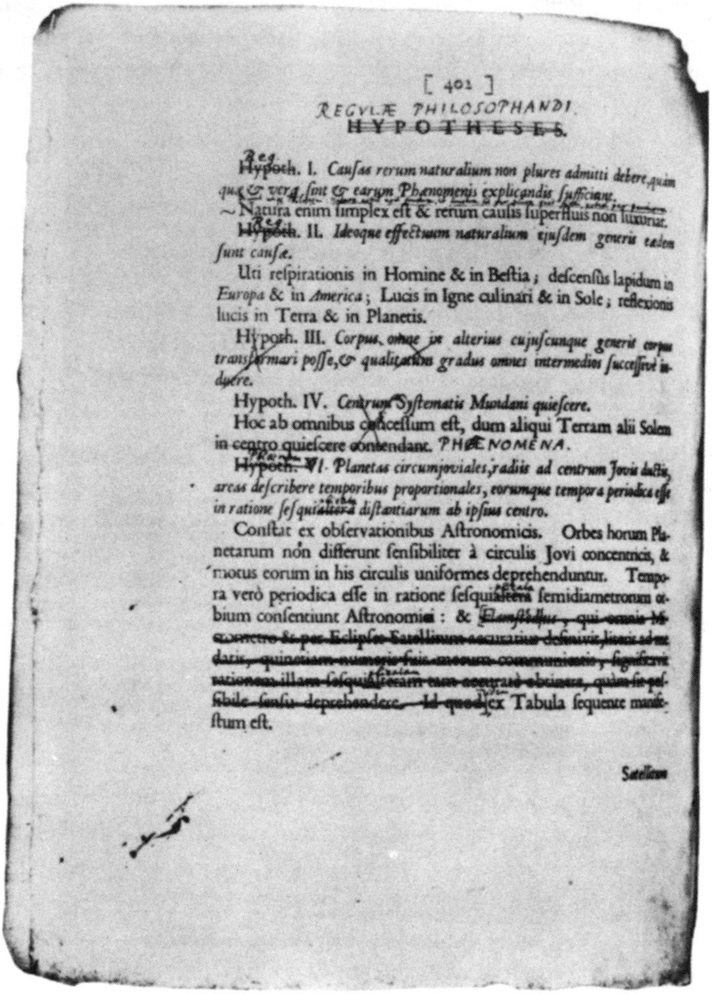

[402]

REGULÆ PHILOSOPHANDI.

~~HYPOTHESES.~~

Reg. ~~Hypoth.~~ I. *Causas rerum naturalium non plures admitti debere, quam quæ & vera sint & earum Phænomenis explicandis sufficiant.*

~ Natura enim simplex est & rerum causis superfluis non luxuriat.

Reg. ~~Hypoth.~~ II. *Ideoque effectuum naturalium ejusdem generis eadem sunt causæ.*

Uti respirationis in Homine & in Bestia; descensûs lapidum in *Europa* & in *America*; Lucis in Igne culinari & in Sole; reflexionis lucis in Terra & in Planetis.

Hypoth. III. *Corpus omne in alterius cujuscunque generis corpus transformari posse, & qualitatum gradus omnes intermedios successivè induere.*

Hypoth. IV. *Centrum Systematis Mundani quiescere.*

Hoc ab omnibus concessum est, dum aliqui Terram alii Solem in centro quiescere contendant. PHÆNOMENA.

~~Hypoth.~~ VI. *Planetas circumjoviales, radiis ad centrum Jovis ductis, areas describere temporibus proportionales, eorumque tempora periodica esse in ratione sesquialtera distantiarum ab ipsius centro.*

Constat ex observationibus Astronomicis. Orbes horum Planetarum non differunt sensibiliter à circulis Jovi concentricis, & motus eorum in his circulis uniformes deprehenduntur. Tempora verò periodica esse in ratione sesquialtera semidiametrorum orbium consentiunt Astronomi: & ~~Flamstedius, qui omnia Micrometro & per Eclipses Satellitum accuratius definivit, literis ad me datis, quinetiam numeris suis mecum communicatis, significavit rationem illam sesquialteram tam accuratè obtinere, quàm sit possibile sensu deprehendere. Id quod~~ ex Tabula sequente manifestum est.

Satellitum

FIGURE 2-4 · Marked-up page from Newton's *Principia*. Reprinted with permission of the Syndics of Cambridge University Library.

the flux and reflux is very small; that is undoubtedly his assumption. But as this assumption is arbitrary, not having been proven, the demonstration that depends on it can be no more than mechanical.

To create a work more perfect than is possible at the moment, Mr. Newton must give us a physics as exact as his mechanics. He will have accomplished that when he is able to substitute the true motions of celestial bodies in place of those he has presupposed.

While the first sentence is laudatory, this review is clearly critical of Newton's *Principia*. It would have no doubt infuriated the book's temperamental author, had he read it. This is very likely, as his library contained a set of the *Journal*. Especially infuriating would be the Frenchman's notion that the Englishman's principles were "arbitrary," that he was not really a "natural philosopher," and that perhaps he will do better next time. Truly, the reviewer seems to be damning with faint praise.

Initially, the French scientific community resisted Newton's theory. French resistance to Newton was overturned only after 1738, through the efforts of three Royal Academy members, Pierre Louis Moreau de Maupertuis, the Marquis de Condorcet, and Alex Clairaut, and those of the great French writer, Voltaire. French resistance was motivated, in part, by their championship of their countryman Descartes's complete, beautiful, and mistaken world system of vortices; it was also motivated by the apparently reasonable but ultimately futile objection to a central limitation of Newton's system: his inability to give an account of the physical realization of such crucial concepts as gravity and action at a distance. While modern physicists are used to such deficiencies in grand theories, the discomfort of his contemporaries with such "incomplete" explanations is understandable.

MEMOIRS OF THE ROYAL ACADEMY OF SCIENCES

Illustrating Botany

Denis Dodart, 1676. "Mémoire pour servir à l'histoire des plantes" (Memoir concerning the history of plants). *Mémoires de l'Académie Royale des Sciences depuis 1666 jusqu'à 1699*, Vol. 4, pp. 121–333.

In contrast to the relatively short English "account" or "narrative" typical of *Philosophical Transactions*, the French *mémoires* run anywhere in length from four or five pages to the length of a full-size book. Indeed, the cited botanical article, well over two hundred pages, first appeared in book form in 1676, then in 1731 as part of the *Mémoires de l'Académie Royale des Sciences depuis 1666 jusqu'à 1699*. The primary "author" was academy member Denis Dodart; however, Dodart notes that "This document is the work of the entire academy. There is no member who has not composed some of it, been a judge of it, and at least contributed some opinions."

Appended to this memoir are thirty-eight engraved plates of plants in black and white (example on p. 56). In contrast to the Royal Society of London, largely financed by membership dues, the French government under Louis XIV financed its scientific society, according to historian of science Alice Stroup, "at a level similar to the annual income of the wealthiest monastery in France." That meant funds were more readily available for the drawing and engraving of illustrations, a very expensive proposition in the seventeenth century.

Although depicting and describing a natural object in a realistic manner for a scientific publication may seem like a relatively straightforward task, it is not. How do you convey the actual size of the object as opposed to its size on the page? For flora and fauna, at what stage of development do you represent them? For plants, do you depict what is below the ground as well as above? And what about color or features not visible except under close inspection? Similar problems exist on what to include and what to omit from the text. Without established conventions for scientific illustration and description, the early practitioners had to improvise. Dodart discussed some of the problems and his solutions at the start of his long memoir. (We have left the botanical names in original French, unless the identification is perfectly clear, despairing of identifying these plants by their proper English names.)

We have taken advantage of this large volume to make the largest plates possible. As a result, we were able to represent plants of middling size in their actual dimensions. When we came upon a plant that was twice the size of our page, or nearly so, one that we could divide in two without rendering it unrecognizable, we ordinarily depicted it in two parts on the same plate.

But because there are many plants larger than this volume, like the *Pancratium*, the *Morelle of Virginia*, and of course, the trees, we thought it appropriate to add to the depiction of the plant one of its components in its actual dimensions, as a basis by which to judge the true dimensions of the whole plant.

This is accomplished in two different ways. Plants that grow not upward but parallel to the ground, both those that sport and do not sport stems, always leave an empty space at the top of the plate, even after we depict them as large as we can, given the limits of the dimensions of our page. In this empty space, we will be able to depict, for example, the spike of the acanthus, actual size, or the disk of the flower of the *Carline,* or indeed, some other part. But for those whose shapes and features are such that they cannot be represented actual size without exceeding the limits of the entire page, as the *Morelle of Virginia*, the *Rose d'Outremer*, the *Belveder*, and all the trees, we will depict in the foreground and at the top of the plate some aspect in its actual size, while the plant as a whole is depicted in a smaller scale in the distance. We will only do this for those trees with a feature particularly worthy of note; for example, the tamarind and all of the conifers.

We add to the principal depiction of each plant that of its seed, either by itself if naked, or with its covering membrane and other components. We also believe we have an obligation to add to the mature plant its juvenile equivalent when that equivalent is different enough to cause confusion. . . .

As we cannot at present print in color and as hand painting is very time consuming, and not always successful, we believe we have made up for this difficulty in some measure by taking care that the gradations of colors will be expressed as well as can be in a black-and-white engraving. Thus we treat dark green and light green differently, white flowers differently from their deeply tinged counterparts.

The Wisdom of Nature Betrayed

Jean Méry, 1693. "Observations de deux foetus enfermés dans une même envelope" (Observations of two fetuses enclosed in the same membrane). *Mémoires de l'Académie Royale des Sciences depuis 1666 jusqu'à 1699*, Vol. 10, pp. 324–25.

On May 15, 1693, Jean Méry addressed the learned members of the Royal Academy on the cause behind the death of twin fetuses. He even brought the fetuses with him for purposes of display. In our selection, Méry lectures on the cause of death, not only the specific case before them, but of twin fetuses in general:

> Though the two infants given by Monsieur Méry in the figure are not abnormal, the manner in which they were enveloped is very rare and consequently very remarkable . . . [The author] provided the meeting of the *Royal Academy of Sciences* with the opportunity of viewing the two fetuses depicted here. There he remarked on the wisdom of nature in its foresight in routinely enclosing each fetus in its own membrane. Thus separated, their umbilical cords cannot interlace; when, instead, the two fetuses are enclosed in a common membrane, they easily intertwine their cords by their movements and consequently suffocate; as actually happened to the fetuses represented here, whose cords obstructed each other's functions, formed a knot that prevented the blood from circulating from the placenta to their vessels, and caused their death.

FIGURE 2-5 · Plant from Denis Dodart's
Memoir concerning the natural history of plants.
Courtesy of Special Collections Research Center,
University of Chicago Library.

The accompanying illustration is a "realistic" drawing in which art is clearly at the service of science. Similar in style to Hooke's flea and the innards of Perrault's chameleon, each of the key anatomical features—the amnion, the chlorion, the placenta, and especially the knotting of the two umbilical chords—is highlighted by means of labeling. The cause of the fetuses' demise—the strangulation discussed in the article—is made visually the most prominent item, placed at the very center of the picture, and marked by the letter "N" for "*nœud.*" Flanking the knotted cord is a grotesque yet accurate rendering of the fetuses, dramatizing the tragic consequences of this medical phenomenon: the wasting and ultimate death of the fetuses resulting from insufficient oxygen and

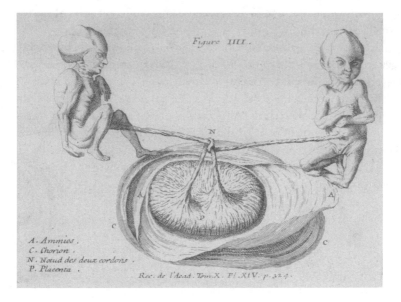

FIGURE 2-6 · Strangled fetuses. Courtesy of Special Collections
Research Center, University of Chicago Library.

nourishment through the cord, the human toll of which is outside the bounds
of normal scientific discourse. The nearly invisible mother is represented in the
picture by her scientifically relevant anatomical features alone. In the text Méry
does not tell us much about this unfortunate Parisian woman, other than she
gave birth to the dead twins at three and a half months into her pregnancy.

Headings, Eighteenth-Century Style

Antoine de Jussieu, 1712. "Description du Coryspermum Hyssopifolium, plante d'un
nouveau genre" (Description of Corispermum Hyssopifolium, plant of a new genus).
Mémoires de l'Académie Royale des Sciences, pp. 185–86.

Printed marginalia—or to use the academic jargon of today, "meta-textual
interventions"—hold a place of no small importance in the early scientific literature. They house the author's (and sometimes editor's) tangential thoughts,
bibliographic information related to a name or title cited nearby, short headings for large swathes of text, small line drawings that illuminate the adjacent
text, digressions, and whatever else a particular author wanted to throw into
an article but felt was not crucial enough for the center stage. Over time, all

these items would become better integrated into different parts of the text and formalized conventions would develop for their presentation.

Early scientific articles (and books as well) employed this presentational device in a haphazard fashion. Some had them in abundance; many did not. The article cited here is two pages in length and has six marginal notes.

The first marginal note, "31 August 1712," gives the date the article officially entered the Academy records. All *mémoires* carried such a marking adjacent to the first line of text, in part for the author's and Academy's protection in any priority disputes (a practice still widely followed in the scientific literature). Immediately below the date comes the word "Etymology," the first of five substantive marginalia. It is accompanied by this sentence in the text:

> The fruits of this plant look so much like a bug in shape and color, I believed I could not give it a name more fitting than that of *Corispermum*, which signifies seed of a bug in Greek.

The next marginalia is "Character," accompanied by this sentence in the text:

> The *Corispermum* is a genus of plant whose flower (1) is without a calyx, and is composed of two opposing petals (2), between which rises to a stamen (3) and a pistil (4), which becomes a rounded fruit at its base, convex on one side (5), a little concave on the other (6), and bordered with a leaf.

About mid-page in the margin is the word "Description," accompanied by this paragraph in the text:

> This plant rises to the height of about a foot: the root is sometimes simple, sometimes branched, other times a little twisted, with a length from two to six inches, covered with some hairy fibers, thickened at the annulus or ring by two or three lines [1/6 to 1/4 inch]. As it grows, the stem divides from the bottom to the top in alternating branches that subdivide into other smaller ones; some are full; some, flexible; some, angular; some, a little grooved along their length; some, sleek; some green; but ordinarily purplish at the base: this color extends sometimes over the whole plant, at which time it begins to subside.

"Taste" introduces the second-to-last paragraph in the text:

> This plant is an annual: chewed, it is pasty and leaves a taste that is a little bitter, biting and disagreeable.

And "Location" sends off the last paragraph in the text:

> It grows in the Lanquedoc region. Mr. Fagon, the first physician to the king, has observed it around Agde in the past. Mr. Nissole, a doctor from Montpellier, and a very skillful botanist, sent the seeds to me the past year.

Marginal notes such as these serve the same purpose as modern headings — summarizing a portion of the article's contents and alerting the reader to sections of possible interest to them.

A Precursor to the Periodic Table

Étienne François Geoffroy, 1718. "Table des différens rapports observés en chemie entre différentes substances" (Table of different relationships observed in chemistry amongst different substances). *Mémoires de l'Académie Royale des Sciences*, pp. 202–12.

In the early eighteenth century, Geoffroy systematically arranged "substances" in a table as a means of better understanding and representing chemical relationships. His is the first in a long line of "affinity tables" ultimately leading to that most famous chemical table of all, the periodic table of elements, almost 150 years later. In his article Geoffroy is careful to avoid the terms "affinity" or "attraction," speaking instead of "relationships" (*rapports* in French). Presumably he wishes to avoid reference to these "occult properties." Still, like the periodic table, this table is animated by theory. It is also an excellent example of the use of tabular presentations for pedagogical and heuristic purposes:

> Concerning the laws of these relationships I have observed that when two substances disposed to unite are combined and others are brought near them or mixed with them, they join to one of them and loosen the grasp of the other. In addition, several others never join with them nor loosen the grasp of the existing combination. From this it appears to me that we can conclude with some certainty that those substances that were joined to one of the two had more relationship of union or disposition to unite to it than the substances which were displaced, and I believe that we can deduce from these observations the following proposition, which is very broad, although we cannot call it general without having examined all possible combinations to assure that there are no genuine exceptions.
>
> *Whenever two substances having some disposition to unite, the one with the other, are found in combination and a third that has a closer relationship with one of the two is added, the third will unite with that one, separating it from the other.*

This proposition is valid for a very great range in chemistry, where we meet, so to speak, some effects of this relationship at each step. It is this property on which depend most of the hidden motions that follow from the intermingling of substances, motions that would be nearly impenetrable without this key to their understanding. But, as little is known about the order these relationships obey, I believe that it would be very useful to identify the principal materials that we are accustomed to work with in chemistry, distinguishing among them, and preparing a table of them, where at a glance, everyone can see the relationships that hold among them.

In this table that I present to you today [August 27, 1718], I display the different relationships that I have collected from many of my own experiments and from the observations of other chemists.

By means of this table, beginners in chemistry will be able quickly to master the relationships that different substances have one with the other, and more experienced chemists will find an easy method for discovering exactly what happens in many of their more complex chemical operations, and to predict the results from the combinations of different mixed substances.

The following pages of the cited article describe the table's contents in great detail. The star of the show, the actual table reproduced on the next page, is relegated to the end of the article, in conformance with eighteenth-century printing practices. It has sixteen columns, each headed with symbols for a chemical reagent defined at the bottom of the table. Below each heading—acid, alkali, metal, or water—are arranged the symbols for those substances that can form combinations with it, in descending order of strength of combination. Many of the symbols are drawn from alchemy; none say anything about the composition of the substances they represent. The table is not about chemical substances at all, but about their transformation.

Heroic Science

Pierre-Louis Moreau de Maupertuis, 1737. "La figure de la terre, determinée par les messieurs de l'Académie Royale des Sciences, qui ont mesuré le degree du Méridien au Circle Polaire." *Mémoires de l'Académie Royale des Sciences*, pp. 389–466. Translation in *The Figure of the Earth, Determined from Observations Made by Order of the French King, at the Polar Circle*. London: T. Cox (1738).

A controversy having arisen in Europe over the shape of the earth—was it flattened at the poles, or at the equator?—the decision was reached to resolve all differences by empirical means: actually to measure the earth at the North

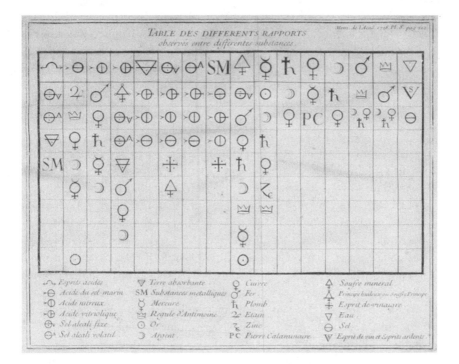

FIGURE 2-7 · Table of different chemical relationships. Courtesy of Special
Collections Research Center, University of Chicago Library.

Pole and at the equator. The French polar expedition to Lapland, a team of
Royal Academy scientists led by Pierre-Louis Moreau de Maupertuis, started
out after the parallel French expedition to Peru, whose goal was the measurement of equatorial girth. Despite its late start, it was Maupertuis's expedition
that returned to France first. Taking advantage of the opportunity afforded
by his early return, Maupertuis determined to publish his results as soon as
possible in book form—the Academy's annual *Mémoires* were routinely several
years behind schedule (those for 1737 did not make it into print until 1740).
As an index of international interest, the book was soon translated into English and Latin. Maupertuis had not only "scooped" the Peru expedition but
successfully trounced his opponent in the controversy, Jacques Cassini, son of
Jean-Dominque, the founder of the Paris observatory, who had opted for an
earth flattened, not at the poles, Maupertuis's position, but at the equator.

Geodesy may seem an unlikely subject for a best-seller. But this hasty judgment seriously underestimates Maupertuis's literary skill. His book is a heroic

travel rollercoaster, a safe but exciting ride through the dangers and discomforts of doing science in the polar regions—the voracious insects in summer, the biting cold in winter. (The technical details of the expedition and their accompanying mathematical graphics are tucked safely into appendices.) We reproduce a typical example of Maupertuis's narrative verve, a brilliant amalgam of exotic anecdote, scientific boasting, incidental polemic, and last but not least, scientific measurements:

> Last Summer we had omitted an Observation of very small moment, and which might have been overlooked in a Country where the making of Observations was less troublesome than here. We had forgot to take the height of an Object that we made use of in measuring, from the *Top of Avasaxa*, the Angle between *Cuitaperi* and *Horrilakero*. And to measure this height, I undertook to go with a Quadrant to the top of the Mountain; so scrupulously careful were we that nothing should be wanting to the perfection of our Work. Imagine a very high Mountain, full of Rocks, that lie hid in a prodigious quantity of Snow, as well as their Cavities, wherein you may sink thro' a Crust of Snow as into an Abyss, and the Undertaking will appear scarce possible. Yet there are two ways of performing it. One is by walking, or rather sliding along, upon two strait Boards eight foot in length, which the *Finlanders* and *Laplanders* use to keep them from sinking into the Snow. But this way of walking requires long practice. The other is by trusting yourself to a *Rain-Deer,* who is used to perform such Journeys.
>
> The Machine which these Animals draw is a sort of Boat scarce large enough to hold the half of one's Body. As this travelling in the Snow is a kind of Navigation, that the Vessel may suffer the less resistance in its Course, it has a sharp Head and a narrow Keel, like an ordinary Boat; and on this Keel it tumbles so from side to side, that if one takes not good care to ballance himself, it will be in danger of over-setting every moment. It is fixt by a thong to the Collar of the Rain-Deer, who, as soon as he finds himself on a firm beaten Road, runs with incredible Fury. If you would stop him, it avails little to pull a sort of Rein which is tied to his Horns. Wild and unmanageable, it will only make him change his Track, or perhaps turn upon you, and revenge himself by kicking. If this happens to a *Laplander,* he turns the Boat over him, and uses it as a Buckler against the Attacks of the Rain-Deer. But as we were Strangers to this Adresse, we might have been killed before we could put ourselves in such a posture of Defence. Our only Defence was a little Stick each of us had got in his hand, by way of Rudder to steer our Course, and keep clear of the Trunks of Trees. In this manner

was I to climb *Avasaxa*, accompanied by the Abbé *Outhier*, two Men and a Woman of the Country, and Mr. *Brunnius* their Curate.

The first part of our Journey was done in a moment; our flight over the plain beaten Road from the Curate's House to the foot of the Mountain can be compared only to that of Birds. And though the Mountain, where there was no track, very much abated the speed of our Rain-deer, they got at length to the top of it; where we immediately made the Observation for which we came. In the mean time our Rain-deer had dug deep holes in the Snow, where they browzed the Moss that covers the Rocks. And the *Laplanders* had lighted a great Fire, at which we presently joined them to warm ourselves. The Cold was so exceeding great, that the Heat of this Fire could reach but to a very small distance. As the Snow just by it melted, it was immediately froze again, forming a hearth of Ice all round.

If our Journey up hill had been painful, our Concern now was lest our return should be too rapid. We had come down a steep, in Conveyances, which, though partly sunk in Snow, slid on notwithstanding, drawn by Animals whose Fury in the Plain we had already try'd, and who, though sinking in the Snow to their Bellies, would endeavour to free themselves by the Swiftness of their flight. We very soon found ourselves at the bottom of the Hill; a moment after, all this great River was crossed, and we back at the Curate's House.

Next day we finished our Survey; and had now no reason to regret the Toils we had gone thro', when we saw what surprizing exactness the measuring upon a Surface of Ice had given us, The difference between the Measures of our two Companies was but four Inches upon a distance of 7406 Toises 5 feet (14,800 yards). An exactness not to be expected, and almost incredible. We look upon it as an effect of Chance, and that there must have been greater differences, but which in the course of our Work had compensated each other; for this small difference of four Inches rose all in our last day's measuring. Each of our Bands had measured every day the same number of Toises; and the difference every day was but one Inch which the one or the other had gained. This coincidence, which was owing partly to the Ice, and partly to our own Accuracy, shewed at the same time the perfect equality of our [measuring] *Rods;* for the smallest difference in their lengths, must, in so great a distance, have been multiplied to a considerable Sum.

We already knew the *Amplitude* of our Arc; and our Figure, every other way determined, wanted only to be applied to its *Scale*, that is, the length of our Base. This we were now Masters of; and immediately found that the length of the Arc of the Meridian intercepted between the two Parallels that pass through the Observatories of *Torneå* and *Kittis* is 55023 1/2 Toises (62.5 miles). That the *Amplitude* of this Arc being 57′ 27″. the Degree of the Meridian at the Polar Circle is greater by 1000 Toises (2000 yards) than it should be according to Mr. *Cassini,* in his Treatise *On the Magnitude and Figure of the Earth.*

In this passage, Maupertuis demonstrates his stylistic agility as he moves deftly from globe trekking to globe measuring.

An Early Conservationist

Comte de Buffon (Georges Louis Leclerc), 1739. "Sur la conservation et le retablissement des forests" (On the conservation and re-establishment of forests). *Mémoires de l'Académie Royale des Sciences*, pp. 140–56.

While hardly a household name except among historians of science, Comte de Buffon was one of the giants in eighteenth-century science. His main scientific product was the 36-volume *Histoire Naturelle*, produced over a forty-year stretch (1749–89). Under that short title appear theory-based descriptions encompassing all that was known about the animal, mineral, and vegetable kingdoms. Steven Jay Gould was not exaggerating when he called it "one of the most comprehensive and monumental efforts ever made by one man (with a little help from his friends, of course) in science or literature." Like Gould himself, Buffon was a prose stylist of the highest order, and his encyclopedic volumes sold comparably with the best-selling strictly literary works of the day.

Although Buffon devoted most of his energies to writing books, he contributed a few *mémoires* to the Royal Academy. In the one cited, he tackles a problem that is not just technical but social: the destruction of French forests and practical interventions for their long-term protection. Evidently, anxiety over society's voracious appetite for natural resources long predates the modern era.

Buffon begins by arguing that a dwindling supply of high-quality timber for shipbuilding and other uses should make woodland management a major concern of the state and, indeed, of every good French citizen. He further asserts that one of the raisons d'être of the state-financed Royal Academy is to solve such important social and economic problems for the public good. In arguing for the public utility of his research project, Buffon is following an argumentative strategy still alive and well today in such diverse fields as particle physics and genetics:

The wood that was formerly very common, now barely suffices for indispensable uses, and we are threatened by its running out in the future; this could be a true loss for the State, which would be obliged to have to resort to neighbors, and to draw from them at great expense what our care and some slight economizing is able to procure for us. But it is necessary to recognize the problem at this time; it is necessary to begin as soon as today; for if our indolence continues, if the urgent desire that possesses us continues to increase our indifference to posterity, and if the policy concerning woodlands is not reformed, it is to be feared that the forests, the most noble part of the domain of our king, will become waste, and that the woodland so essential to our maritime forces will be found to have been consumed and destroyed without hope of renewal in the immediate future.

In this paragraph, Buffon uses a rhetorical tactic unusual for modern scientific prose, parallel syntax for the purpose of emphasis: *"it is necessary* to recognize" and *"it is necessary* to begin" followed by *"if* our indolence," *"if* the urgent desire," and *"if* the policy." (Rhetoricians refer to this pattern as "anaphora" repetition at the start of successive phrases or sentences.) The first half of the sentence warns his readers that deforestation is a critical problem; the second half foreshadows the consequence of the three "ifs": if action is not taken at the present time, the future of France could be bleak.

The second paragraph, after repeating the problem and call for action, turns to another distinguished Academician who had done preliminary work on solving the problem eighteen years earlier and, lastly, to Buffon's own research following up on his colleague's initial efforts:

Those who are in charge of woodland conservation themselves complain of their wasting away; but it is not enough to complain of an evil that we feel already, and that will only increase with time, it is necessary to find a remedy, and every good citizen ought to lay before the public the experiments and reflections that they have made on that account. Such has always been the principal objective of the Academy; public utility is the goal of its work. These considerations have moved Mr. de Reaumur to share with us, in 1721, some excellent observations on the state of the Royal Woods. He states some incontestable facts, he propounds some sound views, and he conveys some experiments that would bring honor to those who carried them out. Moved by these same concerns, and finding myself within reach of these woods, I observed them with particular attention; and finally moved to action by the directives of Count Maurepas, I have conducted several experiments on this subject for 7 to 8 years. Driven by the twin motives of utility and scientific curiosity to make the best use of my little woodland, I have created some nurseries of forest trees, I have sown and planted some large expanses, and having made all these trials on a large scale, I am at a point of reporting some little success concerning several practices that worked on a small scale, and that agricultural authors have commended. It is in this case as in all the other arts, the model that succeeds the best on a small scale, often fails at a large one.

Several pages later, the author elaborates on the problem identified in the last sentence, what a contemporary scientist might call "technology scaleup":

As I had wished to instruct myself thoroughly on the sowing and planting of woodlands, after having read the little that our agricultural authors say on this matter, I became attached to some English authors, such as Evelyn, Miller, etc., who appeared to me to be more grounded in fact and to speak from experience. I had hoped at first to imitate their methods on all points, and I planted and sowed some woods according to their dictates, but after a short time I perceived that this method was extravagantly expensive, and that in following their counsel, the woods, before reaching maturity, would cost ten times more than their value. I recognized then that all their experiments were made on a small scale in gardens, in nurseries, or more often in some parks where they are able to cultivate and care for young trees, but this is not what we look for when we wish to plant woods.... I have thus been obliged to abandon these authors and their methods, and to seek instruction by other means. I have tried lots of different methods, which for the most part, I admit, have not met with much success, but which, at least, have taught me something and put me on the road to success.

The Secret History of the Auvergne

Nicolas Desmarest, 1771. "Sur l'origine et la nature du basalt prismatique" (On the origin and nature of prismatic basalt). *Histoire de l'Académie Royale des Sciences*, pp. 23–25.

The great advances in natural history in eighteenth-century France also included geology. In the *Histoire naturelle* volume of 1774, for example, Buffon made the first experimentally based estimate for the age of the earth ("indeterminably long," but at least 75,000 years). A few years before, a governmental official who conducted geological field research in his spare time, Nicolas Desmarest, wrote two ground-breaking *mémoires* on the volcanic origin of basalt, a key finding because of its eventual impact on geological theory.

This three-page summary of Desmarest's first *mémoire* was read before the Royal Academy in 1771, but neither it nor its *mémoire* appeared in print until three years later. This typically delayed publication is a tribute to the leisurely pace of eighteenth-century science, a pace difficult to imagine today. In his work, Desmarest explores the origin of basalt, a dark volcanic rock whose formation is now attributed to magma, a molten material originating in the earth's upper mantle. At the time, however, basalt was generally attributed to sedimentation, a precipitate from a hypothetical ocean that once covered the entire globe. According to the leading school of geological thought, and to its leader, Abraham Gottlob Werner, all rocks and minerals were a consequence of this sedimentation.

There were serious problems with this theory. It was, for example, difficult to dispose of this hypothesized universal ocean once its work of sedimentation was completed. But far more telling than any theoretical objection was the objection of fact: empirical evidence that all rocks were not sedimentary. This is precisely the evidence that Desmarest sought to gather, focusing on an area of extinct volcanoes in France. His article is the consequence of meticulous fieldwork culminating in a map of the Auvergne region showing the placement and direction of its various lava flows and fields of basalt. The map, a masterpiece of the engraver's art too large to reproduce in this book, gives the reader a God's-eye view of the region with a series of arrows pointing only one way, *away* from the cones of extinct volcanoes. It is a visual argument, a leap from fact to theory difficult to counter, so long as the Auvergne region is typical, a typicality that Desmarest tries to establish with reference to other regions of current or former volcanic activity. He also must explain away apparent anomalies, such as the displacement of fields of basalt from patterns of volcanic flow and the absence of agreed-upon signs of volcanic activity—volcanic cinders and lavas—from some fields of basalt. In doing so, he provides us with a durable explanation of the sculpting of the landscape through the action of running water.

We reproduce here the third-person précis of Desmarest's *mémoire*. This précis brilliantly boils down Desmarest's seventy pages of evidence and argument into a few pages, and closes with a warning to scientists like Werner who combined too much theory with too little fieldwork:

It is not so long since volcanoes were looked upon as isolated phenomena. No attention was paid to the fact that a large part of the earth was covered with the products of volcanic action and that it was necessary to see them as one of the most general causes that act on the surface of our globe. Sponge-like lavas and scoriae [stones that appear porous and cinder-like] were the only substances attributed to volcanic action. There were, however, other products of a very different kind but which had the same origin. These are immense masses composed of large irregular prisms, whose number of facets vary, but are in general uniform in each mass. These prismatic stones are continuous throughout the length of the trails they make, or are divided into articulated segments that are enclosed, the one in the other. They are black in color, their granules tightly packed; they are easily polished, very hard, catch fire if fire is applied to them, and are fusible without the aid of any catalyst. Agricola [a sixteenth-century German mineralogist] has described the stones like this that are found at Stolpen in Saxony and has called them *basalts*. The same name has been given to similar stones, which form the *Giant's Causeway* in Ireland.

Monsieur Desmarest observed the same prismatic stones in the Auvergne, a great many parts of which are almost completely covered with them in an infinite variety of guises. It is in the Auvergne that, by comparing the position of these masses of stones with the directions of the courses of the extinct volcanoes in the region, he hypothesized that these stones were the products of the action of these volcanoes. Other observations confirmed this hypothesis. First, he found that often the base of these prismatic stones rested on matter very like scoriae and sponge-like lavas, while the upper part was covered over also with the same substances. Monsieur Raspe, a government official of Kassel in Germany, observed that the summit of the mountain where the Landgrave waterfalls are found, was formed of scoria and lavas, while the body of the mountain was basalt. In addition, in the quarries in the vicinity of Rome, the upper part furnishes sponge-like stones suitable for millwheels, while the lower part yields a compact stone like basalt with which the Roman streets are paved. Finally, the masses of stones obviously produced by the eruptions of Vesuvius rest by their bases on scoriae and sponge-like lavas, and have characteristics that perfectly coincide with those of prismatic basalt. Sometimes the basalt appears without the lavas that, placed beside it, would indicate its origin. But it is easy to see that this basalt has been dislodged from its original position by rain water or the water of melting snows. At the same time, the lavas and the burning soils were wholly consumed, and the resulting debris, swept up by the action of these forms of water, was deposited far from such other soils and such other stones.

If Monsieur Desmarest's evidence, taken together, isn't sufficiently striking, the pieces of basalt that he presented to the Academy appeared all that could be desired as proof. Time seems to have effaced the vestiges of nature's operations, but here and there nature has left behind traces pregnant with meaning. At the same time, she has produced those with the talent to observe, to recognize these traces and to read her secrets in them. Among the pieces of granite collected in Auvergne beside the masses of prismatic basalt, Monsieur Desmarest was able to see where this granite, composed of fine spar and quartz, was commingled with basalt in the same stone. This is to be expected, since granite was its natural state up to the point of total fusion, of becoming basalt, as can be seen by a scrupulous examination of its fine structure. Thus basalt is a granite that, formed in the volcano's interior, spews forth onto the soil in flaming torrents. When it cools, it takes the form of prisms.

To his research on the origin of basalt, M. Desmarest has joined a description of the different prismatic masses of basalt that he observed or that were described by other naturalists. Almost everywhere basalt is found mixed with scoriae and all known volcanoes are surrounded by masses of basalt. Ireland, the Orkneys, Italy, a large part of Germany, the Auvergne, and Reunion Island are covered with this substance, and offer, at the same time, evidence that it is to volcanoes, active or extinct, that they owe their origin.

In addition to his analysis, Monsieur Desmarest has added a discussion on the nomenclature of this class of minerals. In other articles, Monsieur Desmarest gives us a detailed investigation of the formation of basalt, comparing prismatic basalts with those to which naturalists in the ancient world gave this name. But all of this is only a part of a comprehensive work that Monsieur Desmarest is writing on volcanoes and the products of their activity and which he intends to publish shortly. This piece, which he extracted from it, is well-suited to whet the public's appetite for the work itself. In mineralogy, there has not been a theory as comprehensive and at the same time as firmly based on evidence than that of Monsieur Desmarest.

It would be foolish to insist that Desmarest put the last nail into the coffin of the neptunists, those who believed with Werner that all rocks and minerals were sediments from the universal ocean. Throughout his life, in fact, Desmarest remained a neptunist, believing that basalt was an exception, and that most rocks and minerals were the result of sedimentation. Nor should we put Desmarest on a scientific pedestal: it is simply not true that basalt is composed of formerly molten granite. Nonetheless, Desmarest's meticulous fieldwork and cautious theorizing showed that neptunist theory was in need of serious revision.

A Farewell to Phlogiston

Antoine-Laurent Lavoisier, 1777. "Sur la combustion en général" (On combustion in general). *Mémoires de l'Académie Royale des Sciences*, pp. 592–600.

After years of careful experiments on combustion and respiration and pondering the theoretical significance of his results, Antoine Lavoisier was emboldened to present them before his colleagues at the Royal Academy. This he did at an Academy meeting of November 1777, attended by a special guest from America, Benjamin Franklin.

This famous text begins with Lavoisier's personal philosophy of science — stressing the importance of not only acquiring experimental facts and observations but of offering a theoretical interpretation of them. Moreover, he does not take the simplistic view that experiment and observation should invariably precede theory; rather, the three should work together:

> As dangerous as the systematic spirit is in the physical Sciences, it is also to be feared that in hoarding without order a great multiplicity of experiments, we obscure the Science rather than clarify it; that we make access unnecessarily difficult for those who attempt to enter it; that we obtain at the cost of long and arduous work, only disorder and confusion. Facts, observations, experiments, these are the materials of a great edifice; but it is necessary to avoid, when one is stitching them together, coagulations in the Science; on the contrary it is necessary to classify them, to distinguish that which belongs to each category, to each part of the edifice, and finally, to organize them in advance to form a part of the whole to which they belong.
>
> From this point of view, systems in Science are, properly speaking, methods of approximation which put us on the path to the solution of a problem; these are the hypotheses which, successively modified, corrected & changed to the degree that they are contradicted by experiment, lead us one day, without fail, by dint of exclusions and eliminations, to knowledge of the true laws of Nature.

"Emboldened by these reflections," Lavoisier then goes on to announce his new theory of combustion, one opposed by phlogistonists whose intellectual progenitor was the German chemist, Georg Ernst Stahl. Although phlogistonists differed among themselves, they agreed that, on combustion, phlogiston left the substance being consumed and combined with air. This meant that metallic oxides, such as rust, were simple substances, whereas their metals, such as iron, were combinations of the oxide and phlogiston. There was the well-known problem that the oxides seemed to weigh more than their metals, but these problems were not enough to undermine a fundamental faith in a theory that explained so much.

In his *mémoire*, Lavoisier is not about to undermine completely so central a doctrine of existing chemistry. Phlogiston no longer explains combustion, but it does not disappear; rather, it combines with pure air to form heat and light. In effect, phlogiston is displaced, rather than eliminated. This *mémoire* represents a stage in Lavoisier's full conversion as an anti-phlogistonist, a combination of daring and caution:

> The existence of the matter of fire, of phlogiston in metals, in sulfur, etc., is then really only a hypothesis, a supposition which, once admitted, explains, it is true, some of the phenomena of calcination and combustion; but if I am able to demonstrate that these same phenomena are able to be explained in just as natural a manner by an opposing hypothesis, that is to say without supposing that the matter of fire or phlogiston exists in materials called *combustible*, the system of Stahl will be found to be shaken to its foundations.
>
> Without a doubt it will not be amiss to demand first, what do I mean by the matter of fire? I reply with Franklin, Boërhaave, and some of the philosophers of antiquity that the matter of fire or of light is a very subtle, very rarified, very elastic fluid which surrounds all parts of the planet we inhabit, which penetrates bodies composed of it with greater or less ease, and which tends, when free, to be in equilibrium with everything.
>
> I will add, borrowing from the language of chemistry, that this fluid is the solvent for a great number of bodies; that it combines with them in the same manner as water combines with salt and as acids combine with metals; and that the bodies thus combined and dissolved by the igneous fluid lose in part the properties which they had before the combination and acquire new ones that approach those of the matter of fire. . . .
>
> Pure air, the dephlogisticated air of Mr. Priestley, is then, from this point of view, the true combustible body and, perhaps, the only one in Nature, and we see that there is no longer need, in explaining the phenomena of combustion, of supposing that there exists an immense quantity of fixed fire in all bodies that we call *combustible*, that on the contrary, it is very probable that little of this fire exists in metals, sulfur, phosphorus, and the majority of very solid, heavy, and compact bodies; and perhaps even that the matter of free fire exists only in these substances by virtue of its property of coming into equilibrium with all neighboring bodies.

In the first paragraph, Lavoisier proclaims his independence from the old chemistry in bold language: "the system of Stahl will be . . . shaken to its foundations." In the last paragraph, he explicates his new chemistry in a language more conciliatory, more cautious: "from this point of view . . . perhaps," "it is very probable," "perhaps even." In chapter 7, we will have more to say about the importance of hedging like this in scientific prose.

Geological Explanation in the Late Eighteenth Century

....

Antoine-Laurent Lavoisier, 1789. "Observations générales sur les couches modernes
horizontals qui ont été deposes par la mer et sur les consequences qu'on peut
tirer de leurs dispositions relativement à l'ancienneté du globe terrestre"
(General observations on the modern horizontal beds that have been deposited
by the sea and on the conclusions one is able to draw from their dispositions
relative to the age of the earth). *Mémoires de l'Académie Royale des Sciences*,
pp. 351–71.

Through the nineteenth century, scientists viewed theory with suspicion. In the
Baconian model, one first ran experiments or made empirical observations, then,
maybe, drew some theoretical inferences, after dutifully apologizing for having
strayed from the "facts." Even a scientist of the first rank like chemist Humphrey
Davy followed these conventions, ending his 1821 *Philosophical Transactions*
article reporting his experiments on electricity and magnetism: "I will not
indulge myself by entering far into the theoretical part of this subject; but
a number of curious speculations cannot fail to present themselves to every
philosophical mind, in consequence of the facts developed; such as whether the
magnetism of the earth may not be owing to its electricity."

As we saw in the previous chemical article, Lavoisier had no such trepida-
tions. In the cited geological article here, he even openly admitted to having
formed his theory long before he had gathered sufficient evidence to support
it:

> There are two ways of presenting objects in scientific matters; the first con-
> sists of moving from phenomena to the causes that have produced them;
> the second, to hypothesize the cause, then see if the phenomena presented
> by the observations coincides exactly with the hypothesis. The last method
> is rarely that which we follow in the search for new truths, but it is often
> useful for teaching students: it spares them difficulties and weariness, and it
> is also the method that I believe I ought to adopt in the suite of mineralogical
> memoirs that I propose to give successively to the Academy.

In this case, Lavoisier's hypothesis was that because the earth's oceans had
advanced and retreated through several cycles over long periods of time, this
ebb and flow should be visible in deep vertical sections in the earth. In the late
eighteenth century, geologists had believed that the ancient land masses were
at one time submerged by an ocean, which retreated and left various sediments.
But that theory did not gel with what Lavoisier observed in the field:

Here we found masses of sea shells among which we see some that are thin and fragile; most are neither worn nor abraded; they are precisely in the state where the animal left them when it died: all those with an oblong shape are lying horizontally; nearly all are in a position that is determined by their center of gravity: all the features that surround these shells indicate a completely tranquil environment and, if not an absolute repose, no more than slight movements independent of exterior forces.

A few feet above or below where this observation was made presents a dramatically altered scene; we see there no trace of living or animate beings; we find, instead, rounded pebbles whose angles have been worn smooth by rapid movement over a long period; this is a picture of a raging sea, which shatters against the shore and violently churns a considerable mass of pebbles. How can we reconcile observations so opposed? How can effects so different be explained by the same cause? How can the movement which wears out quartz, rock crystal, the hardest stones, which turns their angles into curves, still preserve the fragile and light shells?

Lavoisier's introductory paragraphs present a paradox almost as dramatic as the clock paradox Einstein presents in his great paper on special relativity (chapter 9). Lavoisier, of course, has an explanation for this dilemma. The dilemma is not real, only apparent:

This contrast of tranquility and agitation, order and disorder, separation and mingling, appeared to be inexplicable at first glance; nonetheless, by dint of examining and re-examining the same objects, in different times and different places, and by dint of combining observations and facts, it seemed to me that we are able to explain these surprising phenomena in a simple and natural manner, and to succeed in determining the principal laws that nature follows in the arrangement of horizontal beds. I meditated a long time on these phenomena before I formed an explanation that seems to have enough consistency and unity for me to present it to the Academy.

Lavoisier's solution involves defining two kinds of beds: "pelagic," a tranquil bed formed in the open ocean; and "littoral," a turbulent bed formed at the expense of coastlines. These two types of bed are the remnants from the rising and falling of the oceans and alternate in horizontal sections of the earth's crust. The above quotations are filled with antitheses ("hardest stones/fragile and light shells," "tranquility/agitation," "order/disorder"), mirroring his scientific theory.

The rest of the article presents Lavoisier's observational evidence and argument for that theory, which according to Stephen Jay Gould, while "a wonderful exemplar of scientific procedure at its best," turned out "wrong in every detail." Even great scientists like Lavoisier take wrong turns now and again. Nonetheless, as Gould also noted, this article was an important advance over the previous theory and a step toward the geological work carried out by James Hutton and others soon thereafter (chapter 4).

3

INTERNATIONALIZATION *and* SPECIALIZATION

In the beginning was the General Scientific Journal. And the General
Scientific Journal begat the Specialty Journal, and the Specialty Journal
begat the Single-Subject Journal, whether according to class of compounds, specific disease, or methodology. And the Single-Subject Journal begat the Interdisciplinary Journal to link up the specialties at an
earlier evolutionary date. And the scientific community saw the journals
were good, and they were fruitful and multiplied.

Eliot M. Berry, in 1981, *New England Journal of Medicine*

FOLLOWING THE LEAD of England and France, other counties
soon launched their own general scientific periodicals (also called journals, essays, acta, commentaries, proceedings, transactions, miscellanea,
and memoirs)—many of which had some strong bond with a scientific society. To give but two notable examples from the eighteenth century, we find
the annual reports of the Royal Academy of Sciences and Belles-Lettres in
Berlin and the transactions from the American Philosophical Society meetings
in Philadelphia.

While numerous learned journals "from many considerable parts of the world" were established outside of England and France during the seventeenth through the eighteenth centuries, we concentrate at the beginning of this chapter on selections from two countries that would later emerge as world powers in science, Germany and America. Before the nineteenth century, scientists in the German states—men like Gottfried Wilhelm Leibniz and Johann Heinrich Lambert—were already doing distinguished science, while Americans like Benjamin Smith Barton and Caspar Wistar were creating the foundations of a scientific community that was not to mature for nearly a century.

By the end of the eighteenth century, scientific periodicals focused on particular areas of science began appearing sporadically in Europe and England. In France we have the *Connaissance des Temps*, an astronomical and meteorological journal, and *Annales de Chimie*, a chemical journal. In Germany we have the *Astronomisches Jahrbuch*, an astronomical yearbook, and the *Bergmännisches Journal*, a mining journal. By the middle of the nineteenth century, specialized journals proliferated. We have the *Monthly Notices* of the Royal Astronomical Society, the *Annalen der Physik*, the *Bulletin* of the Chemical Society of Paris, the *Journal* of the Geological Society of London, the *Botanische Zeitung*, and the *Annales* of the Entomological Society of France. Our excerpts from this specialized literature range widely. They come from chemistry, physics, zoology, botany, and evolutionary biology, and they cover subjects as diverse as an early formulation of the law of the conservation of energy, and the description of a brilliant red flower used in India in funeral processions and as a shoe polish.

In the nineteenth century, increased professionalization accompanied this specialization. According to historian Joseph Ben-David, "the transformation of science into a status approaching that of a professional career and into a bureaucratic, organized activity took place in Germany between 1825 and 1900." Not only in Germany but in other scientifically mature countries as well, professionals working in different branches of science came to the fore while virtuosi and savants with diverse interests faded into the background. As a consequence, we more often find professionals writing for other professionals. During this same period, not coincidentally, the growth curve for the scientific literature shifted from linear to exponential as an intellectual gold rush spread throughout the major cities in Europe. By 1863, there were more than 1,500 scientific periodicals, up from one hundred a century earlier.

Initially at least, this increased professionalization and specialization did not mean homogenization in the writing style. Our selections from this period thus range widely, from the standard neutral scientific prose of Julius Mayer and Rudolph Clausius, to a biting polemical attack by Thomas Huxley, a parody by Friedrich Wöhler, and a vicious diatribe by Hermann Kolbe, all scientists of the first rank.

GERMAN LITERATURE

Useful Fictions and the Birth of the Calculus

....

Gottfried Wilhelm Leibniz, 1684. "Additio ad schedam de dimensionibus figuram inveniendis" (Addition to the article on the calculation of the various dimensions of figures). *Acta Eruditorum*, Vol. 5, pp. 233–36. From the French translation of the Latin by Marc Parmentier, in *G. W. Leibniz: La Naissance du Calcul Différentiel, 26 Articles des* Acta Eruditorum. Paris: Libraire Philosophique J. Vrin (1989), pp. 93–95.

Like the French *Journal des Sçavans*, the German *Acta Eruditorum* began life in the seventeenth century as a general periodical that included scientific material along with many other scholarly topics. It was created by Otto Mencke, the first editor and chief financial backer, and it is considered one of the forefront learned journals in Europe well into the eighteenth century. Among other important discoveries, it published most of the articles in which Gottfried Leibniz announced and articulated his invention of the calculus. Contributions from other well-known men of science included an international cast: Huygens, Newton, Boyle, Leeuwenhoek, the Bernoulli brothers, and Halley. And Latin was the language chosen to appeal to the international marketplace.

Leibniz wrote about fifty articles over three decades for the *Acta Eruditorum*, the most well known being those on calculus, a mathematical method now routinely taught in high school. Leibniz's invention of the calculus is not an achievement of mathematics alone, since modern science and engineering are impossible to imagine without it. For Leibniz, the central insight into the calculus was a "useful fiction." He imagined a circle as a polygon of infinite sides, the successive tangents to any of which represent the instantaneous change in velocity of an accelerating body, the succession of all of which represent the change in this velocity over time:

The very learned mathematician, *Jean Christophe Sturm*, has given us in these *Acta* of last March, a method for establishing the various dimensions of figures established by Euclid and Archimedes, but more directly and in shorter compass than is ordinarily the case: it consists in the reduction of infinite series, in the division of the abscissas (that is to say the portions of the axis) continually in two, and in the construction of parallelograms whose height is the portions of the axis and whose base is the ordinate, a method that differs in its particulars from case to case. As he expressly wished to know how I felt about this solution, along with other mathematicians, I have thought it my duty briefly to give my view of it; I would doubtless have done this already had I paid attention earlier to this passage of the *Acta*. I certainly agree that this way of solving these problems is correct and worthy of praise. It seems to me, however, that one can derive it, as all the other methods that are used, from what seems to me *the general principle for determining the various dimensions of curves, namely, to consider that a curvilinear figure is equivalent to a polygon with an infinity of sides.* It follows from this that everything that can be established concerning such a polygon, one independent of its number of sides, approaches closer to the truth the more sides the polygon has, a multiplication that sees to it that the final error will range within acceptable limits. And this can be said of the curve also.

A polygon of infinite sides is a useful fiction in that it helps us solve problems; according to Leibniz, it "cannot lead to error for it is sufficient to substitute for the infinitely small as small a thing as one may wish, so that the error may be less than any given amount."

Leibniz contributed, not only to mathematical substance, but to mathematical expression, a system of symbols so nicely suited to their tasks that many remain in use to this day: the geometric symbols for congruence and similarity, the sign of equality in proportions, the colon for division, the dot for multiplication, and in the calculus, the signs for the derivative and for integration. Leibniz did not merely invent these signs; he worked hard through correspondence to achieve consensus from other mathematicians in their use, and considered not only the needs of his fellows, but also those of the typesetters. He felt that a perfect system of signs was one that reflected the true nature of the things they described: "In signs one observes an advantage in discovery which is greatest when they express the exact nature of a thing briefly and, as it were, picture it; then indeed the labor of thought is wonderfully diminished."

Breaking the Barrier: A Female Artist-Explorer-Naturalist from Frankfurt

Maria Sibylla Merian, 1705. *Metamorphosis Insectorum Surinamensium* (Metamorphosis of the Insects of Suriname). Amsterdam: Gerardum Valk.

Throughout the first half of our book, the alert reader will note the near absence of female authors. Until well into the twentieth century, women were excluded from the premier scientific societies and other organized scientific activities. Even in the twenty-first century, certain scientific disciplines remain male-dominated. So it does not come as a surprise that shockingly few female names grace the pages of the early scientific literature across all national borders.

Despite this hostile environment, a few industrious women did manage to make significant contributions to science. We have already discussed the astronomical work of Caroline Herschel in England (chapter 1). She was not alone. Surprisingly, during the early eighteenth century, about 14% of German astronomers were women. As far as we know, however, their work did not make it into the scientific literature of the day, at least not under their own names.

Maria Merian was an exception in many respects. It was in the field of natural history that this German-born woman made a name for herself through scientific book publication in the early eighteenth century. In 1699, financed out of her own pocket, resident alien Frau Merian sailed from her home in Amsterdam to a Dutch colony in northern South America, where she observed insect and plant life in the rain forests over a two-year period. Not only did she do the necessary fieldwork and drawing, she wrote the cited volume in Dutch and Latin (likely with some assistant from a translator), then self-published the manuscript in both languages. The resulting elegant volume documenting her observations has sixty figures, similar to the one on display here, and is one of the wonders of early scientific bookmaking.

Merian's Suriname insects all appear in a "natural setting," yet they are arranged in a highly artificial way to tell a story easily grasped by a lay audience. In large part, that story centers upon different insects' metamorphosis from egg to adulthood.

As an example, the next image represents the four main stages in the development of the White Witch moth: egg, caterpillar, cocoon, and adult. To illustrate this evolutionary process in a single picture, Merian drew a drop-shaped moth egg clinging to a rubber plant. Crawling about it is a caterpillar, having emerged from the egg and nibbled away at the plant leaves for nourishment. Below that is the caterpillar's silken cocoon, hanging from the plant like a Christmas tree ornament. Finally appear two versions of the resulting adult moth—one resting on a branch, the other flying—both fully displaying their beautiful wings (which can reach up to a fourteen-inch span). The sitting moth shows one side of its inner wing; the airborne moth, the full span as though mounted for display in a museum collection.

In basic format, Merian's *Metamorphosis Insectorum Surinamensium* anticipates *Curtis's Botanical Magazine* from a century later (example given later in this chapter), with each figure occupying a whole page and accompanied by a page of explanatory text. It also reflects one of the major activities of seventeenth- and eighteenth-century science—collecting typical, abnormal, and exotic specimens of the natural world, naming them, displaying them as part of museum-like collections, and describing them in written form with the aid of engravings like Merian's.

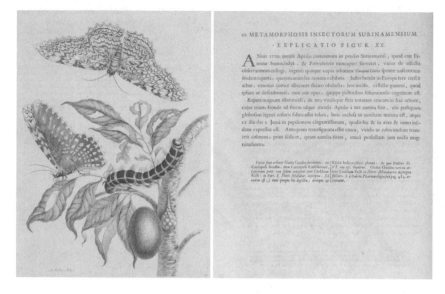

FIGURE 3-1 · Metamorphosis of witch moth from Suriname: drawing and facing text.
Courtesy of Special Collections Research Center, University of Chicago Library.

Visualizing Data, Finding Hidden Trends

Johann Heinrich Lambert, 1769. "Essai d'hygrométrie ou sur la mesure de l'humidité"
(Essay on hygrometry or the measurement of humidity). *Histoire de l'Académie Royale
des Sciences et des Belles-Lettres de Berlin*, pp. 68–127.

The brainchild of Gottfried Leibniz, the Societas Regia Scientiarium in Berlin
began the *Miscellanea Berolinensia* in 1710. This Prussian society floundered
from the start, and its publication had little impact on the international scene,
with a mere seven volumes issued between its founding and 1744. Soon thereafter, the Societas merged with the newly formed Prussian Royal Academy of
Sciences and Belles-Lettres. Closely modeled on the Royal Academy in Paris,
this organization launched its own periodical, employing French instead of the
expected choice of Latin. Its annual publication, *Histoire de l'Académie Royale
des Sciences et des Belles-Lettres de Berlin*, eventually became a major competitor
to its similarly named French predecessor and put Berlin on the map as a center
of scientific excellence.

The selected article by physicist Johann Lambert in the Prussian Royal
Academy's *Histoire* contains the first substantive use ever of a Cartesian graph in
a scientific journal. Scientific articles without graphs? For the present-day reader

that's hard to imagine. Yet the line graph as we know it was not "invented" until the late eighteenth century, a hundred years after the scientific journal made its debut. In this visual form, the author typically plots some independent variable, such as time, on the abscissa (horizontal axis) and some dependent variable, such as a physical or chemical property that changes over time, on the ordinate (vertical axis). This arrangement is ideally suited for communicating multiple data points at a glance, visually representing implied cause and effect, uncovering trends in a large mass of data, and making comparisons among multiple data sets. As astronomer J. F. W. Herschel noted in an article from the 1833 *Memoirs of the Royal Astronomical Society*, blank line-graph charts with pre-drawn axes "are so very useful for a great variety of purposes, that every person engaged in . . . physico-mathematical inquiries of any description, will find his account in keeping a stock of them always at hand."

In the nineteenth and twentieth centuries, graphs have become more and more central as the sciences have moved away from the descriptive and qualitative, and toward the mathematical and quantitative. The graph is now considered the most important form for scientific visual representation, supplanting geometric diagrams and realistic drawings of natural objects and research equipment. Lambert is the first scientist to freely use graphs in his books and articles.

Our selection comes from an article by Lambert on measuring humidity with a hygrometer, a subject of special interest and controversy in late eighteenth-century physics. Lambert's Figure 4 plots the decrease of the height of water in a capillary tube measured over time and varying temperatures (curves DEF and ABC). The horizontal axis is divided into hours on January sixth: from 8 am to 8 pm, with noon represented as 0. The vertical axis measures two quantities, the Réamur temperature (0° to 80°, freezing to boiling) and length in "lines" (one-twelfth of an inch per line). The tangents to DEF give the rate of evaporation. In Figure 4 Lambert gives one example, DEG, tangent to the curve at point E of PH. From his calculations, Lambert was able to derive a graph showing the evaporation rate as a function of temperature. In this second graph, Lambert's Figure 5, the horizontal axis is the temperature; and the vertical, the evaporation rate.

Here is Lambert's discussion of his two graphs:

> In order to display my observations of January 6, I have represented them on a larger scale in the fourth figure. The abscissa is divided into hours, the first hour into ten-minute segments. The ordinate has two scales, the first for the temperature, the second, for evaporation. The curve ABC displays the rise and fall in temperature, or the degree of heat of the water. And the curve DEF gives the drop in the surface of the water in the capillary tubes. I availed myself of a similar figure, but on a larger scale, to compare the rate

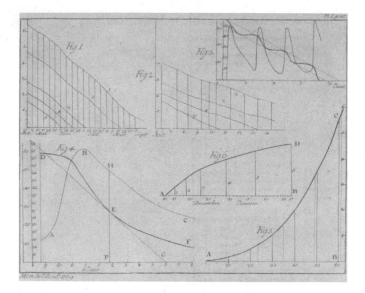

FIGURE 3-2 · Two early Cartesian graphs: evaporation as a function of time and as a function of temperature. Courtesy of Special Collections Research Center, University of Chicago Library.

of evaporation with the degree of heat. To obtain this measure by inference, it is necessary that for each ordinate PH a tangent EG be drawn. As the time PG is to EP, so any period of 24 hours is to a fourth number, which expresses the number of lines that evaporate in the interval of a day when the water's heat is constantly at PH. By this means I find the following correspondences:

HEAT OF	DAILY EVAPORATION
61°	67 lines
60°	65
49°	39 ½
35°	17.2
23°	8.7

These numbers, with a slight correction that needs to be made for the last, form the curve of the fifth figure [where the heat values in the table are represented by *dotted* vertical lines]. The abscissas are divided into degrees of temperature, and the ordinates into the inches and lines of evaporation that correspond. Since the curve's convexity turns towards the axis, it follows that the evaporation proceeds faster than the temperature rises.

Figure 5 allows us to see the steep rise of evaporation as a function of temperature, a rate invisible in the original data and in Figure 4, constructed from that data. Figure 5 allows Lambert to conjecture as to the mechanical cause:

> From this curvature, we can draw conclusions relative to the forces implicated in evaporation. Because, as evaporation follows the law that relates surface area to rate, I have already inferred that the active force must be found in the air. This force works more effectively when the forces of cohesion in the water are lessened, and it is clear that heat contributes to this effect by the expansion that it produces. This expansion lessens the forces of cohesion because water flows apart more readily the hotter it is. The expansion enlarges the spaces found between the water particles, and this allows the particles of air freer access so that they may absorb those of water all the more easily.

Getting Personal

Anonymous, 1790. Review of *Three Letters on Mineralogy* by Johann Jacob Ferber. *Bergmännisches Journal*, Vol. 3, pp. 174–97.

Johann Jacob Ferber, a Swede by birth, had a successful career in mineralogy and geology, publishing extensively and rising professionally until felled at forty-seven by a stroke that proved fatal. Ferber was able to contribute to geological theory as well as to descriptive geology. While in Italy he observed active volcanoes and was able to testify to the volcanic origin of basalt, evidence against the view of the Neptunists that all rocks had an aqueous origin.

Given Ferber's excellent credentials, including Regional Mining Inspector for the King of Prussia, it is difficult to understand the vitriolic attack on his work that the review in this German mining journal represents. The author savages Ferber's character—allegedly in the interest of science—since the writer's cudgel is a social norm: terminological propriety. In the reviewer's mind, Ferber's allegedly sloppy practices constitute a problem; if they are not criticized, if they happen to prevail, geology in the German-speaking lands will suffer badly:

> From the announcement, one might reasonably expect that [the three letters that make up the book on which he is commenting] would contain all sorts of interesting intelligence. But because of many personally abusive remarks, what appears therein is so highly unpleasant to read that we are obliged to censure it. From one so scholarly as Mr. F[erber], it is so unexpected, and it is so thoroughly disagreeable to normal readers to read the intelligence communicated therein—presumed to be science—in *so slovenly a discourse*, so full of *vague and incorrect terms*. In our report, without Herr F.'s being

able legitimately to raise his voice against us for *pedantry* and *quibbling*, we thought it necessary to point to various instances of vagueness and incorrectness, into which he has strayed. Currently scientists have already come so far in *the definition of mineralogical terms* that such uncertainties would create *extraordinary disorder* if they were to run riot and appear in important works; as a consequence, instead of advancing in this science, we would take *several steps* backward. We must conclude this report by not concealing the painful observation imposed upon us by reading this little book repeatedly from cover to cover: that, with increasing years, Herr Ferber *seems more and more to lay bare the weakness of his character*, or more accurately, *of his heart*, for we would rather not speak about defects of mind or knowledge (emphasis in the original).

To add insult to insult, the editor of the journal appends a long footnote to the article summarizing the conclusions of a contributor to the German journal, *Chemischen Annalen*, alleging that Herr Ferber has illegitimately asserted priority of discovery. The author, Herr Höpfner, ends by stating: "This is not the only error that Regional Mining Inspector Ferber makes."

Despite the rapid movement of the sciences such as geology toward professionalization, in no scientific community in the eighteenth century are social tendencies strong enough to completely exclude the intrusion of the very personal, such as expressions like "the weakness of his character" in the Ferber book review. We shall see further examples of this argument *ad personam* later in this chapter.

AMERICAN LITERATURE

Scientific Zoology in America

Benjamin Smith Barton, 1799. "Some account of an American species of Dipus or Jerboa." *Transactions of the American Philosophical Society*, Vol. 4, pp. 114–24.

Across the ocean from England and Europe, scientific societies and their publications took root in America even before the Declaration of Independence. America's first scientific society was the American Philosophical Society, started by Benjamin Franklin, among others, in Philadelphia in 1743 to pursue "all philosophical Experiments that let Light into the Nature of Things." After petering out due to lack of interest after less than four years, the society was revived permanently in 1768. The first volume of its *Transactions* (1771) contains many articles reporting observations and measurements on the transit of Venus across the sun in 1769—used for estimating the distance between the earth and sun—a cutting-edge problem in astronomy at the time. This promising inaugural volume helped establish an international reputation not only for the American Philosophical Society, but for American science in general. But the real strength of early American science lies not in its astronomy, but in its natural history of the largely unexplored American continent.

The selected article here, as well as the next three, come from the very early *Transactions of the American Philosophical Society*. With a versatility not at all unusual for the time, its author, Benjamin Barton, was a botanist, zoologist, linguist, anthropologist, and physician. While his wide ambitions exceeded his accomplishments, these were not insignificant. He is the author of several books reporting "facts, observations, and conjectures" on American natural history and a series of zoological articles from which the present excerpt derives.

Barton's scientific writing contrasts with that of his fellow countryman, the aptly named John Godman, for whom the study of nature was a semi-religious calling. For Godman, "other individuals may, but the enlightened student of nature *never can* forget the omnipresence of Diety—it is every where before his eyes, and in his heart—obvious and palpable;—it is a consciousness, not a doctrine; a reality, not an opinion, identified with his very being, and attested to his understanding by every circumstance of his existence." This sort of passionate nature writing is, of course, still very much with us. In contrast, Dr. Barton's prose represents the objective scientific naturalist, intent on extending the work of the great European scientists, Carolus Linnaeus and Comte de Buffon, to the flora and fauna of America.

In our selected passage, Barton argues for the appropriateness of the name Dipus Americanus by comparing this rodent and a contrasting European species in terms of their external physical features. The resulting prose is serviceable, certainly able to "enlighten the mind," but hardly able to "warm the heart":

From these descriptions, it appears that the Mus longipes of Palles is larger than the Dipus Americanus. This circumstance is farther confirmed by Zimmerman, who says that the size of the first of these animals is between that of the rat and the field-mouse. The colour of the Dipus Americanus is rather of a dark than a tawny, colour. Below the colour in both animals seems to be the same. As far as I can judge from the figure of the Dipus Jaculus, I should think that the head of this animal is more oblong than that of the Dipus Americanus. The ears of the Dipus meridanus are said to be large. Those of the Dipus Americanus are much smaller than the ears of the Dipus Jaculus, or any of the other species of the genus of which I have seen figures. The feet of the Dipus Americanus are not white, but of a reddish or flesh colour. The

FIGURE 3-3 · Dipus Americanus. Courtesy of Special
Collections Research Center, University of Chicago Library.

soles of the feet of the Dipus meridianus are said to be very villous [covered
with hair-like projections]; but the soles of the feet of the Dipus Americanus
are nearly naked. In the Dipus Americanus, the *tuberculum* [bony projec-
tion] of the fore-feet is entirely destitute of a nail. The thighs of the hind
legs of the Dipus meridianus are said to be very thick, or fleshy. Those of
the Dipus Americanus do not appear to be remarkably so. The tail of the
Dipus meridianus is said to be shorter than the body. The tail of the Dipus
Americanus is considerably longer than the body.

Thomas Jefferson Corrects the Comte de Buffon

Thomas Jefferson, 1799. "A memoir on the discovery of certain bones of a quadruped of
the clawed kind in the western parts of Virginia." *Transactions of the American
Philosophical Society*, Vol. 4, pp. 246–60.

It would not be fair to call Thomas Jefferson a professional scientist; it would be equally unjust to label him a rank amateur. Throughout his career he was deeply interested in science and its furtherance in the new Republic. While he did not exclude the speculative or theoretical, he preferred applied science. This is apparent in a paper in which the newly elected President of the American Philosophical Society discoursed on the improvement of the plough, read before the Society in 1798. Such a topic was consistent both with his philosophy and with the philosophy of the Society he led, founded "for promoting useful knowledge."

It might seem that another paper by Jefferson, read the previous year, was the antithesis of the practical and patriotic concerns that motivated both Jefferson and the founders of the American Philosophical Society. This article concerns the bones of a large mammal that Jefferson identifies (mistakenly) as a lion of enormous size. But for Jefferson the bones are evidence for two theses propounded nearly two decades earlier in his *Notes on the State of Virginia*. The first is that extinction is rare or nonexistent in nature; the second, that the great French naturalist, Comte de Buffon, is mistaken when he asserts that the animals of the New World are smaller than those of the Old.

For Jefferson, the answer to the question of extinction would combine science with patriotism, scientific argument with political argument. In his 1797 paper (not published until the 1799 *Transactions*), he cites anecdotal evidence that his enormous "lion" still exists, and quiets doubters with the following argument from possibility (a presage of the Lewis and Clark expedition that Jefferson funded):

> The progress of the new population would soon drive off the larger animals, and the largest first. In the present interior of our continent there is surely space and range enough for elephants and lions, if in that climate they could subsist; and for mammoths and megalonyxes [the "lion" Jefferson has just described] who may subsist there. Our entire ignorance of the immense country to the West and North-West, and of its contents, does not authorize us to say what it does not contain.

Jefferson the paleontologist was stung by Buffon's imputation that the mammals of America were punier than those of the Old World. He thought he saw a contradiction between the large size of the bones he had just analyzed and Buffon's statement that "there is in the combination of constituents and of other physical causes, something that goes against the increase in size of life *in this new world*; there are obstacles to development and perhaps to the formation of germinal elements that conduce to large size." His dander up, Jefferson counters Buffon by arguing the ludicrousness of the reverse position:

Are we then from all this to draw a conclusion, the reverse of that of M. de Buffon, that nature, has formed the larger animals of America, like its lakes, its rivers, and mountains on a greater and prouder scale than in the other hemisphere? Not at all, we are to conclude that she has formed some things large and some things small, on both sides of the earth for reasons which she has not enabled us to penetrate; and that we ought not to shut our eyes upon one half of her facts, and build systems on the other half.

In his final sentence Jefferson makes his larger scientific point: no natural history can be complete or correct without taking American animals into consideration.

Caspar Wistar Corrects Thomas Jefferson

Caspar Wistar, 1799. "A description of the bones deposited, by the President [Thomas Jefferson, president of the American Philosophical Society], in the Museum of the Society, and represented in the annexed plates." *Transactions of the American Philosophical Society*, Vol. 4, pp. 526–31.

Thomas Jefferson was mistaken when he identified the bones of the megalonyx as those of an enormous lion. It was his American Philosophical Society colleague, Caspar Wistar, a Professor of Anatomy at the University of Pennsylvania and author of the first textbook on anatomy published in America, who got the identification right. His correction appears as the last article in the same volume of the *Transactions* carrying Jefferson's article. After examining the bones that Jefferson had deposited in the Museum of the American Philosophical Society, Wistar recognized them as the foot and claw of a bradypus, or giant sloth, that once roamed the New World.

In the selected passage, the expert knowledge possessed by Wistar permits him to confidently extrapolate from a few bones to a large sloth. (In this passage, the technical terms are anatomical and refer to the claw of the sloth and its immediately connecting bones. The "megatherium" Wistar mentions is another species of extinct sloth.) As was the case in the Barton passage, Wistar builds his case by comparison with earlier descriptions by other naturalists:

From the shortness of the metacarpal bone, and the form and arrangement of the other bones of the paw, and also from the form of the solitary metacarpal bone, it seems probable that the animal did not walk on the toes, *it is also evident that the last phalanx was not retracted.* The particular form of No. 2, and its connection with the metatarsal bone, and with No. 3, must have produced a peculiar species of flexion in the toes, which combined with the greater flexion of the last phalanx upon the second, must have enabled the animal to turn the claws under the soals of his feet; from this view of the subject there seems to have been some analogy between the foot of this animal and those of the bradypus—having no specimens of that animal I derive this conclusion from the description of its feet given by M. Daubeton.

Notwithstanding a general resemblance, they differ in some important points—In the sloth the figure of the metacarpal bone was such that M. Daubeton could not determine from it, whether the bone belonged to the metacarpus or the phalanges—but there could be no doubt as to these bones, for they are unequivocally metacarpal or metatarsal—The sloth has but two phalanges in addition to the supposed metacarpal bone, whereas the animal in question had bone No. 2 and the two phalanges besides. The relative size or proportions of the phalanges, must have differed greatly in the two animals. M. Daubeton describes the first phalanx as very long, and the last, or claw bone, as very short in the sloth, but the reverse is the case with these bones— There is however an unguis described by M. Daubeton which is particularly interesting, it was presented by M. De la Condamine as belonging to a large species of sloth, and although not entire, its length measured round the convexity was half a foot, and its breadth, at the base, an inch and a half.

We are naturally led to inquire whether these bones are similar to those of the great skeleton found lately in Paraguay, but for want of a good plate, or a full description we are unable at present to decide upon that subject—If however any credit be due to the representation given in the Monthly Magazine for Sept. 1796 published in London (the only plate I have seen) these bones could not have belonged to a skeleton of that animal—for according to that representation the lower end of the ulna is much larger, and articulated with a larger portion of the foot in the megatherium than in the megalonix—The upper end of the radius also is much larger than the lower in that figure, whereas the reverse is the case with the megalonix, and the difference in the claw bones is still greater, as will appear to every one who compares the two.

Despite Wistar's excellent detective work, the French naturalist Anselme Desmarest, son of the geologist Nicolas (see chapter 2), honored Jefferson by naming the extinct species *Megalonyx Jeffersoni*.

A Schizophrenic Style of Communicating Science

Thomas Say, 1818. "A monograph of North American insects of the genus Cicindela." *Transactions of the American Philosophical Society*, new series, Vol. 1, pp. 401–26.

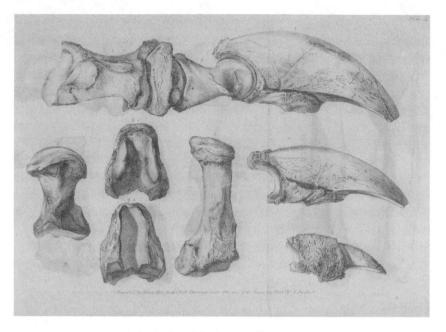

FIGURE 3-4 · Bones from Megalonyx Jeffersoni. Courtesy of Special
Collections Research Center, University of Chicago Library.

The early American journals are filled with new flora and fauna found
throughout North America, and brought to life in both verbal descriptions
and artistic engravings. John James Audubon's work on birds is the most
widely known example. But many others, both native and foreign visitors,
also contributed to the enormous enterprise of systematically documenting
the splendors of the various species of plant and animal inhabiting the new
world.

    This selection, from the father of American entomology, describes a North
American beetle—it is the verbal equivalent of Audubon's famous water-color
prints. The article has two distinct parts. The first is a general description of the
insect's habits and characteristics, meant for the "general observer" and written
with a skill any novelist would envy. It contrasts sharply with the humdrum style
of Barton and Wistar. The second part reports elaborate technical details aimed
at the specialist. This starling juxtaposition reveals the vast differences between a
prose constructed out of everyday language and a technical vocabulary. Thomas
Say's use of sentence fragments in the second part further heightens its semantic
density. A lengthy sampling of the first style follows:

It will perhaps be thought necessary, previous to entering into a technical detail of the characters of the genus Cicindela, and of the indigenous individuals which are comprehended by it, that some account of the manners of this sprightly tribe should be given, and of such circumstances, relating to them, as may serve to present them to the recollection of the general observer. I shall accordingly proceed to state, that these insects usually frequent arid, denudated soils; are very agile, run with greater celerity than the majority of the vast order to which they belong; and rise upon the wing, almost with the facility of the common fly. They are always to be seen, during the warm season, in roads or pathways, open to the sun, where the earth is beaten firm and level. At the approach of the traveller, they fly up suddenly to the height of a few feet, pursuing then a horizontal course, and alighting again at a short distance in advance, as suddenly as they arose. The same individual may be roused again and again, but when he perceives himself the object of a particular pursuit, he evades the danger by a distant and circuitous flight, usually directed towards his original station. It is worthy of observation, as a peculiarity common to the species, that when they alight, after having been driven from their previous position, they usually perform an evolution in the air near the earth, so as to bring the head in the direction of the advancing danger, in order to be the more certainly warned of its too near approach.

They lead a predatory life, and as it would appear, are well adapted to it, by their swiftness, and powerful weapons of attack. The beaten path, or open sandy plain, is preferred, that the operations of the insects may not be impeded by the stems and leaves of vegetables, through which, owing to their elongated feet, they pass with evident difficulty and embarrassment. They prey voraciously upon the smaller and weaker insects, upon larvæ and worms, preferring those whose bodies are furnished with a membranaceous cuticle, more readily permeable to their *instrumenta cibaria*.

The same rapacity is observable in the larva, or imperfect stage of existence, of these insects, that we have occasion to remark in the parent; but not having been endowed by nature with the same light and active frame of body, they are under the necessity of resorting to stratagem and ambuscade for the acquisition of the prey, which is denied to their sluggish gait. The remark is, I believe, generally correct, though liable to many signal exceptions, that carnivorous animals display more cunning, industry, and intelligence, than those whose food is herbs, for the acquisition of which, fewer of the mental attributes are requisite; we see throughout the animated creation, that the development of these qualities, as well as of the corporeal functions, are in exact correspondence with their necessities; and that where a portion of the one is withheld, an additional proportion of the other is imparted. This larva has a very large head, elongated abdomen, and six short feet placed near the head; when walking, the body rests upon the earth, and is dragged forward slowly by the feet. Notwithstanding these disadvantages they contrive means to administer plentifully to an appetite, sharpened by a rapid increase of size. A cylindrical hole is dug in the ground to a considerable depth, by means of the feet and mandibles, and the earth transported from it, on the concave surface of the head; this cell is enlarged and deepened, as the inhabitant increases in size, so that its diameter is always nearly equal to that of the head. At the surface of the earth they lay in wait for their prey, nicely closing the orifice of the hole by the depressed head, that the plain may appear uninterrupted; when an incautious or unsuspecting insect approaches sufficiently near, it is seized by a sudden effort of the larva, and hurried to the bottom of the dwelling, to be devoured at leisure. These holes we sometimes remark, dug in a footpath; they draw the eye by the motion of the inhabitant retreating from the surface, alarmed at the approach of danger.

Say brings his beetle to life for general readers by the beauty and gracefulness of his descriptive prose, while in this next short passage, he exploits a highly technical vocabulary to establish for specialists this insect's place within the large body of zoological knowledge at the time. We quote a brief segment from a much longer technical description:

Order V.—COLEOPTERA.

Section I. PENTAMERA. —*Family I.* ENTOMOPHAGA. —*Tribe I.*

CICINDELETÆ.

Genus CICINDELA.

Cicindela. *Linn. Fabr. Latr.*
Buprestes. *Geoff.*

Essential Character.

Maxillæ monodactyle.
Mentum trifid, inner division scarcely shorter.
Intermediate and *posterior palpi* subequal, filiform.
Tibia simple.

Artificial Character.

Antennæ filiform.
Clypeus shorter than the labrum.
Maxillæ with two very distinct palpi, of which the exterior one, is nearly equal to the labial palpi, penultimate joint of the latter hairy.

An American Faraday?

....

Joseph Henry, 1832. "On the production of currents and sparks of electricity from magnetism." *American Journal of Science and Arts*, Vol. 22, pp. 403–8.

We now turn from natural history to physics, a discipline that American science would come to dominate in the early twentieth century. Joseph Henry, most famous as the first Secretary of the Smithsonian Institution, was the first American physicist of European stature after the death of Benjamin Franklin. He was capable of doing science on the level of Michael Faraday at his finest. Did that make him an American Faraday? Not exactly. Faraday was a genius; for Faraday, the creative lightning struck over and over. Henry was not so blessed. Still, in his case this lightning did strike twice in one article, the one excerpted here. Henry reports his discovery of two forms of electrical induction. Mutual induction, the transfer of magnetic to electrical force, is a phenomenon absolutely central to the electrification of America and the world: it is at the heart of the electric generator. Self induction, a less important phenomenon, is the consequence of a potential difference that appears within a circuit when a connection is broken; a current surges briefly in the opposite direction, creating a vivid spark. The electrical unit of measure "henry" commemorates his early contributions to American science.

Unfortunately for Henry, Faraday also discovered both forms of induction, discoveries virtually simultaneous to those of his American competitor. In the end, Faraday was given primary credit for the more important discovery of mutual induction, Henry for the less important discovery of self induction. This rough justice is more a matter for scientists avid for credit, than for science, which is perfectly indifferent to such matters.

Even Henry's discovery of self induction had some international competition, from a Scottish scientist. We reproduce the final paragraphs of this ground-breaking article, recounting Henry's anxiety over priority for self induction followed by his lucid explanation of the cause of that phenomenon:

> It appears from the May No. of the Annals of Philosophy, that I have been anticipated in this experiment of drawing sparks from the magnet by Mr. James D. Forbes who obtained a spark [from a natural magnet] on the 30th of March; my experiments being made during the last two weeks of June. A simple notification of his result is given, *without any account of the experiment*, which is reserved for a communication to the Royal Society of Edinburgh; my result is therefore entirely independent of his and was *undoubtedly obtained by a different process*.

I have made several other experiments in relation to the same subject, but which *more important duties* will not permit me to verify in time for this paper. I may however mention one fact which I have not seen noticed in any work and which appears to me to belong to the same class of phaenomena as those before described: it is this; when a small battery is moderately excited by diluted acid and its poles, which must be terminated by cups of mercury, are connected by a copper wire, not more than one foot in length, no spark is perceived when the connection is either formed or broken: but if a wire thirty or forty feet long be used, instead of a short wire, though no spark will be perceptible when the connection is made, yet when it is broken by drawing one end of the wire from its cup of mercury a vivid spark is produced. If the action of the battery be very intense, a spark will be given by the short wire; in this case it is only necessary to wait a few minutes until the action partially subsides and until no more sparks are given from the short wire; if the long wire be now substituted a spark will again be obtained. The effect appears somewhat increased by coiling the wire into a helix; it seems also to depend in some measure on the length and thickness of the wire; I can account for these phaenomena only by supposing the long wire to become charged with electricity which by its reaction on itself projects a spark when the connection is broken [italics added].

In terms of priority, there are two key phrases: Forbes failed to give "any account of the experiment," and Henry "undoubtedly obtained [his result] by a different process." In this passage as well as other excerpts in the chapter, we see scientists openly discussing possible competition for the same discovery at about the same time, making a preemptive argument for their priority.

Henry's passage also tells us something about science as a profession in the nineteenth century. While Faraday had the leisure to pursue his research in his laboratory at the Royal Institution in London, Henry was busy working seven hours a day as a science and math teacher at the Albany Academy, his "more important duties" mentioned in the quoted passage (second paragraph). We also see something about nineteenth-century scientific style: the personal has not yet been banished.

Applying Medical Science to Medical Practice

Oliver Wendell Holmes, Sr., 1843. "Contagiousness of puerperal fever." *New England Quarterly Journal of Medicine*, pp. 503–30.

Joseph Lister, 1867. "On the antiseptic principle in the practice of surgery." *British Medical Journal*, Vol. 2, pp. 246–54.

Besides being father of a famous U.S. supreme court justice, Oliver Wendell Holmes, Sr., is best known for the rousing poem "Old Ironsides," a plea to save the retired battleship *Constitution* from demolition, and for a single article on child-bed fever, an often fatal infection of the female genital tract. Holmes begins his article with a litany of reports of women who died, having been attended by physicians and midwives who moved from patient to patient without washing their hands or changing their clothes. This was a world in which physicians shifted blithely from autopsy to accouchement. In one case, a physician carried "the pelvic viscera to the classroom. . . . The same evening he attended a woman in labor without previously changing his clothes." Unsurprisingly, the patient died.

Holmes's rhetoric in the passage below is more Ciceronian than scientific. After page upon page of anecdotal accounts, he ends his article with a plea for reform, cast in a style more suitable for the courtroom than the examining room:

I have no wish to express any harsh feeling with regard to the painful subject which has come before us. If there are any so far excited by the story of these dreadful events that they ask for some word of indignant remonstrance to show that science does not turn the hearts of its followers into ice or stone, let me remind them that such words have been uttered by those who speak with an authority I could not claim. It is as a lesson rather than as a reproach that I call up the memory of these irreparable errors and wrongs. No tongue can tell the heart-breaking calamity they have caused; they have closed the eyes just opened upon a new world of love and happiness; they have bowed the strength of manhood into the dust; they have cast the helplessness of infancy into the stranger's arms, or bequeathed it, with less cruelty, the death of its dying parent. There is no tone deep enough for regret, and no voice loud enough for warning. The woman about to become a mother or with her newborn infant upon her bosom, should be the object of trembling care and sympathy wherever she bears her tender burden or stretches her aching limbs. The very outcast of the streets has pity upon her sister in degradation when the seal of promised maternity is impressed upon her. The remorseless vengeance of the law, brought down upon its victim by a machinery as sure as destiny, is arrested in its fall at a word which reveals her transient claim for mercy. The solemn prayer of the liturgy singles out her sorrows from the multiplied trials of life, to plead for her in the hour of peril. God forbid that any member of the profession to which she trusts her life, doubly precious at that eventful period, should hazard it negligently, unadvisedly, or selfishly!

Holmes's high-flown peroration is followed by some down-to-earth advice:

There may be some among those whom I address who are disposed to ask the question, What course are we to follow in relation to this matter? The facts are before them, and the answer must be left to their own judgment and conscience. If any should care to know my own conclusions, they are the following; and in taking the liberty to state them very freely and broadly, I would ask the inquirer to examine them as freely in the light of the evidence which has been laid before him.

1. A physician holding himself in readiness to attend cases of midwifery should never take any active part in the post-mortem examination of cases of puerperal fever.

2. If a physician is present at such autopsies, he should use thorough ablution, change every article of dress, and allow twenty-four hours or more to elapse before attending to any case of midwifery. It may be well to extend the same caution to cases of simple peritonitis.

3. Similar precautions should be taken after the autopsy or surgical treatment of cases of erysipelas [an acute skin disease], if the physician is obliged to unite such offices with his obstetrical duties, which is in the highest degree inexpedient.

4. On the occurrence of a single case of puerperal fever in his practice, the physician is bound to consider the next female he attends in labor, unless some weeks at least have elapsed, as in danger of being infected by him, and it is his duty to take every precaution to diminish her risk of disease and death.

5. If within a short period two cases of puerperal fever happen close to each other, in the practice of the same physician, the disease not existing or prevailing in the neighborhood, he would do wisely to relinquish his obstetrical practice for at least one month, and endeavor to free himself by every available means from any noxious influence he may carry about with him.

6. The occurrence of three or more closely connected cases, in the practice of one individual, no others existing in the neighborhood, and no other sufficient cause being alleged for the coincidence, is prima facie evidence that he is the vehicle of contagion.

7. It is the duty of the physician to take every precaution that the disease shall not be introduced by nurses or other assistants, by making proper inquiries concerning them, and giving timely warning of every suspected source of danger.

8. Whatever indulgence may be granted to those who have heretofore been the ignorant causes of so much misery, the time has come when the existence of a private pestilence in the sphere of a single physician should be looked upon, not as a misfortune, but a crime; and in the knowledge of such

occurrences the duties of the practitioner to his profession should give way to his paramount obligations to society.

Despite his goodwill and his heartfelt pleas, Holmes failed to persuade the majority of his fellow physicians to avoid contagion. Why? Perhaps it was because his recommended procedures were based on anecdote and conjecture, not on hard science. This was, of course, not Holmes's fault: there was no science available regarding the cause of common diseases. In the late 1840s similar claims based upon clinical observations made by a physician working in Vienna, Ignác Semmelweis, met a similar fate. But a change was in the offing. Writing in England two decades after Holmes, Joseph Lister exemplifies the revolution in medicine that Pasteur had begun in France (see chapter 4). For Lister, the avoidance of contagion was based on a germ theory of disease backed by clinical research:

> In the course of an extended investigation into the nature of inflammation, and the healthy and morbid conditions of the blood in relation to it, I arrived several years ago at the conclusion that the essential cause of suppuration in wounds is decomposition brought about by the influence of the atmosphere upon blood or serum retained within them, and, in the case of contused wounds, upon portions of tissue destroyed by the violence of the injury.
>
> To prevent the occurrence of suppuration with all its attendant risks was an object manifestly desirable, but till lately apparently unattainable, since it seemed hopeless to attempt to exclude the oxygen which was universally regarded as the agent by which putrefaction was effected. But when it had been shown by the researches of Pasteur that the septic properties of the atmosphere depended not on the oxygen, or any gaseous constituent, but on minute organisms suspended in it, which owed their energy to their vitality, it occurred to me that decomposition in the injured part might be avoided without excluding the air, by applying as a dressing some material capable of destroying the life of the floating particles. Upon this principle I have based a practice of which I will now attempt to give an account.

SPECIALIZED LITERATURE: BIOLOGY

Specialization without Professionalization

Anonymous, 1791. "Hibiscus Rosa Sinensis. China-Rose Hibiscus." *Botanical Magazine; or Flower-Garden Displayed*, Vol. 5, No. 158.

During the later decades of the eighteenth century, scientists along with their societies and publications became more specialized as a means of coping with the expanding frontiers of scientific knowledge. Overall, the general desire for higher professional standards in science led to an influx of individual articles primarily aimed at subject matter experts. As François Rozier eloquently, if brusquely, put it in the preface to the first issue of a French specialized journal he founded, "We will not offer to idle amateurs purely agreeable works or the sweet illusion of believing themselves to be initiated to the truly knowledgeable." He further asserted that his journal would "reject everything that is nothing more than undigested compilation and that is wanting in new and useful views."

This international trend toward specialization, however, did not always translate into specialization of the readers or impenetrability of the prose to outsiders. Such is the case with *Botanical Magazine; or Flower-Garden Displayed*, founded by William Curtis in 1787 "for the use of ladies, gentlemen, and gardeners, as wish to become scientifically acquainted with the plants they cultivate," and still in existence today as *Curtis's Botanical Magazine*. Each issue contains a series of vibrant watercolor depictions of plants rendered in lovingly exact detail. (Until 1948, the magazine manufactured these realistic images by mass-producing a line drawing by a professional artist or the author, then having groups of women and children hand color each illustration.) Accompanying each illustration—at least in the early days of *Botanical Magazine*—is a page or two of verbal description, including information on the plant's habitat, color, scent, size, flower placement, method of pollination and cultivation, practical uses, and incidental observations such as the following:

> Rumphius in his *Herbarium Amboinense* gives an excellent account of this beautiful native of the East-Indies, accompanied by a representation of it with double flowers, in which state it is more particularly cultivated in all the gardens in India, as well as China; he informs us it grows to the full size of our hazel, and that it varies with white flowers.
>
> The inhabitants of India, he observes, are extremely partial to whatever is red, they confer it as a colour which tends to exhilarate; and hence they not only cultivate this plant universally in their gardens, but use its flowers on all occasions of festivity, and even in their sepulchral rites: he mentions also an oeconomical purpose to which the flowers are applied, little consistent with their elegance and beauty, that of blacking shoes, hence their name of *Rose calceolariae*; the shoes, after the color is imparted to them, are rubbed with the hand, to give them a gloss, and which thereby receives a bluish tinge, to discharge which they have recourse to lemon juice.

FIGURE 3-5 · China-Rose Hibiscus.
Courtesy of Special Collections Research Center,
University of Chicago Library.

With us it is kept in the stove, where it thrives and flowers readily during most of the summer; the single blossoms last but a short time, yet their superiority arising from the curious and beautiful structure of the interior parts of the flower, compensates for the shortness of their duration.

It is usually increased by cuttings.

Huxley on the Horror of Chambers

T. H. Huxley, 1854. Review of *Vestiges of the Natural History of Creation*. In *British and Foreign Medico-Chirurgical Review*, Vol. 13, pp. 425–39.

Book reviews were a common ingredient of scientific journals from their very first years. These started out typically as an expanded table of contents, then evolved into summaries with critical evaluation of the content. This selection comes from one of the most vituperative critical reviews ever appearing in a learned scientific journal. In this medical journal, Huxley lambastes an anonymous and immensely popular book proposing evolution by a master program built into primordial forms of life. This book, posthumously revealed to have been written by a successful book publisher, Robert Chambers, riled both religious conservatives and scientists like Huxley. The former felt threatened by the idea that humans had evolved over time, and the latter mainly objected to the many flaws in its biological, geological, and astronomical science.

Huxley apparently was also appalled that this "mass of pretentious nonsense" had gone through no less than ten editions in ten years. This fifteen-page review goes into technical detail primarily of interest to the specialist. In the first few paragraphs, however, Huxley makes perfectly clear his negative opinion of the book under review:

> In the mind of any one at all practically acquainted with science, the appearance of a new edition of the "Vestiges" at the present day, has much the effect that the inconvenient pertinacity of *Banquo* had upon *Macbeth*. "Time was, that when the brains were out, the man would die." So time was, that when a book had been shown to be a mass of pretentious nonsense, it, too, quietly sunk into its proper limbo. But these days appear, unhappily, to have gone by, and the same utter ignorance of the public mind as to the methods of science and the criterion of truth, which were evidenced to a Faraday by the greedy reception of the table-turning folly, have encouraged the author of the "Vestiges" to venture upon a *tenth* edition, "with extensive additions and emendations." We doubt not that this edition—very pretty and well got up it is—will be as greedily swallowed by those to whom it is offered, as any of the other nine, to the great glory and no small profit, of its modest and ingenious author. We grudge no man either the glory or the profit to be obtained from charlatanerie, and we can hardly expect that those who are so ignorant of science as to be misled by the "Vestiges," will read what we have to say upon the subject; but a book may, like a weed, acquire an importance by neglect, which it could have attained in no other mode; and, therefore, it becomes our somewhat unpleasant duty to devote a few of our pages to an examination of some of the leading points of this once attractive and still notorious work of fiction: indeed we feel the more called upon to undertake this criticism at present, since, as we shall see, the "Vestigiarian"

has bolstered up his case by the use of names and authorities, which, were it justifiable, might give a certain value to his statements.

It would be no less wearisome than unprofitable to go into a detailed examination of all the blunders and mis-statements of the "Vestiges"—to drag to light all the suggestions of the false and suppressions of the true, which abound in almost every page, and which, in a work of such pretension, of such long elaboration, and so filled with whining assertions of sincerity, are almost as culpable, if they proceed from ignorance, as if they were the result of intention. We propose, therefore, to confine our attention to the fundamental proposition of the book and to some one or two of those matters connected with the biological sciences, which come more particularly within the province of this review.

Huxley's opening reference to Shakespeare assumes that his reader is not just a scientist, but a cultivated person, one able to appreciate a rapier-like assault accompanied by a firm moral purpose: Chambers has the effrontery to dress his meretricious opinions in the garb of scientific authority. While Huxley's reference to Shakespeare is obvious, his reference to Michael Faraday may be less so. It is to two articles in which the great physicist showed that the popular belief that spirits could turn tables during séances lacked scientific foundation.

Charles Darwin Is Forced to Publish

Charles Lyell and J. D. Hooker, 1858. "Summary of papers communicated by Charles Darwin and Alfred Russel Wallace to Linnean Society of London, July 1, 1858." *Journal of the Proceedings of the Linnean Society*, Vol. 3, pp. 45–46.

As a stopgap measure to ensure Charles Darwin's magisterial book in progress was not scooped by Alfred Russel Wallace's superb, but far less detailed scientific article on natural selection, geologist Charles Lyell and botanist Joseph Hooker quickly organized the somewhat unusual reading of papers on the subject written by both authors at a meeting of the Linnean Society, followed shortly thereafter by publication in its journal. In the rush to publish something that would establish his priority, Darwin contributed a short extract on natural selection from his work in progress, along with a somewhat abbreviated form of a letter he sent to the American botanist Asa Gray, in which he concluded "This sketch is *most* imperfect; but in so short a space I cannot make it better. Your imagination must fill up very wide blanks." For some scientific discoveries only the commodiousness of a book will do.

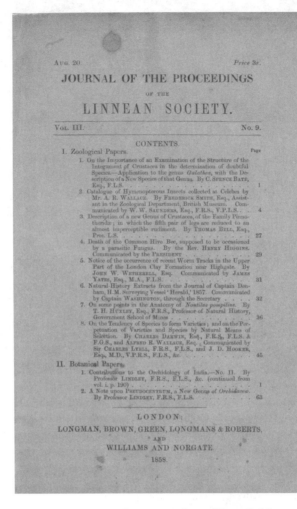

Aug. 20   Price 3s.

JOURNAL OF THE PROCEEDINGS

OF THE

LINNEAN SOCIETY.

Vol. III.   No. 9.

CONTENTS.

I. Zoological Papers.   Page

1. On the Importance of an Examination of the Structure of the
   Integument of Crustacea in the determination of doubtful
   Species.—Application to the genus *Galathea*, with the De-
   scription of a New Species of that Genus. By C. SPENCE BATE,
   Esq., F.L.S. 1

2. Catalogue of Hymenopterous Insects collected at Celebes by
   Mr. A. R. WALLACE. By FREDERICK SMITH, Esq., Assistant in the Zoological Department, British Museum. Communicated by W. W. SAUNDERS, Esq., F.R.S., V.P.L.S. . . . 4

3. Description of a new Genus of Crustacea, of the Family Pinno-
   theridæ; in which the fifth pair of legs are reduced to an
   almost imperceptible rudiment. By THOMAS BELL, Esq.,
   Pres. L.S. 27

4. Death of the Common Hive Bee, supposed to be occasioned
   by a parasitic Fungus. By the Rev. HENRY HIGGINS.
   Communicated by the PRESIDENT 29

5. Notice of the occurrence of recent Worm Tracks in the Upper
   Part of the London Clay Formation near Highgate. By
   JOHN W. WETHERELL, Esq. Communicated by JAMES
   YATES, Esq., M.A., F.L.S. 31

6. Natural-History Extracts from the Journal of Captain Den-
   ham, H.M. Surveying Vessel 'Herald,' 1857. Communicated
   by Captain WASHINGTON, through the Secretary 32

7. On some points in the Anatomy of *Nautilus pompilius*. By
   T. H. HUXLEY, Esq., F.R.S., Professor of Natural History,
   Government School of Mines 36

8. On the Tendency of Species to form Varieties; and on the Per-
   petuation of Varieties and Species by Natural Means of
   Selection. By CHARLES DARWIN, Esq., F.R.S., F.L.S. &
   F.G.S., and ALFRED R. WALLACE, Esq. Communicated by
   Sir CHARLES LYELL, F.R.S., F.L.S., and J. D. HOOKER,
   Esq., M.D., V.P.R.S., F.L.S., &c 45

II. Botanical Papers.

1. Contributions to the Orchidology of India.—No. II. By
   Professor LINDLEY, F.R.S., F.L.S., &c. (continued from
   vol. i. p. 190) 1

2. A Note upon PSEUDOCENTRUM, a New Genus of *Orchideæ*.
   By Professor LINDLEY, F.R.S., F.L.S. 63

LONDON:

LONGMAN, BROWN, GREEN, LONGMANS & ROBERTS,

AND

WILLIAMS AND NORGATE.

1858.

FIGURE 3-6 · Cover to 1858 issue of *Journal of the
Proceedings of the Linnean Society* with first papers on natural
selection. Courtesy of Special Collections Research
Center, University of Chicago Library.

Neither author was among the approximately 30 who attended the now-
famous meeting: Darwin remained home mourning the death of his youngest
child, while Wallace was in the field collecting specimens. Many members
attending the July 1 meeting when the papers were read apparently did not
understand they had witnessed a historic event. At the beginning of the new

year, the Society's President observed that the year 1858 had not "been marked by any of these striking discoveries which at once revolutionize, so to speak, [our] department of science." It was not until publication of Darwin's *Origin of the Species* in 1859 that a fierce public controversy was ignited, a controversy lasting to the present day.

The selection below, in the form of a short letter addressed to the Linnean Society's secretary, reproduces the Lyell-Hooker preface to the papers by Darwin and Wallace:

My Dear Sir,—

The accompanying papers, which we have the honour of communicating to the Linnean Society, and which all relate to the same subject, viz. the Laws which affect the Production of Varieties, Races, and Species, contain the results of the investigations of two indefatigable naturalists, Mr. Charles Darwin and Mr. Alfred Wallace.

These gentlemen having, independently and unknown to one another, conceived the same very ingenious theory to account for the appearance and perpetuation of varieties and of specific forms on our planet, may both fairly claim the merit of being original thinkers in this important line of inquiry; but neither of them having published his views, though Mr. Darwin has for many years past been repeatedly urged by us to do so, and both authors having now unreservedly placed their papers in our hands, we think it would best promote the interests of science that a selection from them should be laid before the Linnean Society.

Taken in the order of their dates, they consist of:—

1. Extracts from a MS. work on Species*, by Mr. Darwin, which was sketched in 1839, and copied in 1844, when the copy was read by Dr. Hooker, and its contents afterwards communicated to Sir Charles Lyell. The first Part is devoted to "The Variation of Organic Beings under Domestication and in their Natural State;" and the second chapter of that Part, from which we propose to read to the Society the extracts referred to, is headed, "On the Variation of Organic Beings in a state of Nature; on the Natural Means of Selection; on the Comparison of Domestic Races and true Species."

2. An abstract of a private letter addressed to Professor Asa Gray, of Boston, U.S., in October 1857, by Mr. Darwin, in which he repeats his views, and which shows that these remained unaltered from 1839 to 1857.

3. An Essay by Mr. Wallace, entitled "On the Tendency of Varieties to depart indefinitely from the Original Type." This was written at Ternate [Indonesia], in February 1858, for the perusal of his friend and correspondent

Mr. Darwin, and sent to him with the expressed wish that it should be forwarded to Sir Charles Lyell, if Mr. Darwin thought it sufficiently novel and interesting. So highly did Mr. Darwin appreciate the value of the views therein set forth, that he proposed, in a letter to Sir Charles Lyell, to obtain Mr. Wallace's consent to allow the Essay to be published as soon as possible. Of this step we highly approved, provided Mr. Darwin did not withhold from the public, as he was strongly inclined to do (in favour of Mr. Wallace), the memoir which he had himself written on the same subject, and which, as before stated, one of us had perused in 1844, and the contents of which we had both of us been privy to for many years. On representing this to Mr. Darwin, he gave us permission to make what use we thought proper of his memoir, &c.; and in adopting our present course, of presenting it to the Linnean Society, we have explained to him that we are not solely considering the relative claims to priority of himself and his friend, but the interests of science generally; for we feel it to be desirable that views founded on a wide deduction from facts, and matured by years of reflection, should constitute at once a goal from which others may start, and that, while the scientific world is waiting for the appearance of Mr. Darwin's complete work, some of the leading results of his labours, as well as those of his able correspondent, should together be laid before the public. We have the honour to be yours very obediently,

Charles Lyell
Jos. D. Hooker

* This MS. work was never intended for publication, and therefore was not written with care. — C. D. 1858.

The obvious purpose behind this skillfully crafted letter is to head off what could have turned into an "ungentlemanly" controversy over priority. One could easily imagine its mediating authors slaving over the exact phrasing and arrangement of paragraphs to employ in this delicate matter. The underlying argument at work here would appear to be that Darwin concocted his theory of evolution first, but only dared share it with a few friends in private letters and conversations, and only now grudgingly publishes something imperfect, "not written with care," before the release of his complete work "matured by years of reflection," while Wallace was the first to transform his theory into a scientific article for public consumption. Thus, both men deserve credit for a "very ingenious theory" and can "fairly claim the merit of being original thinkers in this important line of inquiry." In the end, Darwin won the most

On the Tendency of Species to form Varieties; and on the Per-
petuation of Varieties and Species by Natural Means of
Selection. By CHARLES DARWIN, Esq., F.R.S., F.L.S., &
F.G.S., and ALFRED WALLACE, Esq. Communicated by Sir
CHARLES LYELL, F.R.S., F.L.S., and J. D. HOOKER, Esq.,
M.D., V.P.R.S., F.L.S., &c.

[Read July 1st, 1858.]

London, June 30th, 1858.

MY DEAR SIR,—The accompanying papers, which we have the
honour of communicating to the Linnean Society, and which all
relate to the same subject, viz. the Laws which affect the Production of Varieties, Races, and Species, contain the results of the
investigations of two indefatigable naturalists, Mr. Charles Darwin
and Mr. Alfred Wallace.

These gentlemen having, independently and unknown to one
another, conceived the same very ingenious theory to account for
the appearance and perpetuation of varieties and of specific forms
on our planet, may both fairly claim the merit of being original
thinkers in this important line of inquiry; but neither of them
having published his views, though Mr. Darwin has for many
years past been repeatedly urged by us to do so, and both authors
having now unreservedly placed their papers in our hands, we
think it would best promote the interests of science that a selection from them should be laid before the Linnean Society.

Taken in the order of their dates, they consist of:—

1. Extracts from a MS. work on Species*, by Mr. Darwin, which
was sketched in 1839, and copied in 1844, when the copy was read
by Dr. Hooker, and its contents afterwards communicated to Sir
Charles Lyell. The first Part is devoted to "The Variation of
Organic Beings under Domestication and in their Natural State;"
and the second chapter of that Part, from which we propose to
read to the Society the extracts referred to, is headed, "On the
Variation of Organic Beings in a state of Nature; on the Natural
Means of Selection; on the Comparison of Domestic Races and
true Species."

2. An abstract of a private letter addressed to Professor Asa
Gray, of Boston, U.S., in October 1857, by Mr. Darwin, in which

* This MS. work was never intended for publication, and therefore was not
written with care.—C. D. 1858.

FIGURE 3-7 · First page of Lyell-Hooker summary of
evolutionary pages in 1858 issue of *Journal of the Proceedings
of the Linnean Society*. Courtesy of Special Collections
Research Center, University of Chicago Library.

recognition, at least in part because of the powerful argument he formulated a year later in book form, buttressing his theory with a mountain of evidence.

SPECIALIZED LITERATURE: PHYSICS

Mayer Discovers the Conservation of Energy—or Does He?

Julius Robert Mayer, 1842. "Bemerkungen über die Kräfte der unbelebeten Natur" (Remarks concerning the forces of inorganic nature). *Annalen der Chemie und Pharmacie*, Vol. 42, pp. 233–40. Modified from translation by G. C. Foster, *Philosophical Magazine* (1862), Vol. 24, pp. 371–77.

This publication led to a priority dispute over the discoverer of the law of conservation of energy, the principle that its total amount does not vary. If I cause a bowling ball to be lifted, I have done work; I have effected its elevation by the expenditure of energy. But where did the energy go? It has not disappeared, but is now contained in the bowling ball as *potential energy*. Potential energy is real enough, as you will see if I subsequently drop the ball on your foot.

The German physician Julius Robert Mayer thought that he was the first to articulate this important principle in print; but so did James Joule, the English physicist. Reading over Mayer's article today, it seems to us that he discovered only a special case of energy conservation, the convertibility of heat and work. (In physics, "work" is the product of the force applied to an object and the distance the object moves. According to physics, writing this book is not work.)

In this article, Mayer does not speak of the conservation of energy, but rather the equivalence of "falling force and motion" and "heat." He derives this equivalence from the general principle of the equivalence of cause and effect:

> If **falling force and motion** are equivalent to **heat**, **heat** must also naturally be equivalent to **motion and falling force**. Just as heat appears as an *effect* of the diminution of bulk and of the cessation of motion, so also does heat disappear as a *cause* when its effects are produced in the shape of motion, expansion, or raising of weight [bold added; italics in original].

Mayer conveys his new principle in the form of a sentence structure called "antimetabole" in classical rhetoric texts, where two terms in the first part of the sentence are repeated in reverse order in the second part: "If **falling force and motion** are equivalent to **heat**, **heat** must also naturally be equivalent to **motion and falling force**." The next sentence works similarly but is a little more elaborate, involving both repetition and antithesis in which the terms in the first part of the sentence have a corresponding term in the second part ("heat/heat," "appear/disappear," "effect/cause," "diminution of bulk/expansion, or raising of weight," "cessation of motion/shape of motion").

Having stated this interesting partnership among heat, force, and motion, Mayer transforms his abstract formulation into real-world terms. He instantiates the first equivalence (motion into heat) with a very old technology (water-wheels), and the second equivalence (heat into motion) with a new technology in the early nineteenth century (steam engines in locomotives):

> In water-mills, the continual diminution in bulk which the earth undergoes, owing to the fall of the water, gives rise to motion, which afterwards disappears again, continually calling forth a great quantity of heat; and inversely, the steam-engine serves to turn heat back into motion or the raising of weights. A locomotive engine with its train may be compared to a distilling apparatus; the heat applied under the boiler turns into motion, and this is turned back to heat at the axles of the wheels.

Finally, Mayer gives an example showing the practical application of his abstract principle. He calculates the mechanical equivalent of heat, the amount of heat needed to raise a given quantity of water one degree centigrade. He then applies this thermodynamic calculation to steam engines to show how inefficient they are in converting heat to work, arguing that there is much room for improvement. Mayer suggests switching to electricity. Here is how he presents this example, making a strong argument for the utility of his research:

> Our claim follows as a necessary consequence of the principle "causa aequat effectum" [the cause equals the effect], and is in perfect accord with all the phenomena of nature. We close with a practical deduction. The solution of the equations for falling force and motion requires that the space fallen through in a given time, e.g. the first second, should be experimentally determined; in like manner, the solution of the equations subsisting between falling force and motion on the one hand and heat on the other, requires an answer to the question, How great is the amount of heat corresponding to a certain quantity of falling force or motion? For instance, we must ascertain how high a given weight requires to be raised above the ground in order that its falling force may be equivalent to the raising of the temperature of an equal weight of water from 0° to 1° C. The attempt to show that such an equation is the expression of a physical truth may be regarded as the substance of the foregoing remarks.

By applying the principles that have been set forth to the relations between the temperature and the volume of gases, we find that the sinking of a mercury column by which a gas is compressed is equivalent to the quantity of heat set free by the compression; it follows from this that if the ratios of the heat capacities of air under constant pressure and constant volume equals 1.421, then the warming of a given weight of water from 0° to 1° C corresponds to the fall of an equal weight from a height of 365 meters. If we compare with this result the working of our best steam-engines, we see how small a part only of the heat applied under the boiler is really transformed into motion or the raising of weights; and this may serve as justification for the attempts at the profitable production of motion by some other method than the expenditure of the chemical difference between carbon and oxygen—more particularly by the transformation into motion of electricity obtained by chemical means.

The Nature of Heat

Rudolf Clausius, 1857. "Über die Art der Bewegung welche wir Wärme nennen" (On the nature of the motion we call heat). *Annalen der Physik*, Vol. 100, pp. 353–80. Modified from the translation in *Philosophical Magazine* (1857), Vol. 14, pp. 108–27.

As a consequence of the Industrial Revolution, and especially the widespread use of James Watt's invention the steam engine, English and German physics focused on the nature of heat and its conversion to work. In England we have Michael Faraday, James Prescott Joule, and James Clerk Maxwell; in Germany and Austria, Julius Robert Mayer, Rudolf Clausius, Hermann von Helmholz, and Ludwig Bolzmann. It was a glorious time for the development of thermodynamics, the science of heat.

Although Clausius's thermodynamic paper covers a great deal of ground, it nevertheless focuses on the molecular motion of gases induced by heat. Clausius's compatriot, August Karl Krönig, had posited a translatory motion for these molecules, one in which they moved in three dimensions: up, down, and sideways, in a straight line until they hit an obstacle—a wall of a vessel or another molecule; then the molecules bounced off in another direction, something like a billiard ball. Clausius agreed—to a point. He dissented in that he felt that these motions were not enough to account for the work potential of the gas. Clausius accounts for this difference by postulating two additional motions: the molecules rotate, and their atomic and subatomic constituents vibrate. He constructed this remarkable explanation almost a half century before the existence of molecules had been confirmed experimentally by Jean Perrin (see chapter 5).

In the section we excerpt, Clausius states his amazingly prescient theory in plain language and also addresses a priority issue regarding Krönig's earlier research:

A memoir has lately been published by Krönig, under the title, "Essential features of a theory of gases," in which I recognize some of my own views. Seeing that Krönig has arrived at these views just as I have, independently, and has published them before me, all claim to priority on my part is of course out of the question; nevertheless, the subject having once been mooted in this memoir, I feel myself induced to publish those parts of my own views which I have not yet found in it. For the present, I shall confine myself to a brief indication of a few principal points, and reserve a more complete analysis for another time.

Krönig assumes that the molecules of a gas do not oscillate about definite positions of equilibrium, but that they move with constant velocity in straight lines until they strike against other molecules, or against some surface which forms an impenetrable barrier. I share this view completely, and I also believe that the expansive force of the gas arises from this motion. On the other hand, I am of the opinion that this is not their only motion.

In the first place, the hypothesis of a rotary as well as a progressive motion of the molecules at once suggests itself; for at every impact of two bodies, unless it happens to be central and rectilinear, a rotary as well as a translatory motion ensues.

I am also of an opinion that vibrations take place within these masses as they move forward. Such vibrations are conceivable in several ways. Even if we limit ourselves to the consideration of the atomic masses, and regard these as absolutely rigid, it is still possible that a molecule, which consists of several atoms, may be not absolutely rigid in structure, but that within it the several atoms are movable, at least to a certain extent, and thus may be capable of oscillating with respect to each other.

I may also remark, that by thus ascribing a movement to the atomic masses themselves, we do not exclude the hypothesis that each atomic mass may consist in part of a quantity of finer matter, which, without separating from the atom, may still be moveable within its vicinity.

By means of a mathematical investigation given at the end of the present memoir, it may be proved that the *vis vita* [work in the sense of force times distance] of the translatory motion alone is too small to represent the whole heat present in the gas; so that unless we argue from probabilities, we are compelled to assume one or more motions of another kind. According to this calculation, the excess of the whole *vis vita* over that of the translatory motion alone is particularly important in gases of a complicated chemical constitution, in which each molecule consists of a great number of atoms.

Clausius's argument bears a close resemblance to William Harvey's for the circulation of the blood from two centuries earlier. Just as Harvey infers that the ebb and flow of the blood is insufficient to account for the rapidity of the pulse, so Clausius infers that the motions Krönig posits are insufficient to account for the *vis vita*. In subsequent sections, he describes molecular behavior in exquisite detail; in his final section, he derives his conclusion mathematically. The time had come when physics could not exist separate from its mathematical realization.

SPECIALIZED LITERATURE: CHEMISTRY

Argument by Parody

Anonymous (Friedrich Wöhler), 1839. "Das enträthselte Geheimniss der geistigen Gährung" (Solving the riddle of alcoholic fermentation). *Annalen der Pharmacie*, Vol. 29, pp. 100–104.

Although this article is anonymous, historians believe that its author was Friedrich Wöhler, who had the blessing of the journal editor, Justus Liebig. It reports bogus experiments meant to ridicule recently published research on alcoholic fermentation by Theodore Schwann and Charles Cagniard-Latour. Several decades after publication, the prominent English scientist and essayist T. H. Huxley called it "the most surprising paper that ever made its appearance in a grave scientific journal." In it, Liebig and Wöhler—who made many important contributions to the then-new field of organic chemistry—had the misfortune to dismiss with contempt the correct view that yeast is a living organism capable of converting sugar water into alcohol. Instead, they believed that yeast acted as an inanimate catalyst in a chemical reaction. We quote a large segment from this parody—a literary concoction worthy, in Huxley's estimation (and ours), of Jonathan Swift:

> By the simplest means that can be imagined, a new theory of alcoholic fermentation has been revealed to me. I now regard this process, so mysterious

up to now, as fully understood. It shows how simply Nature brings forth its wonders. I made this discovery thanks to a remarkable microscope constructed, according to the plan of the renowned Ehrenberg, by none other than the distinguished artisan Pistorius.

Under this microscope, beer yeast, suspended in water, decomposes into a swarm of small pellets, scarcely 1/800 of a line [0.00125 inches] across, and numerous fine threads, consisting, unmistakably, of protein. When these small pellets are transferred to sugar water, we see them for what they are—eggs. These swell and burst, giving rise to tiny animals that multiply with a rapidity unparalleled among living creatures. These animals look like no other species on the face of the earth. They are shaped like Beindorf distillation flasks without their cooling apparatus. A tube projecting from their helmet-shaped heads forms a kind of snout lined with fine bristles 1/2000 of a line [0.0005 inches] long. No teeth or eyes are evident, but a stomach, intestines, and an anus—a rose-red spot—are clearly distinguishable, as is an organ devoted to urinary secretion. From the moment they spring forth from their eggs, these creatures extract sugar from the solution that surrounds them; it can be seen very plainly as it arrives in their stomachs. The sugar is immediately digested, a process simultaneous with, though easily distinguishable from the evacuation of waste that follows. In a word, these creatures turn sugar into alcohol and carbon dioxide. When filled, their urinary bladders are shaped like a champagne bottle; when empty, they resemble a small button. The process of evacuation may be observed: a bubble of gas multiplies in diameter until it is ten times its original size; at this point it is released into the surrounding solution by means of a spiral rotation of the creature's exterior ring-shaped muscles. I consider compelling the parallel between this action and that of galvanic electricity conducted through a wire. In the opinion of renowned physicists, magnetism is generated under such conditions, magnetism that follows a stern dictate of Nature in taking a spiral course. I cite this as evidence in support of Döbereiner and Schweigger's contention that fermentation and electro-magnetism are inextricably intertwined. These observations show that their insight was correct; they discovered the truth, though they lacked the assistance of my powerful microscope.

Let us lay this electro-magnetic hypothesis aside for a moment and return to the physiology of these newly discovered animals. From their anuses, a characteristic fluid arises, lighter than the surrounding medium, and from their truly enormous sexual organs a stream of carbon dioxide is squirted out at very short intervals... Quantitative data and illustrations of this new organism will follow in a more detailed monograph.

This article creates its effect by combining a deadpan style, typical of science, with a description that is progressively more raucous and ribald. The final stroke in the quoted passage is the parallelism asserted between the rotatory muscle action of these imaginary organisms and electromagnetism.

Finally, a Chemistry of Real Structures

Archibald Scott Couper, 1858. "On a new chemical theory." *Philosophical Magazine*, Vol. 16, pp. 104–16.

This chemical article appeared in the same year in the two leading journals for the physical sciences in England and France, *Philosophical Magazine* and *Annales de Chimie et de Physique*. The publication of this brilliant paper, however, had been apparently delayed due to the slowness of Couper's mentor in shipping it from his desk to a publisher's. As a consequence, another young man, August Kekulé, successfully claimed priority in the discovery of the tetravalency of carbon and the ability of this element to form carbon chains by linking with other of its exemplars. It was this self-binding power that led directly to Kekulé's impressive solution to the problem of the structure of benzene—his famous benzene ring consisting of six carbons bound to each other and to six neighboring hydrogens. This insight formed a keystone in modern organic chemistry. Kekulé went on to a brilliant career as a chemist; Couper did not.

Before Kekulé and Couper, the chemists Charles Gerhardt and Alexander Williamson had proposed a theoretical program in which organic compounds were classified by types, clusters of elements from which new compounds were created by substitution. Gerhardt and Williamson did not claim that these types actually existed in nature. They only insisted that such types were a useful means of classification.

In his paper, Couper finds these types objectionable: not only are they imaginary, they stifle research into the real principles by which organic compounds form. Those principles, Couper insists, will be elucidated only when the binding forces of the chemical elements, especially the ubiquitous element carbon, are elucidated. He argues by analogy "to the common events of life." He imagines someone who claims that there is a cluster of special words that together form the foundation of language. New words come from the combination and recombination of these words. (In analogy to the chemical theory he is criticizing, Couper speaks of their "substitution and double decomposition.") Couper says that this is absurd. The only sensible way to solve the problem of word formation is to start with the letters as the fundamental building blocks whose rules of combination are to be elucidated. (Couper must be taken to mean the letters not of the actual, but of a phonetic alphabet.) The following is Couper's polemic against the type theory:

> Should the principle which is therein adopted be applied to the common events of life, it will be found that it is simply absurd. Suppose that some one were to systematize the formation of letters into words that formed the contents of a book. Were he to begin by saying that he had discovered a *certain word which would serve as a type, and from which by substitution and double decomposition all the others are to be derived,* — that he by this means could not only form new words, but new books, and books almost *ad infinitum,* — that this word also formed an admirable point of comparison with all the others, — that in all this there were only a few difficulties, but that these might be ingeniously overcome, — he would state certainly an empirical truth. At the same time, however, his method would, judged by the light of common sense, be an absurdity. But a principle which common sense brands with absurdity, is philosophically false and a scientific blunder . . . .
>
> The sure and invincible method at arriving at every truth which the mind is capable of discovering is always one and the same. It is that, namely, of throwing away all generalization, of going back to first principles, and of letting the mind be guided by these alone. It is the same in common matters. It is the same in science. To reach the structure of words we must go back, seek out the undecomposable elements, viz. the letters, and study carefully their powers and bearing. Having ascertained these, the composition and structure of every possible word is revealed. It would be well to call to recollection the parallelism of chemical research with that of every other search after truth; for it has been in overlooking this that in chemistry false and vacillating theories have been advocated and a wrong route so often pursued. In mathematics, the starting-point is not generalizations, but axioms, ultimate principles. In metaphysics Descartes led the way to progress by analysing till he thought he could reach some ultimate elements beyond which it was impossible for him to go, then studying their force and power, and proceeding synthetically. The recognition of this method wrought the regenerations of science and philosophy.
>
> On the other hand, look where Gerhardt's generalization of Williamson's generalization leads him, and legitimately too, — a fact which his logical spirit clearly discerned. He is led not to explain bodies according to their composition and inherent properties, but to think it necessary to restrict chemical science to the arrangement of bodies according to their decomposition,

and to deny the possibility of our comprehending their molecular constitution. Can such a view tend to the advancement of science? Would it not
be only rational, in accepting this veto, to renounce chemical research altogether?

These reflections naturally lead to the inquiry after another theory more
adequate to satisfy the just demands which can be made upon it. There is one
which, as it is supported by many distinguished chemists, cannot be passed
over altogether unnoticed. It is that of the theory of certain combinates in
organic chemistry which are to be viewed as analogous to, "playing the part
of," inorganic elements. These are denominated radicals, and are supposed
to be contained in all organic chemical products.

In addition to this, and also in connexion with it, there is a doctrine
describing many combinates to be copulated, conjugated, by addition.

It is impossible here to enter upon any extensive criticism of this theory.
I can only remark that it is not merely an unprofitable figure of language,
but is injurious to science, inasmuch as it tends to arrest scientific inquiry by
adopting the notion that these quasi elements contain some unknown and
ultimate power which it is impossible to explain. It stifles inquiry at the very
point where an explanation is demanded, by putting the seal of elements, of
ultimate powers, on bodies which are known to be anything but this.

Science demands the strict adherence to a principle in direct contradiction
of this view. The first principle, without which research cannot advance a
step, dare not be ignored; namely, that a whole is simply a derivative of its
parts. As a consequence of this, it follows that it is absolutely necessary to
scientific unity and research to consider these bodies as entirely derivative,
and as containing no secret ultimate power whatever, and that the properties
which these so-called quasi elements possess are a direct consequence of the
properties of the individual elements of which they are made up.

Nor is the doctrine of bodies being "conjugated by addition" a whit in
advance of that which I have just been considering. This doctrine adopts
the simple expedient of dividing certain combinates, if possible, into two
imaginary parts, of which one or both are bodies already known. Then it
tells us that these two parts are found united in this body. But how are
they united, or what force binds them together, it does not inquire. Is this
explication arbitrary? Is it instructive? Is it science?

Couper's is a powerful analogy, accompanied by a wicked polemic. It is not
entirely unlikely that this combination of evident power and equally evident
hostility contributed to the unfortunate fate of his neglected paper.

No Holds Barred

....

Hermann Kolbe, 1871. "Moden der modernen Chemie" (Fashions in "modern" chemistry). *Journal für praktische Chemie, Neue Folge*, Vol. 4, pp. 241–71.

In the previous selection, we glimpsed a young chemist trying to make his mark and upturn the status quo. In this next selection, one of the leading chemists of his day, the 53-year-old Hermann Kolbe, launches the opening salvo in a polemical campaign against August Kekulé's theory of structural chemistry. It is arguably one of the most extreme attacks on a promising new theory, and on its proponents, in the history of science. The question at issue in the debate is the extent to which chemists could specify the structure of molecules, the actual position of their atoms relative to one another in three-dimensional space. An organic compound, such as benzene, has a chemical formula of C_6H_6—that is, it consists of six carbon and six hydrogen atoms. This is an empirical formula; it says nothing about the relative positions of the carbon and hydrogen atoms in space. About these, Kolbe insisted, nothing *could* be said.

To make his point about the enormous disparity between the evidence for chemical structure and the highly speculative conclusions of structural chemists, Kolbe resorts to a series of biting rhetorical questions followed by a prank meant to illustrate the absurdity of his opponents' position:

> Do Herr Schützenberger [eminent French chemist then at Ecole Supérieure des Sciences], and do "modern" chemists who so easily go astray in exactly the same manner, fail to realize that, to construct these so-called constitution formulas, they need hardly be fully awake? Do they not realize that these formulas say nothing whatever about the true constitution of compounds? Above all, do they not realize the utter impossibility of turning the few facts we have about the chemical constitution of compounds into anything more than a vague conjecture concerning any of the three unknown bonds of the compound in question?
>
> I wanted to reveal once and for all the completely impermissible elasticity of the structural formulas of so-called bonding chemistry. So, as soon as I had read through Schützenberger's treatise, I derived three empirical formulas from his analysis:

$$CO \, PtCl_2$$
$$C_2O_2 \, PtCl_2$$
$$C_3O_3 \, Pt_2Cl_4$$

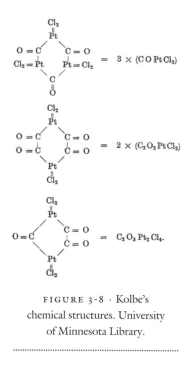

FIGURE 3-8 · Kolbe's
chemical structures. University
of Minnesota Library.

....

Then I said to my copyist, to whom the subject was terra incognita, and to whom I communicated only that the bonds must be between platinum chloride and carbonic oxide: Turn these three empirical formulas into structural ones. In just half an hour, unencumbered by any specialist knowledge, he spread before me three beautiful examples of his speculative enterprise:

As we can easily see, the last schema coincides with that developed by Schützenberger; in the case of the first two, Kekulé's six-element benzene ring spontaneously occurred to the copyist. If I were a modern chemist, I must confess, this formulation would please me more than Schützenberger's, especially as no fact of chemistry bars either the tripling or the doubling of the formulas.

Herr Kolbe's polemic here—and his even more personal attacks that followed—in favor of theories that the chemical community largely regarded as outdated, or at least under serious question, did not go over well. Many leading German chemists interpreted Kolbe's behavior as a loss of intellectual control, a loss evidenced by the tendency of his strong feelings to overwhelm the conceptual points he wanted to make. One of his many critical targets, the young Dutch chemist Jacobus Hendricus Van't Hoff, responded in a chemistry journal "I can only say that such behavior fortunately is not a sign of the times, but rather must be regarded as a contribution to understanding a single individual."

4

SELECT
PRE-MODERN
CLASSICS

THE SCIENTIFIC LITERATURE from the late eighteenth through the nineteenth century is increasingly dominated by a passion for factual precision, largely in the form of quantifications and a technical vocabulary, and by somewhat less timidity about theorizing to explain the acquired facts. Furthermore, in no sense do the various sciences appear to be advancing toward the impossible dream of a single theory that would explain the whole of nature; rather, each science is developing explanatory structures appropriate to its enterprise. As the sciences develop, they diverge rather than converge. Chemists are concerned with the nature of the elements, and the various ways in which they combine to form compounds; geologists, with the description and causal history of the physical features of the earth; and medical researchers, with the fundamental cause of various diseases. Physics focuses more and more on fundamental entities and fundamental forces and their mathematical description. Biology becomes, not a single unified discipline, but a federation of disciplines: very diverse enterprises from taxonomy to physiology to evolutionary theory. What the scientific literature reflects is not the division of labor within a single enterprise, science, but a loose coalition of enterprises with less and less conceptually in common.

In this chapter, we begin by sampling classic theory-driven articles by James Hutton, Alfred Russel Wallace, and Gregor Mendel, the first representing the inauguration of a revolution in geology, the second in evolutionary biology, the third in genetics—each opening a new research field cultivated with great fecundity right up to the present day. Also from the life sciences, we sample a trio of medical classics from Louis Pasteur, Rudolf Virchow, and Robert Koch, whose insights into the microscopic phenomena behind "certain common illnesses" helped revolutionize medical research and practice throughout the world.

The remaining articles in our menagerie reflect the rapid maturity of chemistry and physics during the nineteenth century. In chemistry, we read about Jöns Jacob Berzelius's advance in terminology; Gustav Kirchhoff and Robert Bunsen's, in element identification and the composition of luminous objects in outer space; August Kekulé's, in organic chemistry and the spatial representation of molecular structure; and as a crowning achievement, Dimtri Ivanovich Mendeleev's periodic table. In physics, the achievements are no less impressive: James Clerk Maxwell's mathematization of Michael Faraday's experimental insights in electromagnetism, Wilhelm Konrad Röntgen's extension of the spectrum to X-rays, and Pierre and Marie Curie's discovery of radioactive elements.

EARTH SCIENCE

Bringing Obscure Scientific Prose into the Light

James Hutton, 1788. "Theory of the earth; or an investigation of the laws observable in the composition, dissolution, and restoration of land upon the globe." *Transactions of the Royal Society of Edinburgh*, Vol. 1, pp. 209–304.

John Playfair, 1805. "Biographical account of the late Dr. James Hutton." *Transactions of the Royal Society of Edinburgh*, Vol. 5, pp. 39–99.

James Hutton was the first to propose a comprehensive theory in which the earth's age extended beyond the biblical calculation of 6,000 years. In his poetic words, the geologic evidence suggests "no vestige of a beginning,—no prospect of an end." Hutton's theory was that the present earth was formed by a cyclic process involving, as he put it in the title, "the composition, dissolution, and restoration of land upon the globe." Over unimaginably vast stretches of time, the environmental conditions dissolved the earth's continents into the oceans, where the land eventually compressed into solid matter and formed new continents. This cyclic process of decay and rebirth was not directly observable or measurable, so Hutton had to deduce it by studying present rock formations and their chemical composition, inferring the gradual geological forces at work, then extrapolating the very distant past from the present. But the power of Hutton's theory is not matched by the power of his prose:

> LET us now consider how far the other proposition, of strata being elevated by the power of heat above the level of the sea, may be confirmed from the examination of natural appearances.
>
> THE strata formed at the bottom of the ocean are necessarily horizontal in their position, or nearly so, and continuous in their horizontal direction or extent. They may change, and gradually assume the nature of each other, so far as concerns the materials of which they are formed; but there cannot be any sudden change, fracture or displacement naturally in the body of a stratum. But, if these strata are cemented by the heat of fusion, and erected with an expansive power acting below, we may expect to find every species of fracture, dislocation and contortion, in those bodies, and every degree of departure from a horizontal towards a vertical position.
>
> THE strata of the globe are actually found in every possible position: For from horizontal, they are frequently found vertical; from continuous, they are broken and separated in every possible direction; and, from a plane, they are bent and doubled. It is impossible that they could have originally been formed, by the known laws of nature, in their present state and position; and the power that has been necessarily required for their change, has not been inferior to that which might have been required for their elevation from the place in which they had been formed. . . .
>
> WE are now to conclude, that the land on which we dwell had been elevated from a lower situation by the same agent which had been employed in consolidating the strata, in giving them stability, and preparing them for the purpose of the living world. The agent is matter actuated by extreme heat, and expanded with amazing force.

After Hutton's death (1797), John Playfair, one of the leaders of early geological research based on the stratagraphic evidence and editor of the Royal Society of Edinburgh *Transactions*, clarified, extended, and promoted Hutton's theory. In an 1805 *Transactions* article describing Hutton and his work, Playfair vividly re-created an expedition the two of them made to the North Sea coast of Scotland, where they viewed Devonian Old Red Sandstone lying on top of grey sandstone from a different geological epoch, combined in a distorted pattern explainable by Hutton's theory. Unlike Hutton, Playfair is a literary artist as well as a scientist. He brings to life what Hutton dryly describes in abstract terms:

On us who saw these phenomena for the first time, the impression made will not easily be forgotten. The palpable evidence presented to us, of one of the most extraordinary and important facts in the natural history of the earth, gave a reality and substance to those theoretical speculations which, however probable, had never till now been directly authenticated by the testimony of the senses. We often said to ourselves, what clearer evidence could we have had of the different formation of these rocks, and of the long interval which separated their formation, had we actually seen them emerging from the bosom of the deep? We felt ourselves necessarily carried back to the time when the schistus on which we stood was yet at the bottom of the sea, and when the sandstone before us was only beginning to be deposited (in the shape of sand or mud) from the waters of a superincumbent ocean. An epocha still more remote presented itself, when even the most ancient of these rocks, instead of standing upright in vertical beds, lay in horizontal planes at the bottom of the sea and was not yet disturbed by that immeasurable force which has burst asunder the solid pavement of the globe. Revolutions still more remote appeared in the distance of this extraordinary perspective. The mind seemed to grow giddy by looking so far into the abyss of time; and while we listened with earnestness and admiration to the philosopher [Hutton] who was now unfolding to us the order and series of these wonderful events, we became sensible how much farther reason may sometimes go than imagination can venture to follow.

BIOLOGICAL SCIENCES

Metaphor in Evolutionary Biology
....

Alfred Russel Wallace, 1858. "On the tendency of varieties to depart indefinitely from the original type." *Journal of the Proceedings of the Linnean Society (Zoology)*, Vol. 3, pp. 53–62.

While bedridden with malaria during a field trip in the Malaysia Archipelago, Alfred Russel Wallace conceived in a flash of insight the central tenet of biological evolution—that its driving force was a winnowing process operating on variation of species in an environment of limited resources over immense swathes of time. Darwin had arrived at this theory some twenty years earlier, but refrained from publishing until he had amassed sufficient evidence to withstand the intense critical scrutiny he knew would surely come.

While Darwin's name is far more closely associated with evolutionary theory than Wallace's, both men had virtually the same revolutionary idea, and in both cases it arrived as a eureka moment, a mental flash in which the scattered pieces of an intellectual puzzle suddenly fit perfectly together. In his *Autobiography*, Darwin writes: "I can remember the very spot in the road whilst in my carriage, when to my joy the solution occurred to me; and this was long after I had come to Down. The solution, as I believe, is that the modified offspring of all dominant and increasing forms tend to become adapted to many and highly diversified places in the economy of nature." Similarly, Wallace wrote: "it suddenly flashed upon me [that] in every generation the inferior would be inevitably killed off and the superior would remain—that is, *the fittest would survive.*"

The following selection includes Wallace's use of philosopher Herbert Spencer's famous phrase "survival of the fittest" (Wallace never cared for the term "natural selection"). In this passage, Wallace uses a mechanical metaphor for the evolutionary process: it acts like "the centrifugal governor of the steam engine, which checks and corrects any irregularities almost before they become evident."

There is a difference in the metaphors for evolution used by Darwin and Wallace. In his *Notebooks* Darwin speaks of evolution in terms of "the warring of species" and of "a hundred thousand wedges trying to force every kind of adapted structure into the gaps in the oeconomy of Nature." This difference seems to indicate that, while Darwin sees intra- and inter-species competition as primary, Wallace sees the scarcity of environmental resources as the essential factor. This difference is also a difference in the *use* of metaphors. For Darwin, his metaphors helped him think about science; for Wallace, on the other hand, his metaphor helps us understand his argument:

> One of the strongest arguments which have been adduced to prove the original and permanent distinctness of species is, that *varieties* produced in a state of domesticity are more or less unstable, and often have a tendency, if left to themselves, to return to the normal form of the parent species; and this instability is considered to be a distinctive peculiarity of all varieties, even of those occurring among wild animals in a state of nature, and to constitute a provision for preserving unchanged the originally created distinct species.
>
> In the absence of scarcity or facts and observations as to *varieties* occurring among wild animals, this argument has had great weight with naturalists, and has led to a very general and somewhat prejudiced belief in the stability of species. . . .
>
> But it is the object of the present paper to show that this assumption is altogether false, that there is a general principle in nature which will cause many *varieties* to survive the parent species, and to give rise to successive variations departing further and further from the original type, and which also produces, in domesticated animals, the tendency of varieties to return to the parent form.

The life of wild animals is a struggle for existence. The full exertion of all their faculties and all their energies is required to preserve their own existence and provide for that of their infant offspring. The possibility of procuring food during the least favourable seasons, and of escaping the attacks of their most dangerous enemies, are the primary conditions which determine the existence both of individuals and of entire species. These conditions will also determine the population of a species; and by a careful consideration of all the circumstances we may be enabled to comprehend, and in some degree to explain, what at first sight appears so inexplicable—the excessive abundance of some species, while others closely allied to them are very rare. . . .

It appears evident, therefore, that so long as a country remains physically unchanged, the numbers of its animal population cannot materially increase. If one species does so, some others requiring the same kind of food much diminish in proportion. The numbers that die annually must be immense; and as the individual existence of each animal depends upon itself, those that die must be the weakest—the very young, the aged, and the diseased,—while those that prolong their existence can only be the most perfect in health and vigour—those who are best able to obtain food regularly, and avoid their numerous enemies. It is, as we commenced by remarking, "a struggle for existence," in which the weakest and least perfectly organized must always succumb. . . .

We have also here an acting cause to account for that balance so often observed in nature,—a deficiency in one set of organs always being compensated by an increased development of some others—powerful wings accompanying weak feet, or great velocity making up for the absence of defensive weapons; for it has been shown that all varieties in which an unbalanced deficiency occurred could not long continue their existence. The action of this principle is exactly like that of the centrifugal governor of the steam engine, which checks and corrects any irregularities almost before they become evident; and in like manner no unbalanced deficiency in the animal kingdom can ever reach any conspicuous magnitude, because it would make itself felt at the very first step, by rendering existence difficult and extinction almost sure to follow.

Choosing a Research Organism

Gregor Mendel, 1866. "Versuche über Pflanzenhybriden" (Experiments in plant hybridization). *Verhandlungen des Naturforschenden Vereines in Brünn*, Vol. 4, pp. 3–47. Translated by Eva R. Sherwood, in Curt Stern and Eva R. Sherwood, eds., *The Origins of Genetics: A Mendel Source Book*. San Francisco: W. H. Freeman (1966), pp. 1–48.

The Darwinian revolution was successful only in retrospect. Even with the help of allies, most famously Thomas Henry Huxley, Darwin's theory faced considerable opposition, not only on religious, but also on scientific grounds. For example, Darwin's theory presupposed heritability, but his original theory did not satisfactorily address this crucial issue; moreover, his later theory of heredity, pangenesis, was unsatisfactory as well. Less than a decade after the Darwin and Wallace articles and the *Origin* were published, so was a solution to Darwin's problem. Gregor Mendel's first article on heritability in the common pea was published in 1866. Darwin himself had a copy of that particular *Proceedings* in his library. And as indicated by the uncut pages in that issue, he likely never read Mendel's article. It had no immediate influence and had to be "rediscovered" early in the following century.

In launching his experimental project, Mendel had to first solve an important methodological problem: choice of research organism. He starts with a methodological discussion of the characteristics a plant should have if we are to "determine the number of different forms in which hybrid progeny appear, permit classification of these forms in each generation with certainty, and ascertain their numerical interrelationships." The plant, Mendel tells us, should possess quantifiable features that will occur in every generation, be protected from contamination by foreign pollen during flowering, and yield hybrids as fertile as the parents. Mendel argues that the garden pea meets all requirements.

Here is how Mendel explained the advantages of the pea's floral structure for his breeding experiments; this passage is a direct ancestor of the materials and methods section of the modern scientific article (see chapter 6):

> Selection of the plant group for experiments of this kind must be made with the greatest possible care if one does not want to jeopardize all possibility of success from the very outset.
>
> The experimental plants must necessarily
>
> 1. Possess constant differing traits.
>
> 2. Their hybrids must be protected from the influence of all foreign pollen during the flowering period or easily lend themselves to such protection.
>
> 3. There should be no marked disturbances in the fertility of the hybrids and their offspring in successive generations.
>
> Contamination with foreign pollen that might take place during the experiment without being recognized would lead to quite erroneous conclusions. Occasional forms with reduced fertility or complete sterility, which occur among the offspring of many hybrids, would render the experiments very difficult or defeat them entirely. To discover the relationships of hybrid forms to each other and to their parental types it seems necessary to observe *without exception all* members of the series of offspring in each generation.

From the start, special attention was given to the *Leguminosae* because of their particular floral structure. Experiments with several members of this family led to the conclusion that the genus *Pisum* had the qualifications demanded to a sufficient degree. Some quite distinct forms of this genus possess constant traits that are easily and reliably distinguishable, and yield perfectly fertile hybrid offspring from reciprocal crosses. Furthermore, interference by foreign pollen cannot easily occur, since the fertilizing organs are closely surrounded by the keel [a longitudinal floral ridge], and the anthers burst within the bud; thus the stigma is covered with pollen even before the flower opens. This fact is of particular importance. The ease with which this plant can be cultivated in open ground and in pots, as well as its relatively short growth period, are further advantages worth mentioning. Artificial fertilization is somewhat cumbersome, but it nearly always succeeds. For this purpose the not yet fully developed bud is opened, the keel is removed, and each stamen [which consists of a stem or filament and on which perches the anther or pollen-bearing part of the plant] is carefully extracted with forceps, after which the stigma [which receives the pollen] can be dusted at once with foreign pollen.

From several seed dealers a total of 34 more or less distinct varieties of peas were procured and subjected to two years of testing. In one variety a few markedly deviating forms were noticed among a fairly large number of like plants. These, however, did not vary in the following year and were exactly like another variety obtained from the same seed dealer; no doubt the seeds had been accidentally mixed. All other varieties yielded quite similar and constant offspring; at least during the next two test years no essential change could be noticed. Twenty-two of these varieties were selected for fertilization and planted annually throughout the entire experimental period. They remained stable without exception.

Once the reliability of a natural object has been proven for experimental study within the scientific community at large, as Mendel had done, then subsequent articles need not justify its choice. The choice of *Pisum* is recapitulated in the choice for later genetic research of such historically important organisms as *Drosophila* (fruit fly) and *Escherichia coli* (bacteria).

MEDICAL SCIENCE

The Anatomy of Pathology
....

Rudolf Virchow, 1855. "Cellular-Pathologie." *Archiv für pathologische anatomie und physiologie und für klinische medizin*, Vol. 8 (1), pp. 3–39.

At the end of the fourth decade of the nineteenth century, "medical science" was considered an oxymoron along the lines of "military intelligence" and "dark light." In a medical textbook from 1839, *Elements of Pathological Anatomy*, the famous American surgeon Samuel Gross complained:

> Of the essence of disease, very little is known; indeed, nothing at all; nor can the utmost ingenuity hope to remove the veil which still envelops the subject, until the physiology and pathology of the muscular and nervous systems shall be better understood. The proximate cause of morbid action, and the immediate cause of life in the healthy state, are as inscrutable to the human mind as the cause of gravitation, of attraction, and repulsion. All we can boast of is, that we know something of their effects; beyond this, it is extremely problematical whether we shall be able to penetrate. With this, indeed, every philosophical inquirer after truth should be contented, remembering that the secrets of nature are not easily detected, and that to God alone belongs the knowledge of the intrinsic properties of things.

Unknown to Gross, the solution to at least some of his mysteries was at hand: in Rudolf Virchow, medical science possessed a vanguard of formidable investigative and persuasive powers. In the following passage, Virchow emphasizes the inadequacy of traditional medicine, based on what he calls "practical experience" derived from treating patients, when compared with scientific medicine, based on close reasoning coupled with careful microscopic examination. To do so, Virchow draws his audience into the discourse with a sequence of rhetorical questions that work their way by turning his hearers into his collaborators.

Armed with a microscope—in his hands a virtual scalpel—Virchow showed that the cell is the primary building block of life, that cells come only from other cells. From this principle, he infers that the "natural-historical standpoint" of such physician-researchers as Thomas Sydenham is to be rejected. Diseases are not particular entities but cellular malfunctions. It is these malfunctions that ought to be the object of scrutiny in medical science:

> The practitioner will always put diagnosis first, and who can blame him? But when, for example, a tumor appears before him, how does he actually make his diagnosis? What questions does he ask? In my experience, he asks about the configuration of the malignancy before him. But malignancy isn't a matter of shape; it is a property that certain tumors have, and once we know, for example, that we are dealing with a cancer, then we also know that it is malignant. You've got to know what cancer is and how, malignancy aside, you can tell a cancer from other tumors. It's not enough for you to say

that because the cancerous growth is malignant, everything that is malignant must be a cancer: that's just arguing in a circle, and a vicious one at that. You may take umbrage at this, if you will, but you have to establish exactly what this tumor is, in other words, its histology or physiology.

Despite all his experience, even the busy surgeon, if he wants to delve deeper than his predecessors, must have recourse to histology and, therefore, to the microscope. Surgery is now exactly where internal medicine once was concerning the knowledge of its chosen field of study. Clinicians of the older generation had recourse to their pneumonias and their rheumatisms, and many who are still practicing continue to believe that they can cure diseases better than those who use such 'newfangled' methods as percussion and auscultation. But who believes that their treatments match the diseases they purport to cure? And who is not persuaded that many illnesses are presented to them of whose existence they haven't a clue? What is the point of all this prattle about experience, if you can't identify an illness that is staring you in the face! But if everything that is malignant must be cancerous, and everything that is benign or nearly so isn't—if you already have your solution before you address the problem—there is no use trying to reason with you.

Unfortunately, matters are not resolved so easily. Herr Bennett, whose book on cancerous and cancer-like tumors did not end the debate on the subject, a short time ago induced the Edinburgh Physiological Society to form a Cancer Committee whose sole mission was to produce a report that would provide some closure. But despite many meetings over a considerable period, the Committee dissolved without reaching consensus (*Monthly Journal* 1854. Nov. p. 468). Neither has the discussion of the Paris Academy advanced the subject. How do things stand now? It seems to me, simply put, that because the matter has been treated too superficially, the heart of the question has not been touched; in particular, no one has been able to rid himself of an outmoded point of view: the classification of pathological end-products must still be made by means of the old model of natural history, according to which certain specific properties of these end-products are presupposed.

Concerning cancer, this causal analysis is right on target. But the roster of infectious diseases that was the focus of Virchow's most famous successors, Louis Pasteur and Robert Koch, had a different cause altogether: alien microorganisms. That Virchow distanced himself from these discoveries, that he opposed germ theory, must be seen in the context of Joseph Priestley's opposition to the new chemistry, and Albert Einstein's, to the new quantum dynamics: mistaken only in retrospect, and never in the slightest irrational. In the final analysis, it takes nothing from Virchow's pioneering insight into the cellular foundation of life that he mistook what became the next step in medical science: not the fight against such cellular malfunctions as cancer, but rather against alien invasion—the bacilli.

The Birth of Germ Theory

Louis Pasteur, 1880. "De l'extension de la théorie des germes á l'étiologie de quelques maladies communes" (On the extension of the germ theory to the etiology of certain common diseases). *Comptes Rendus de l'Académie des Sciences*, Vol. 90, pp. 1033–44.

Robert Koch and Louis Pasteur, the two founding fathers of the germ theory of disease, were not friends or collaborators but antagonists. In part, this was because one was German, the other, French; but for the most part, this was because by Koch's standard, Pasteur was an amateur, a sloppy experimentalist who did not heed the three postulates implicit in his study of anthrax and tuberculosis, the subject of the next selection:

1. The parasite occurs in every case of the disease.
2. The parasite does not occur in other diseases or without causing disease.
3. After being fully isolated and repeatedly grown in pure culture, the parasite can induce the disease by being introduced into the healthy animal.

Pasteur certainly does not fulfill any of these three conditions in the following paper—not even the first. He begins with a case of boils appearing on the skin of his assistant, Charles Chamberland. Once the pus from these boils is cultured, Pasteur identifies a form of staphylococcus:

When I began the studies now occupying my attention, I was attempting to extend the germ theory to certain common illnesses. When will I be able to return to that work? In my desire to see it carried on by others, I take the liberty of presenting it to the public just as it is.

I. Concerning Boils. In May 1879, one of the workers in my laboratory had a number of boils, appearing at short intervals, sometimes on one part of his body, and sometimes on another. Coincident with my constant interest in the immense role microscopic organisms play in nature, I inquired whether the pus in the boils might not contain one of these organisms whose presence, development, and transportation here and there in the body after a breach had been effected would provoke a local inflammation and pus formation, and might explain the short or long term recurrence of the illness. It was easy enough to subject this idea to the test of observation.

First observation.—June 2, a puncture was made at the base of the small cone of pus at the apex of a boil on the nape of the neck. The fluid from the puncture was at once seeded into a culture in the presence of air freed of impurities—of course taking the precautions necessary to exclude any other micro-organisms, either at the time of puncture, at the time of seeding, or during a stay in the oven, which was kept at the constant temperature of about 35°C. The next day, the liquid culture lost its limpidity and contained a single organism, consisting of small spherical points arranged in pairs, sometimes in fours, but frequently in irregular masses. Two fluids were preferred in these experiments—chicken and yeast bouillon. According as one or the other of these fluids was used, development exhibited a little variation. This can be described. With the yeast water, the pairs of minute granules are distributed throughout the liquid, which is uniformly clouded. But with the chicken bouillon, the granules are collected in little masses which line the walls of the flasks while the body of the fluid remains clear, unless it is shaken: in this case it becomes uniformly clouded by the breaking up of the small masses on the bottom of the flasks.

Second observation.—June 10, a new boil appeared on the right thigh of the same person. Pus could not yet be seen under the skin, but this was already prominent and red over a surface the size of a franc (about 1/2 inch). The inflamed part was washed with alcohol, and dried with blotting paper sterilized by being passed through the flame of an alcohol lamp. A puncture at the prominent portion of the boil enabled us to secure a small amount of lymph mixed with blood, which was seeded in a culture at the same time as some blood was taken from a finger. The following days, the blood from the finger remained absolutely sterile: in contrast, that obtained from the center of the forming boil exhibited an abundant growth of the same small organism as just before.

Third observation.—June 14, a new boil appeared on the neck of the same person. The same examination, the same result, that is to say the development of the microscopic organism previously mentioned and complete sterility of the blood in general circulation, taken this time at the base of the boil outside of the inflamed area.

At the time of making these observations I had the occasion to speak to Dr. Maurice Raynaud, who was good enough to send me a patient who had had furuncles [boils] for more than three months. On June 13, I made cultures of the pus of one of this man's boils. The next day there was a general cloudiness of the liquid cultures, consisting entirely of the parasite previously identified, and of this alone.

Fourth observation.—June 14, the same individual showed me a voluminous boil newly forming in the left armpit: the prominence was wide-spread and the skin red, but no pus was yet evident. An incision at the center of the prominence exhibited a small quantity of pus mixed with blood. After seeding, rapid growth for twenty-four hours and the appearance of the same organism [occurred]. Blood from the arm at a distance from the boil remained completely sterile.

June 17, the examination of a fresh boil on the same individual gave the same result, the development of a pure culture of the same organism.

Fifth observation.—July 21, Dr. Maurice Raynaud informed me that there was a woman at the Lariboisière hospital with multiple boils. As a matter of fact her back was covered with them, some in active suppuration, others in the ulcerating stage. I took pus from all of these boils that had not opened. After a few hours, this pus exhibited abundant growth in cultures. It was always the same organism, always uniquely that organism, with no added element. Blood from the inflamed base of the boil, seeded in a culture in its turn, remained sterile.

In brief, it appears certain that every boil contains an aërobic microscopic parasite [one that requires air to live], and it is to this parasite that the local inflammation and the pus formation that follows may be attributed.

Pasteur's case relies on argumentative sleight of hand: the relentless progress from experiment to experiment conceals a shaky design—his conclusions are based on three patients! From Koch's point of view, Pasteur's study would count as preliminary. Pasteur is entitled only to the narrow conclusion that every boil he examined contained a parasite that was coincident with its formation.

Pasteur concludes this paper with a restatement of the article's title as a rhetorical question, some artful hedging regarding the implied positive answer, and a challenge, not without a certain belligerence, to imagined critics or doubters:

Was I sufficiently justified in calling this Communication "On the extension of germ theory to the etiology of certain common diseases"? I have exposed the facts as they have appeared to me and I have hazarded my interpretations of them; but I cannot deny that, in the medical field, it is difficult to support oneself wholly on intuition. Nor can I forget that, for me, medicine and veterinary practice are alien territory. I wish all my contributions to be subject to judgment and criticism. Intolerant of frivolous or biased opposition,

contemptuous of that vulgar skepticism that doubts everything on principle, I welcome with open arms the targeted skepticism that turns doubt
into a method, a skepticism whose rule of conduct has as its motto: "More
light."

The Spread of Germ Theory

Robert Koch, 1882. "Die Aetiologie der Tuberculose" (The etiology of tuberculosis).
Berliner klinische Wochenschrift, Vol. 19, pp. 221–30.

On the evening of March 24, 1881, Robert Koch gave a lecture at the Berlin
Physiological Society that left his audience "spellbound." Years later, Koch
protégé and eventually Nobel Laureate Paul Ehrlich wrote that "that evening
has remained my greatest experience in science." The persuasiveness of Koch's
lecture stems largely from its organization: it consists of a series of experiments
that interlock to form an argument leading irrevocably to the conclusion that
tuberculosis is an instance of the more general germ theory of disease.

For the first time, a human disease—a nineteenth-century scourge—had been
traced to its origin: tuberculosis was caused by the bodily invasion of a specific
bacillus. Koch knew this for a fact because he had isolated the bacillus, cultivated
it in the laboratory and, injecting it into disease-free animals, recreated the
disease. "On the ground of many observations," he says,

> I regard my many experiments as proofs that all tubercular infections, whe-
> ther in men or animals, always have the tuberculosis bacillus as their cause,
> a bacillus that differs in characteristic ways from all other micro-organisms.
> Merely from the fact that they mysteriously coincide, a causal relationship
> between tubercular infection and the presence of this bacillus cannot be
> unproblematically inferred, although a not untrivial degree of probability
> derives from the circumstance that the bacillus is habitually found just where
> the tubercular process is establishing itself, or is in progress, and that it is no
> longer evident whenever the illness comes to a halt.

> To prove beyond the shadow of a doubt that tuberculosis is caused by
> the invasion of the bacillus and that the disease is inextricably linked to
> the growth and reproduction of this parasite, the bacillus must be isolated
> from living animals and cultivated in pure cultures until those cultures are
> clearly uncontaminated by the products of illness from those animals. Finally,
> tubercular symptoms must be produced by injection of these cultured bacilli
> in animals, symptoms indistinguishable from those of naturally-occurring
> tuberculosis.

The proof involved a carefully planned series of experiments meticulously carried out with a single conclusion in mind. Of the great experimenter with electricity, Alessandro Volta, Derek de Solla Price says that he had his "brain in his fingertips"; the same may be said of Koch:

> Again and again the tubercules from inoculation and injection from cultures were found to be identical with tubercules that arose spontaneously or arose after inoculation with tubercular matter from these animals. These tubercules possessed exactly the same configuration of cellular elements and frequently contained in giant cells, just like those of tuberculosis bacilli that arise spontaneously. Moreover, the bacilli whose origin was cultural were isolated again in pure cultures and, as with those produced by inoculation experiments, they acted exactly like those from human and animal tubercules. In addition, the tubercules derived from infection with cultures behaved exactly like their natural counterparts.
>
> Looking back on these numerous experiments with the cultured bacillus, we see that the experimental animals became tubercular without exception, no matter how they were infected, whether by simple inoculation in subcutaneous cell tissue, through injection into the abdominal cavity, or into the anterior chamber of the eye, or directly into the blood-stream. Indeed, it was not a single tubercule that formed, but rather an extraordinary amount of tubercules that corresponded to the number of nuclei of infection that were introduced. In other animals a trace amount of the bacillus introduced through inoculation in the anterior chamber of the eye succeeded in producing the same tubercular iritis as that obtained in the well-known and decisive experiments of Cohnhein, Salomonsen, and Baumgarten, experiments that produced tuberculosis only by means of matter that was genuinely tubercular.
>
> A confusion of our results with spontaneous tuberculosis or with an accidental and unintended infection of the experimental animals with tuberculosis-virus is excluded on the following grounds. First, neither spontaneous tuberculosis nor an accidental infection can cause so massive an eruption in so short a time. Secondly, the control animals, which were otherwise treated exactly like their infected counterparts, remained healthy. Third, numerous guinea pigs and rabbits treated in exactly the same way but inoculated and injected with other substances in no case exhibited the typical symptoms of miliary tuberculosis, which can originate only when the body is suddenly, as it were, overwhelmed with infected nuclei.
>
> Taken together, all of these facts give us grounds for saying that the bacilli that occur in tuberculous matter do not merely accompany tubercular processes; they cause them. In other words, when we have the bacillus before us, we have before us tuberculosis itself.

Near the conclusion of his 1882 paper on tuberculosis, Koch looks forward from diagnosis to therapy:

Until now, tuberculosis has been ordinarily perceived as an expression of social wretchedness and hope for its eradication has been based on an improvement of these conditions. In fact, public health measures are still not directed against the disease itself. But in the future struggle against this terrible plague, humankind will grapple, not with an unknown assailant, but with a tangible parasite whose life-cycle is known for the most part and that is at this moment in the process of being elucidated. The fact that this parasite is found only in animals, and not like the anthrax bacillus, in the general environment as well, gives us hope that the prospects for success in this battle against tuberculosis will be especially favorable.

It is a standard rhetorical move for a conclusion: a promise of a better, more productive future. It must have galled Koch that it was the "sloppy" Pasteur who helped to secure that future by providing vaccines for anthrax, chicken cholera, and swine erysipelas, and a cure for rabies, while his own "cure" for tuberculosis turned into an embarrassing fiasco. Worse yet, Koch's further research led to his abandonment of his second and third postulates, and, finally, to his failure to mention his postulates at all because in his own work he could not meet these often unrealizable standards.

CHEMISTRY

New Symbols for the New Chemistry

Jöns Jacob Berzelius, 1814. "Essay on the cause of chemical proportions, and on some circumstances relating to them: Together with a short and easy method of expressing them." *Annals of Philosophy*, Vol. 4, pp. 51–62.

It would be difficult to imagine chemistry without its strings of symbols arrayed in equations, like $2H + O = H_2O$. These equations rest on a system of abbreviations for the chemical elements, one that the great Swedish chemist, Jöns Jacob Berzelius, invented and promoted. Earlier, far-from-intuitive geometric figures abbreviated the names for elements and compounds. For example, in *A New System of Chemical Philosophy*, John Dalton represented elements by means of marked-up circles: an empty circle symbolized oxygen; a filled circle, carbon; a circle with a center dot, hydrogen; a circle with a vertical line, nitrogen; a circle with a C inside, copper; and so forth. The trouble with such systems is

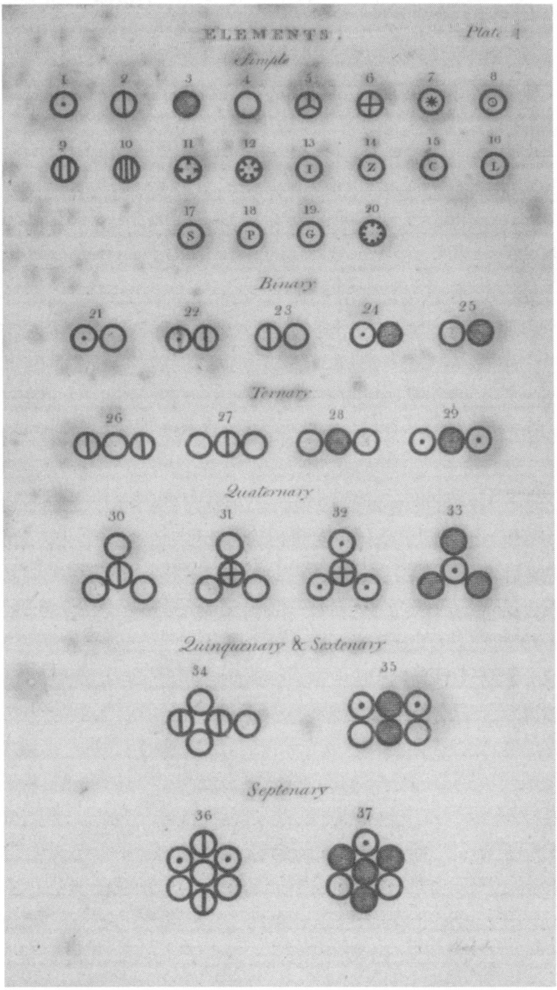

FIGURE 4-1 · Dalton's table of elements and their
combination. Courtesy of Special Collections Research
Center, University of Chicago Library.

that they cannot be used in journal articles unless a special font is created, just
for chemistry. Lavoisier and his colleagues had created a new chemistry and a
new language with which to represent its substances and interactions; it was
Berzelius who created a new symbolism for this new language.

Berzelius recognized that symbolizing the elements by their first letter or two,
without any circles, would be much easier on the typesetter: for example, C for

carbon and Ca for calcium. To ensure international acceptance and uniformity of this system, Berzelius abbreviated the Latin names of the elements. That notation established, he devised a system for representing the proportions of elements in a compound as well as the proportions of elements and compounds in combination. We now know these as the number of atoms in a molecule and the number of molecules in a compound. (Since the computer will not easily permit the placement of superscripts on top of the letters for the elements, where Berzelius put them in this early version of his system, we have transcribed the passage with the superscripts placed just after the elements, where Berzelius put them eventually). In one of a series of essays on how chemical reactions can be translated into mathematical formula, Berzelius wrote:

> when we endeavour to express chemical proportions, we find the necessity for chemical signs. Chemistry has always possessed them, though hitherto they have been of very little utility. They owed their origin, no doubt, to the mysterious relation supposed by the alchemists to exist between the metals and the planets, and to the desire which they had of expressing themselves in a manner incomprehensible to the public. The fellow-laborers in the antiphlogistic revolution published new signs founded on a reasonable principle, the object of which was, that the signs, like the new names, should be definitions of the composition of the substances, and that they should be more easily written than the names of the substances themselves. But, though we must acknowledge that these signs were very well contrived, and very ingenious, they were of no use; because it is easier to write an abbreviated word than to draw a figure, which has but little analogy with letters, and which, to be legible, must be made of a larger size than our ordinary writing. In proposing new chemical signs, I shall endeavour to avoid the inconveniences which rendered the old ones of little utility. I must observe here that the object of the new signs is not that, like the old ones, they should be employed to label vessels in the laboratory: they are destined solely to facilitate the expression of chemical proportions, and to enable us to indicate, without long periphrases, the relative number of volumes of the different constituents contained in each compound body. By determining the weight of the elementary volumes, these figures will enable us to express the numeric result of an analysis as simply, and in a manner as easily remembered, as the algebraic formulas in mechanical philosophy.
>
> The chemical signs ought to be letters, for the greater facility of writing, and not to disfigure a printed book. Though this last circumstance may not appear of any great importance, it ought to be avoided whenever it can be done. I shall take, therefore, for the chemical sign, the *initial letter of the Latin*

name of each elementary substance: but as several have the same initial letter, I shall distinguish them in the following manner: —1. In the class which I call *metalloids* [a class of combustible elements], I shall employ the initial letter only, even when this letter is common to the metalloid and some metal. 2. In the class of metals, I shall distinguish those that have the same initials with another metal, or a metalloid, by writing the first two letters of the word. 3. If the first two letters be common to two metals, I shall, in that case, add to the initial letter the first consonant which they have not in common: for example, S = sulphur, Si = silicium, St = stibium (antimony), Sn = stannum (tin), C = carbonicum, Co = cobaltum (cobalt), Cu = cuprum (copper), O = oxygen, Os = osmium, &c.

The chemical sign expresses always one volume of the substance. When it is necessary to indicate several volumes, it is done by adding the number of volumes: for example, the *oxidum cuprosum* (protoxide of copper) is composed of a volume of oxygen and a volume of metal; therefore its sign is Cu + O. The *oxidum cupricum* (peroxide of copper) is composed of 1 volume of metal and 2 volumes of oxygen; therefore its sign is Cu + 2O. In like manner, the sign for sulphuric acid is S + 3O; for carbonic acid, C + 2O; for water 2H + O, &c.

When we express a compound volume of the first order, we throw away the +, and place the number of volumes above the letter: for example, $CuO + SO^3$ = sulphate of copper, $CuO^2 + 2SO^3$ = persulphate of copper. These formulas have the advantage, that if we take away the oxygen, we see at once the ratio between the combustible radicals. As to the volumes of the second order, it is but rarely of any advantage to express them by formulas as one volume; but if we wish to express them in that way, we may do it by using the parenthesis, as is done in algebraic formulas: for example, alum is composed of 3 volumes of sulphate of aluminia and 1 volume of sulphate of potash. Its symbol is $3(AlO^2 + 2SO^3) + (Po^2 + 2SO^3)$. As to the organic volumes it is at present very uncertain how far figures can be successfully employed to express their composition. We shall have occasion only in the following pages to express the volume of ammonia. It is $6H + N + O$ or H^6NO.

Dalton was aware of Berzelius's system but criticized it because it could not represent the arrangement of the atoms in the molecule, as his system could. And the English were not very friendly initially because Berzelius was not an Englishman. Moreover, Berzelius undermined his own system somewhat by adding "refinements" that made formulas harder to set in ordinary type and, arguably, less transparent to the reader. But all of this changed in the fourth decade of the nineteenth century: in 1831 Edward Turner adopted the symbols of Berzelius for his textbook, *Elements of Chemistry*. Two years later the British Association said it regarded their adoption as "imperative."

Discovering a New Element in a New Way

Gustav Kirchhoff and Robert Bunsen, 1860. "Chemische Analyse durch Spectralbeobachtungen" (Chemical analysis by observations of spectra). *Annalen der Physik und Chemie*, Vol. 110, pp. 161–89. Translation from E. Farber, ed., *Milestones of Modern Chemistry*. New York: Basic Books (1966).

Everyone knows the great theory builders and experimentalists of science: Newton, Lavoisier, Davy, Faraday, Liebig, Darwin, Einstein, Wegener, Heisenberg, Pauling, Watson and Crick, and Hawking come readily to mind. Few, however, know the great inventors of new research methods and equipment. And yet, such invention drives the advance of science every bit as much as new theories and experiments. As Randall Collins put it in *The Sociology of Philosophies*:

> The rapid movement of research equipment from one modification to the next is key to the mode of rapid discovery in which scientists take so much confidence; they feel that discoveries are there to be made along a certain angle of research because the previous generation of equipment has turned up phenomena which are suitable for the intellectual life of the human network.... The genealogy of equipment is carried along by a network of scientific intellectuals, who cultivate and cross-breed their technological crops in order to produce empirical results that can be grafted onto an ongoing lineage of intellectual arguments.

(With due apologies to Randall Collins, we would substitute the more general word "methods" for "equipment.")

The article we excerpt reports a new method that would have a huge impact on analysis of chemical structures and measurements of celestial behavior. This revolutionary method employed a remarkably simple apparatus consisting of a sample holder, a Bunsen burner for igniting the sample, a prism to disperse the light produced, and two telescopes—one for sending the light from the sample to the prism, the other for viewing the resulting spectrum. A third telescope (not shown) is aimed at the mirror (G), so that the analyst can read a horizontal scale in the spectrum. The apparatus's effectiveness is based on the observation that each element produces a unique spectrum when a chemical sample is burned.

In the decades leading up to this article, new elements had been discovered by electrochemical separation. In contrast, Kirchhoff and Bunsen's apparatus

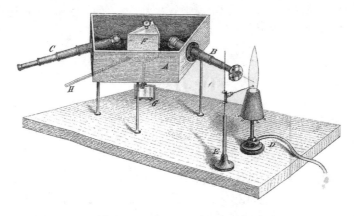

FIGURE 4-2 · First spectroscope. University of Minnesota Library.

identified elements present in small samples—samples too small for electrochemical separation—by means of their unique spectrum when ignited. If that
accomplishment were not enough, the authors announce a new element identified by this apparatus (cesium), and propose the use of a modified version of
it in determining the chemical composition of the stars, heralding a new era in
astronomical research:

> Several substances are known in which contact with a flame produces a spec-
> trum with definite bright lines. A method of qualitative analysis can be based
> on these lines, widening considerably the field of chemical investigations and
> solving hitherto inaccessible problems. We limit ourselves here to develop-
> ing the method for alkali and earth-alkali metals and demonstrating its value
> by some examples.
>
> The lines show up the more distinctly the higher the temperature and the
> lower the luminescence of the flame. The gas burner described by one of us
> (Bunsen, these Ann. 100, p. 85) has a flame of very high temperature and
> little luminescence and is, therefore, particularly suitable for experiments on
> the bright lines that are characteristic of these substances. . . .
>
> For the discovery of hitherto unknown elements, spectrum analysis
> should become a not unimportant resource. If there are substances so rare in
> nature that our present means of analysis fail either to recognize or to separate
> them, then by simply observing the flame spectra of quantities of them so
> small that their analysis eludes our usual methods, we may determine them
> with certainty. We have already had occasion to convince ourselves that there

are such elements, now unknown. Supported by unambiguous results of the spectral-analytical method, we believe we can state right now with complete certainty that there is a fourth metal in the alkali group besides potassium, sodium, and lithium. This element has a simple characteristic spectrum like lithium; displaying only two lines in our apparatus: a faint blue one, almost coinciding with the strontium line $Sr\ \delta$, and another blue one a little further toward the violet end of the spectrum whose intensity and definitiveness competes with the lithium line.

Spectrum analysis, which, as we believe we have shown, offers a wonderfully simple means for discovering the smallest traces of certain elements in terrestrial substances, also opens to chemical research a hitherto completely closed region extending far beyond the limits of the earth and even of the solar system. Since in this analytical method it is sufficient *just to see* the glowing gas to be analyzed, it can easily be applied to the atmosphere of the sun and the brighter fixed stars. However, a modification is needed because of the light emitted by these bodies. In "On the Relationship between the Capacities of Bodies to Emit and Absorb Heat and Light" (Kirchhoff, these Ann. 109, p. 275), one of us has proved theoretically that the spectrum of a glowing gas is *reversed*; i.e., the bright lines are converted into dark ones, if it has behind it a light source of sufficient intensity, one that sends out a continuous spectrum. We can conclude that the spectrum of the sun with its dark lines is just a reversal of the actual spectrum of the sun's atmosphere. Therefore, the chemical analysis of the sun's atmosphere requires only the search for those substances that produce the bright lines that coincide with the dark lines of the solar spectrum.

The relative simplicity of the language in this passage matches the relative simplicity of the apparatus the authors had created. And their argument, for its importance and utility, moves with great celerity from the earth to the heavens.

Thinking in Pictures

August Kekulé, 1865. "Sur la constitution des substances aromatiques" (On the composition of aromatics). *Bulletin de la Société Chimique de France*, Vol. 3, pp. 98–110.

In his old age, August Kekulé claimed that the idea of the carbon chain occurred to him in the summer of 1854 when he was drowsing on top of a London omnibus, where he had a vision of gamboling atoms joining together and forming chains. The important point is not whether the story is true in fact, but whether it represents a psychological truth about the role of visualization in certain scientific discoveries, the sense that scientists *see* solutions, and that rational processes follow rather than precede this vision. In a reply to the French mathematician Jacques Hadamard, Albert Einstein supported this point of view: "The words or the language . . . do not seem to play any role in my mechanism of thought. The psychical entities which seem to serve as elements in thought are certain signs and more or less clear images which can be 'voluntarily' reproduced and combined." This may very well have been what Kekulé had in mind.

By 1858, with the benefit of his insight into the carbon chain, Kekulé had solved the structure of what he called "fatty bodies," and what modern chemists call aliphatic compounds, hydrocarbons forming open-ended chains. In 1865, in the paper we excerpt here, he solved the structure of the much more difficult class of hydrocarbons, the aromatics, those with a family resemblance to benzene (in this passage, he speaks of "affinity" where modern chemists would speak of "valence," the combining capacity of an atom or radical):

> In all aromatic substances there is one and the same atom group, a nucleus which consists of six carbon atoms. Within this nucleus the carbon atoms are found in a form far more compact than in the case of fatty substances. To this nucleus, more carbon atoms are added in the same way and according to the same laws as in the case of fatty bodies.
>
> Before we proceed, we must attempt to give an account of the atomic constitution of this nucleus. The simplest hypothesis is the following: It derives naturally from the tetratomicity of carbon, an assumption that, at this point, needs little defense.
>
> When carbon atoms unite with one another, *one* affinity unit of each atom can bind with *one* affinity unit of the neighboring atom. As I have shown earlier, this explains homology [membership in the same group] and in general the constitution of the fatty bodies.
>
> We can suppose that many carbon atoms combine by *two* affinity units. We can also suppose that they combine alternately by first *one* and then by *two* affinity units. These two ways of combining can be exemplified by the following cycles:
>
> $$1/1, 1/1, 1/1, 1/1 \text{ etc.}$$
> $$1/1, 2/2, 1/1, 2/2 \text{ etc.}$$
>
> If the first cycle explains the constitution of the fatty bodies, the second explains the constitution of aromatic substances, or at least of the nucleus that is common to all these substances.
>
> In effect, when six carbon atoms combine according to this law of symmetry, a group is obtained which, as *an open chain*, still contains *eight*

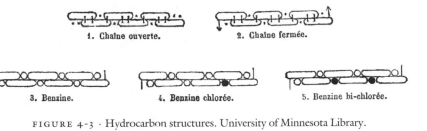

FIGURE 4-3 · Hydrocarbon structures. University of Minnesota Library.

non-saturated affinity units (1. Chaîne ouverte). If another assumption is made, that the two carbon atoms at the ends of chain are linked by one affinity unit, we have *a closed chain* that still contains *six* free affinity units (2. Chaîne fermé).

From this closed chain are derived all the substances that are usually called "aromatic compounds." The open chain occurs in quinone, in chloranil, and in the few substances that stand in close relation to both.

In this diagram, the sausages represent tetravalent carbon atoms, the circles, what Kekulé refers to as monovalent "affinity units." In the top two structures, the carbons are tied together by alternating double and single bonds (double or single vertical lines). The three aromatic structures in the second line, benzene and two variations, all form closed chains. The monovalent atoms for these three are hydrogen (open circles) and chlorine (solid circles). This diagram is an important step in the direction of the current visual representation of organic compounds. Kekulé not only thought in pictures, he communicated in them. And chemists from this point onwards would do so too with ever increasing regularity.

A Natural Order for the Chemical Elements

Dimitri Ivanovich Mendeleev, 1869. "Ueber die Beziehungen der Eigenschaften zu den Atomgewichten der Elemente" (Concerning the relationships of the properties to the atomic weights of the elements). *Zeitschrift für Chemie*, Vol. 12, pp. 405–6.

In three 1869 articles and the book *Osnovy Khimii* (*Principles of Chemistry*), Mendeleev published the first versions of his famous periodic table, which organized the sixty-five known chemical elements into rows and columns guided by the increasing order of their atomic weights and chemical properties. According to Mendeleev, "in the elements there is one accurately measurable property, which is subject to no doubt—namely, that property which is expressed in the atomic weights. Its magnitude indicates the relative mass of the atom, or, if, *we avoid the conception of the atom*, its magnitude shows the relation between the masses, forming the chemical and independent individuals or elements. And according to the sense of all our physico-chemical data, the mass of a substance is that property on which all the remaining properties must be dependent, because they are all determined by similar conditions or by those forces which act in the weight of a substance, and this is directly proportional to the mass of a substance." We added the italics to emphasize Mendeleev's concession to the continuing skepticism about the atom in the latter half of the nineteenth century.

The selected article, from an early specialized chemical journal in Germany, organizes the elements into six columns, each according to its atomic weight. When this is done, the columns display periodicity, that is, they change systematically, while the rows display elements with analogous properties. From this crude first table and subsequent refinements, Mendeleev was able to forecast that gaps in the table would be filled by the discovery of "many new elements" with predictable properties. That prediction was rapidly confirmed with the discovery of gallium in 1875, scandium in 1879, and germanium in 1885.

In reading the Mendeleev selection, bear in mind that "J" is the German for iodine instead of "I," while "typical" does not mean "representative." It refers rather to type theory; it means that "typical" elements are the most important building blocks of compounds. We also now know some of his atomic weights are wrong; for example, Ur (uranium) is 238, not 116. Finally, one also needs to be aware that in European number systems commas substitute for periods: thus, an atomic weight of 35,5 = 35.5.

The following is Mendeleev's article, after having been translated from Russian to German to reach a wider audience in the nineteenth century, then German to English to reach a wider audience in the twentieth century:

By ordering the elements according to increasing atomic weight in vertical rows so that the horizontal rows contain analogous elements, still ordered by increasing atomic weight, one obtains the following arrangement, from which a few general conclusions may be derived.

1. The elements, if arranged according to their atomic weights, exhibit a phased alternation of properties.

2. Chemically analogous elements have either similar atomic weights (Pt, Ir, Os), or weights which increase by equal increments (K, Rb, Cs).

3. The arrangement according to atomic weight corresponds to the valence of the element and to a certain extent the difference in chemical behavior, for example Li, Be, B, C, N, O, F.

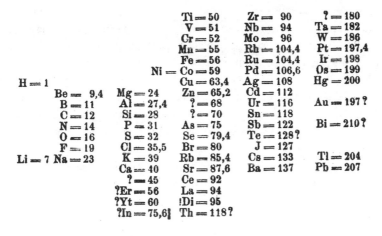

FIGURE 4-4 · Periodic table of elements: first version.
University of Minnesota Library.

4. The elements distributed most widely in nature have small atomic weights, and all such elements are marked by the distinctness of their behavior. They are the typical elements; and so the lightest element H is rightly chosen as the most typical.

5. The magnitude of the atomic weight determines the properties of the element. Therefore, in the study of compounds, not only the quantities and properties of the elements and their reciprocal behavior are to be taken into consideration, but also the atomic weight of the elements. Thus the compounds of S and Te, Cl and J, display not only many analogies, but also striking differences.

6. One can predict the discovery of many new elements, for example, analogues of Si and Al with atomic weights of 65–75.

7. A few atomic weights will probably require correction; for example, Te cannot have the atomic weight 128, but rather 123–126.

8. From the above table, some new analogies between elements are revealed. Thus Ur appears as an analogue of B and Al, as is well known to have been long established experimentally.

Mendeleev's table reflects two enduring tabular properties: an otherwise hidden natural order is discovered and, in the best possible case, the door is opened to further discoveries. Mendeleev's text complements the table by formatting the individual paragraphs into a numbered list of points about the chemical elements and their arrangement—a stylistic manifestation of the drive to quantify almost everything capable of being quantified, even the paragraphs.

In the twentieth century, there was a significant change in the principle underlying the organization of the periodic table, a change from atomic weight to atomic number. In 1913, Henry Moseley reported that the number of protons in a given element (atomic number) worked much better than atomic mass as a basis for arranging the table, a key discovery in its continued evolution. To this day, new elements continue to be discovered with properties consistent with the patterns in Mendeleev's and Moseley's tables. The total number stands at over 115. We think it fair to claim that never has a single table or other visual representation proved so fruitful for science.

PHYSICS

Physics and Fiction

James Clerk Maxwell, 1855. "On Faraday's lines of force." In W. D. Niven, ed., *The Scientific Papers of James Clerk Maxwell*. Cambridge: Cambridge University Press (1890), Vol. 1, pp. 155–229. Reprinted from *Transactions of the Cambridge Philosophical Society*, Vol. 10, Part 1.

Although physics and fiction are supposed to be strangers, Albert Einstein speaks of "the purely fictitious character of the fundamentals of scientific theory." In doing so, he is not being entirely original; he is echoing Maxwell, who showed that physics and fiction can be fruitful partners in the discovery of natural laws.

The great experimentalist Michael Faraday had demonstrated that magnetic fields create lines of force, an effect we can see in action when we sprinkle iron filing over a piece of paper under which a magnet rests. In an article read before the Cambridge Philosophical Society in 1855–56, when he was only twenty-four, Maxwell translates Faraday's observations into a wholly imaginary system of tubes carrying a completely conjectural incompressible fluid. On the basis of this fictional mechanical analogy, he generates a series of equations designed to describe the motions of a particular version of his system, one that obeys Newton's inverse-square law of gravitational attraction. Maxwell makes no claim for the truth of his system or his equations:

> In these six laws I have endeavoured to express the idea which I believe to be the mathematical foundation of the modes of thought indicated in the *Experimental Researches* [of Faraday]. I do not think it contains even

the shadow of a true physical theory; in fact its chief merit as a temporary instrument of research is that it does not, even in appearance, *account for* anything.

His only claim is that his system and his equations may be fruitful in the discovery of natural laws. This next excerpt is, simultaneously, a lesson in physics and in the philosophy of physics, and a fragment of a brilliant argument, written in prose that no editing could improve upon:

> It is by the use of analogies . . . that I have attempted to bring before the mind, in a convenient and manageable form, those mathematical ideas which are necessary to the study of the phenomena of electricity. The methods are generally those suggested by the processes of reasoning which are found in the researches of Faraday, and which, though they have been interpreted mathematically by Prof. Thomson and others, are very generally supposed to be of an indefinite and unmathematical character, when compared with those employed by the professed mathematicians. By the method which I adopt, I hope to render it evident that I am not attempting to establish any physical theory of a science in which I have hardly made a single experiment, and that the limit of my design is to shew how, by a strict application of the ideas and methods of Faraday, the connexion of the very different orders of phenomena which he has discovered may be clearly placed before the mathematical mind. . . .
>
> I have in the first place to explain and illustrate the idea of [Faraday's] "lines of force."
>
> When a body is electrified in any manner, a small body charged with positive electricity, and placed in any given position, will experience a force urging it in a certain direction. If the small body be now negatively electrified, it will be urged by an equal force in a direction exactly opposite.
>
> The same relations hold between a magnetic body and the north or south poles of a small magnet. If the north pole is urged in one direction, the south pole is urged in the opposite direction.
>
> In this way we might find a line passing through any point of space, such that it represents the direction of the force acting on a positively electrified particle, or on an elementary north pole, and the reverse direction of the force on a negatively electrified particle or an elementary south pole. Since at every point of space such a direction may be found, if we commence at any point and draw a line so that, as we go along it, its direction at any point shall always coincide with that of the resultant force at that point, this curve will indicate the direction of that force for every point through which it passes,

and might be called on that account a *line of force*. We might in the same way draw other lines of force, till we had filled all space with curves indicating by their direction that of the force at any assigned point.

We should thus obtain a geometrical model of the physical phenomena, which would tell us the *direction* of the force, but we should still require some method of indicating the *intensity* of the force at any point. If we consider these curves not as mere lines, but as fine tubes of variable section carrying an incompressible fluid, then, since the velocity of the fluid is inversely as the section of the tube, we may make the velocity vary according to any given law, by regulating the section of the tube, and in this way we might represent the intensity of the force as well as its direction by the motion of the fluid in these tubes. This method of representing the intensity of a force by the velocity of an imaginary fluid in a tube is applicable to any conceivable system of forces, but it is capable of great simplification in the case in which the forces are such as can be explained by the hypothesis of attractions varying inversely as the square of the distance, such as those observed in electrical and magnetic phenomena. . . .

I propose, then, first to describe a method by which the motion of such a fluid can be clearly conceived; secondly to trace the consequences of assuming certain conditions of motion, and to point out the application of the method to some of the less complicated phenomena of electricity, magnetism, and galvanism [electricity generated by chemical action]; and lastly to shew how by an extension of these methods, and the introduction of another idea due to Faraday, the laws of the attractions and inductive actions of magnets and currents may be clearly conceived, without making any assumptions as to the physical nature of electricity, or adding anything to that which has already been proved by experiment.

By referring everything to the purely geometrical idea of the motion of an imaginary fluid, I hope to attain generality and precision, and to avoid the dangers arising from a premature theory professing to explain the cause of the phenomena. If the results of mere speculation I have collected are found to be of any use to experimental philosophers, in arranging and interpreting their results, they will have served their purpose, and a mature theory, in which physical facts will be physically explained, will be formed by those who by interrogating Nature herself can obtain the only true solution of the questions which the mathematical theory suggests.

Maxwell begins "On Faraday's lines of force" by describing how he intends to approach Faraday's ideas: not by the use of a physical theory, for then one sees the phenomenon only through a medium; nor by use of a purely mathematical formula, for then one loses sight of the phenomenon to be explained, but by the use of a *physical analogy* representing lines of force as "fine tubes of variable section carrying an incompressible fluid." In this way, he is not confined to the sterility of mathematics, but is able to form in the mind a clear physical conception "without being committed to any theory founded on the physical sciences from which that conception is borrowed."

X-rays: A Rapid Shift from Science to Technology

Wilhelm Conrad Röntgen, 1895. "Über eine neue Art von Strahlen" (Concerning a new kind of ray). *Sitzungsberichte der Physikalischen-medizinishen Gesellschaft zu Würzburgol*, Vol. 87, pp. 132–41. Reprinted in *W. C. Röntgen's Grundlegende Abhandlungen über die X-Strahlen (Seminal Papers concerning X Rays)*. Verlag von Curt Kabitzsch: Würzburg (1915). Modified from the translation by Arthur Stanton, *Nature* (1896), Vol. 53, pp. 274–76.

In the January 1904 *Century Magazine*, Marie Curie noted "how limited is our direct perception of the world which surrounds us, and how numerous and varied may be the phenomena which we pass without a suspicion of their existence until the day when a fortunate hazard reveals them." The discovery of X-rays by Wilhelm Röntgen, and their stunning representation in photographs, is one such "fortunate hazard."

Röntgen was working at his laboratory one November evening in 1895, studying cathode rays in a Hittorf-Crookes tube, a glass tube in which electron beams are generated by a current passed from an anode to a cathode in a vacuum. As a prelude to viewing the cathode rays, he covered the tube with a shield of black paper. By chance, on a nearby bench was a screen treated with barium platinocyanide. When the tube was in operation, Röntgen observed that the screen fluoresced. Then came the eureka moment. Since cathode rays could not have penetrated the covered tube, Röntgen concluded that he must have discovered a new kind of ray. In subsequent experiments, he found that these rays passed through various objects as though they were not even there. When he had his wife interpose her hand in the beam, he saw her bones as a shadowy image on the screen.

The selection below is Röntgen's very straightforward description of his X-ray discovery and followup experiments, presented in the form of an ordered list:

(1) A discharge from a large induction coil is passed through a Hittorf's vacuum tube, or through a well-exhausted Crookes' or Lenard's tube. The

tube is surrounded by a fairly close-fitting shield of black paper; it is then possible to see, in a completely darkened room, that paper covered on one side with barium platinocyanide lights up with brilliant fluorescence when brought into the neighborhood of the tube, whether the painted side or the other be turned towards the tube. The fluorescence is still visible at two metres distance. It is easy to show that the origin of the fluorescence lies within the vacuum tube.

(2) It is seen, therefore, that some agent is capable of penetrating black cardboard which is quite opaque to ultra-violet light, sunlight, or arc-light. It is therefore of interest to investigate how far other bodies can be penetrated by the same agent. It is readily shown that all bodies possess this same transparency, but in very varying degrees. For example, paper is very transparent; the fluorescent screen will light up when placed behind a book of a thousand pages; printer's ink offers no marked resistance. Similarly the fluorescence shows behind two packs of playing cards; a single card between the apparatus and the screen does not visibly diminish the brilliancy of the light. So, again, a single thickness of tinfoil hardly casts a shadow on the screen; several have to be superposed to produce a marked effect. Thick blocks of wood are still transparent. Boards of pine two or three centimeters thick absorb only very little. A piece of sheet aluminum, 15 mm. thick, still allowed the X-rays (as I will call the rays, for the sake of brevity) to pass, but greatly reduced the fluorescence. Ebonite several centimeters thick still allowed the rays to pass through. [Ebonite, also called vulcanite, is a preparation of india-rubber and sulphur hardened by exposure to intense heat.] Glass plates of similar thickness behave variously, depending on their lead content; those with lead content were far more opaque. If a hand is held before the fluorescent screen, the shadow shows the bones clearly with only faint outlines of the surrounding tissues. . . .

(14) The justification of the term "rays," applied to the phenomena, lies partly in the regular shadow pictures produced by the interposition of a more or less permeable body between the source and a photographic plate or fluorescent screen.

Nowhere to be found in the article is even an inkling of the serendipity behind Röntgen's discovery: the typical scientific report represents a rational reconstruction of scientific work—bearing scant relationship with what actually happened in the laboratory or field or observatory on a day-by-day basis. As long ago as the early seventeenth century, Francis Bacon wrote that "never any knowledge was delivered in the same order it was invented."

News of the discovery of X-rays spread like wildfire through the scientific community. It quickly caught the attention of physicians, as evidenced by this excerpt from an article in *Nature*, which appeared in the same year as the English translation of his seminal article:

Medical science seems likely to benefit much by the application of Prof. Röntgen's discovery. The *British Medical Journal* thinks, as an aid to the diagnosis of obscure fractures and internal lesions generally, the new photography will be of great value. From our contemporary we note that already a beginning has been made in this direction, and Prof. Mosetig, of Vienna, has taken photographs which showed with the greatest clearness and precision the injuries caused by a revolver-shot in the left hand of a man, and the position of a small projectile.

On the following page is Röntgen's X-ray photograph showing the hand of his assistant wearing a ring. It was first displayed during his legendary lecture before the Würzburg Physical-Medical Society on January 23, 1896. It also appears in the English translation of his article published in *Nature* that same year. This is one of those instances in scientific communication, infrequent but not all that uncommon, where the images overshadow the accompanying text. These and other photographic images made Röntgen a worldwide celebrity as the man who could make the previously invisible visible.

Arguing a New Radioactive Element into Place

Pierre Curie, Marie Sklodowska Curie, and G. Bémont, 1898. "Sur une nouvelle substance fortement radio-active, contenue dans la pechblende" (On a new, strongly radioactive substance, contained in pitchblende). *Comptes Rendus*, Vol. 127, pp. 1215–17.

One could make a convincing case that Marie Curie was the first woman to publish scientific articles of major significance. In 1898, aged thirty, she published two important articles along with her husband and another man, each article announcing the discovery of a different radioactive element detected in the uranium ore called "pitchblende." The first article reported on polonium, named after Curie's native country; and the second, radium, named from the Latin word for "ray."

Our selection concerns the newly discovered radioactive element radium, a tiny fraction of which lies buried in pitchblende (1 gram of radium is typically found in 7 tons of pitchblende). By labor-intensive chemical separations, the

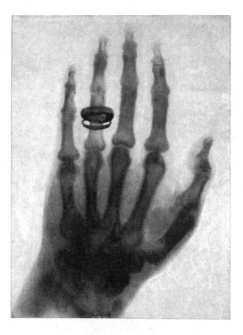

FIGURE 4-5 · X-ray showing hand with ring.
University of Minnesota Library.

Curies had isolated a radioactive fraction of the pitchblende composed pre-
dominantly of barium. The selected passage is not a simple observation of a
previously unknown constituent of nature, but an argument in which the authors hope to convince their readers that radium is an element, is radioactive,
is not the same as the radioactive elements already known (uranium, thorium,
polonium), and has chemical properties similar to barium:

> We believe . . . that this substance [extract from pitchblende], although con-
> stituted for the most part of barium, contains one more new element that
> imparts radioactivity to it and that, besides, is very near to barium in its
> chemical properties.
>
> The following are the reasons that argue in favor of this view:
>
> 1. Barium and its compounds are not ordinarily radioactive, but one of us
> has shown that radioactivity seems to be an atomic property, persisting in all
> the chemical and physical states of the matter. From this point of view, the
> radioactivity of our substance is not due to barium, and must be attributed
> to another element.

2. The first substances that we have obtained, in the form of the hydrated chloride, have a radioactivity 60 times stronger than that of metallic uranium (the radioactive intensity being determined by the magnitude of the conductivity of the air in our apparatus). In dissolving these chlorides in water and precipitating a part of the solute with alcohol, we find that the part precipitated is much more active than that remaining dissolved. On the basis of this finding, we were able to conduct a series of fractionations that yielded more and more radioactive chlorides. We eventually obtained chlorides having an activity 900 times greater than that of uranium. We have been stopped by the lack of material, and . . . we anticipate that the activity could have been much greater, if we had been able to continue. These results may be understood in terms of the presence of a radioactive element whose chloride in a water solution of alcohol is less soluble than that of barium.

3. Mr. Demarçay had wanted to examine the spectrum of our substance, for which we thank him very much. The results of this examination are set forth in a special Note following ours. Mr. Demarçay has found in the spectrum a line that does not appear to belong to any known element. This line, barely visible because the chloride is 60 times more active than uranium, becomes notable when the chloride is enriched by fractionation up to 900 times the activity of uranium. The intensity of this line increases at the same time as the radioactivity, and . . . , we believe, this is a very strong reason for attributing it to the radioactive part of our substance.

The diverse reasons that we have just enumerated lead us to believe that the new radioactive substance harbors a new element, to which we propose to give the name of *radium*.

The argument makes the conclusion inescapable: the element is not barium because it is radioactive; it is not radioactive uranium because it is far more radioactive; it is like no previously discovered radioactive element because it displays a unique spectrum. Calling this new element "radium" is therefore justified. The conclusion snaps together with the satisfactory click of a mathematical proof.

5

EQUATIONS, TABLES, *and* PICTURES

I N   T H I S   C H A P T E R you will find equations, tables, and pictures that changed the way scientists think about the natural world. In the earlier chapters, we presented and discussed visuals that had appeared in the scientific literature through the nineteenth century: Boyle's schematic of his air pump, Hooke's flea, Perrault's chameleon, Maria Merian's Suriname insects, *Botanical Magazine's* plants, Lambert's Cartesian graph, and Mendeleev's periodic table. As important as these examples are to the history of science and its communication, it must be said that such visuals in the typical scientific article are quantitatively relatively scarce into the nineteenth century, in part because of the expense involved in drawing and publishing them. It is not uncommon to find, for example, all the illustrations for an annual volume of a journal huddled together in a slim section at the end.

Today, the scientific article is as much a visual as a verbal document. Our own research has shown that pictures and tables occupy an average of about a quarter of the typical twentieth-century article. Few pages are without them, seamlessly integrated into the verbal text. This fusion of words and visuals overcomes the limitations of each in disclosing and explaining relevant aspects of the natural world. Even in earlier centuries when such visuals were not nearly as plentiful, the fact that any appeared at all, given their expense in reproduction, is a tribute to their communicative utility.

Over the last four centuries, the pictures of science have altered not only in their quantity, but in their character as well. More and more, they depict not the things of nature and the laboratory by themselves, but theories and data trends. In a groundbreaking scholarly article on scientific illustrations in *History of Science*, for example, Martin Rudwick argues convincingly that, in the period from 1760 to 1840, geological drawings of nature became more and more theory-based: what he calls "highly abstract statements in a visual language." Today, pictures from all scientific disciplines are often steeped in theory. In this chapter, we present modern examples regarding the formation of the earth, the behavior of the subatomic world, the chemical structure of a key biological molecule, the origin of life on earth, and a map of the universe.

By far the most common types of modern visual, however, are the Cartesian graph and table of data. Their prominence has to do with the increased quantifications driven by the invention of ever more precise instruments and methods for rapidly measuring and calculating properties. Both tables and graphs overcome the limitations of the verbal in conveying relationships among and within complex arrays of data. Written languages are one-way streets; numbers must be displayed in strict linear sequences not easily compared. Tables and graphs greatly facilitate comparisons: they guide the eye by employing alignment, vertical and horizontal lines, and white space. Our selected tables include one that empirically established the reality of molecules for the first time, and another that ignited a decade-long, bitter conflict in molecular biology. Our representative graph supports the verbal argument that a fundamental constant of physics has changed infinitesimally in magnitude over the past 13 billion years.

Also showing a marked quantitative upturn in the twentieth century are equations, a visual shorthand for communicating mathematical relationships among physical, chemical, or biological properties. Like pictures and tables, these meaningful strings of symbols and numbers concisely capture what can only be described clumsily with words. Instead of writing out "energy equals mass times the speed of light squared" all we need is "$E = mc^2$." We begin this chapter with Einstein's famous equation, a corollary from his newly formed theory of relativity presented in a paper a few months earlier. That example is followed by that of a pure mathematician, G. H. Hardy, applying a "simple" mathematical equation to a problem in Mendelian genetics.

EQUATIONS

An Inspired Afterthought: Einstein's Energy Equation

Albert Einstein, 1905. "Ist die Trägheit eines Körpers von seinem Energiegehalt abhängig?" (Does the inertia of a body depend on its energy content?). *Annalen der Physik*, 4th series, Vol. 18, pp. 639–41. Translated by John Stachel, *Einstein's Miraculous Year: Five Papers that Changed the Face of Physics*. Princeton: Princeton University Press (1998), pp. 161–64.

At the heart of the special theory of relativity lies a set of equations regarding the relative motion of bodies in the three dimensions of space and one of time. These equations appear in Einstein's first and most important paper on special relativity (see chapter 9). In a short follow-up paper published the same year, 1905, Einstein derived the equation every high school physics student now knows: $E = mc^2$, the equivalence of mass and energy.

This simple mathematical expression is directly in accord with Einstein's philosophy of science. According to this philosophy, physics is the search for laws that describe relationships among fundamental entities that completely determine physical laws. "I still believe in the possibility of a model of reality," wrote Einstein in 1934, "that is to say, of a theory which represents things themselves."

In the following passage, Einstein derives the equivalence of mass and energy as a corollary to his theory of special relativity:

The results of an electrodynamic investigation recently published by me in this journal lead to a very interesting conclusion, which will be derived here.

I based that investigation on the Maxwell-Hertz equations for empty space, together with Maxwell's expression for the electromagnetic energy of space, and also the following principle:

The laws according to which the changes of states of physical systems occur do not depend on which one of the two coordinate systems (assumed to be in uniform parallel-translational motion relative to each other) is used to describe these changes (the principle of relativity).

Based on this foundation, I derived the following result...

Neglecting magnitudes of the fourth and higher orders, we can get

$$K_0 \text{-} K_1 = L / V^2 v^2 / 2$$

[That is, the kinetic energy of a body with respect to the one system minus the kinetic energy of a body with respect to the other system equals L, the energy, divided by the square of the velocity of light, multiplied by the square of the velocity of the new coordinate system divided by two.] From this equation, one immediately concludes:

If a body emits the energy L in the form of radiation, its mass decreases by L/V^2. Here it is inessential that the energy taken from the body turns into radiant energy, so we are led to the more general conclusion:

The mass of a body is a measure of its energy content; if the energy changes by L, the mass changes in the same sense by $L/9 \times 10^{20}$, if the energy is measured in ergs and the mass in grams [and the speed of light in centimeters per second].

It is not excluded that it will prove possible to test this theory using bodies whose energy content is variable to a high degree (e.g., radium salts).

If the theory agrees with the facts, then radiation carries inertia between emitting and absorbing bodies.

In this passage, Einstein first reformulates his principle of relativity from his "recently published" article (first three paragraphs), then spins out a new set of equations. From his final displayed equation, shown above, Einstein concludes that energy lost from a body in the form of electromagnetic radiation will decrease in mass by L/V^2 (in modern notation, m = E/c^2). He then takes a creative leap forward, generalizing from this limited case to all forms of energy: that is to say, "The mass of a body is a measure of its energy content," regardless of the particular kind of energy involved.

Up to this point Einstein's argument confidently marches forward with mathematical rigor. The implication is that anyone who can follow his proof must acknowledge its cogency, its imperviousness to rational challenge. No such certainty attaches to the suggested test of his theory, which is only a remote possibility: "It is not excluded that it will prove possible . . ." But this tentativeness casts no shadow on Einstein's conclusion that the mass/energy/speed of light relation, if it "agrees with the facts," is "very interesting." The convertibility of a small mass into an enormous amount of energy (and vice versa) is a startling consequence of his relativity postulates, though Einstein at this time had no way of knowing just how startling it would turn out to be, the atomic bomb being its most notorious manifestation.

A Mathematician Tackles a Scientific Problem: No Apologies Needed

G. H. Hardy, 1908. "Mendelian proportions in a mixed population." *Science*, Vol. 28, pp. 49–50.

Equations have been a key ingredient of theoretical physics and astronomy at least since the time of Newton's *Principia*. Not so the other sciences; as a prominent example, not a single equation interrupts Darwin's *Origin*. In the wake of Mendel's theory of inheritance, that mathematical aversion changed dramatically for evolutionary biology in the early twentieth century.

The selected article is one of the earliest examples. Its author is Godfrey H. Hardy, a world-famous Cambridge University don and mathematician. His brief letter to the editor of *Science* sets out a founding equation of population genetics, based on the results of Mendel's experimental program. Hardy modestly begins by admitting his outsider status: he is a mathematician with no expert knowledge of biology. And yet, he intends to express a "simple point" apparently missed by biological experts who questioned the mathematical viability of Mendelian genetics.

The problem Hardy tackled was why doesn't the larger number of dominant traits (allele A) in a given population wipe out the recessive ones (allele a) after a few generations? As the legend goes, Reginald Punnett gave a lecture on Mendelian genetics, where he was asked a question that stumped him: "Why it was that, if brown eyes were dominant to blue, the population was not becoming increasingly brown eyed, yet there was no reason for supposing such to be the case." He took this mathematical brain teaser to his friend and fellow cricket enthusiast, G. H. Hardy, who reportedly solved it in short order. (In one of those many instances of simultaneous discovery in science, in the same year, Wilhelm Weinberg, a German physician, published an article reaching the same conclusion by essentially the same mathematical route.)

Here is how mathematician Hardy presented his argument, opening with his apology for entering the alien realm of genetics:

> I am reluctant to intrude in a discussion concerning matters of which I have no expert knowledge, and I should have expected the very simple point which I wish to make to have been familiar to biologists. However, some remarks of Mr. Udny Yule, to which Mr. R. C. Punnett has called my attention, suggest that it may still be worth making.
>
> In the *Proceedings of the Royal Society of Medicine* (Vol. I., p. 165) Mr. Yule is reported to have suggested, as a criticism of the Mendelian position, that if brachydactyly is dominant "in the course of time one would expect, in the absence of counteracting factors, to get three brachydactylous persons [shortened bones and hands and feet, an inherited condition] to one normal."
>
> It is not difficult to prove, however, that such an expectation would be quite groundless. Suppose that Aa is a pair of Mendelian characters, A being

dominant, and that in any given generation the numbers of pure dominants (AA), heterozygotes (Aa), and pure recessives (aa) are as p : 2q : r. Finally, suppose that the numbers are fairly large, so that the mating may be regarded as random, that the sexes are evenly distributed among the three varieties, and that all are equally fertile. A little mathematics of the multiplication-table type is enough to show that in the next generation the numbers will be as

$$(p + q)^2 : 2(p + q)(q + r) : (q + r)^2$$

or as $p_1 : 2q_1 : r_1$, say.

The interesting question is—in what circumstances will this distribution be the same as that in the generation before? It is easy to see that the condition for this is $q^2 = pr$. And since $qr^2 = p_1r_1$, whatever the values of p, q and r may be, the distribution will in any case continue unchanged after the second generation.

Suppose, to take a definite instance, that A is brachydactyly, and that we start from a population of pure brachydactylous and pure normal persons, say in the ratio of 1:10,000. Then p = 1, q = 0, r = 10,000 and p_1 = 1, q_1 = 10,000, r_1 = 100,000,000. If brachydactyly is dominant, the proportion of brachydactylous persons in the second generation is 20,001:100,020,001, or practically 2:10,000, twice that in the first generation; and this proportion will afterwards have no tendency whatever to increase. If, on the other hand, brachydactyly were recessive, the proportion in the second generation would be 1:100,020,001, or practically 1:100,000,000, and this proportion would afterwards have no tendency to decrease.

In a word, there is not the slightest foundation for the idea that a dominant character should show a tendency to spread over a whole population, or that a recessive should tend to die out.

One need not follow Hardy's mathematics in detail here. His basic argument is clear. First, Hardy makes some reasonable simplifying assumptions about inheritance, without which his model building would have been hopelessly complex. Next, he develops a mathematical law for making the necessary genetic calculations, leading to the conclusion that the distribution of genes in his simplified case will "continue unchanged after the second generation." He follows that up with a specific example that makes his mathematics more concrete. Dropping any pretense of modesty or politeness, Hardy then repeats his conclusion for emphasis: "there is not the slightest foundation for the idea that a dominant character should show a tendency to spread over a whole population, or that a recessive should tend to die out."

TABLES

Tracking Molecular Movements
....

Jean Perrin, 1909. "Mouvement Brownien et réalité moléculaire" (Brownian movement and molecular reality). *Annales de Chimie et de Physique*, Vol. 18, pp. 5–114. Translated by F. Soddy, *Brownian Movement and Molecular Reality*. London: Taylor and Francis (1910).

French physicist Jean Perrin brings the natural phenomenon known as Brownian motion sensually to life. First he describes the behavior of a large object dropped into a glass of water; then he contrasts that with the amazing behavior of microscopic objects in water, where the effects of gravity are virtually absent:

> When we consider a fluid mass in equilibrium, for example, some water in a glass, all the parts of this mass appear completely motionless to us. If we put into it an object of greater density it falls and, if it is spherical, it falls exactly vertically. The fall, it is true, is the slower the smaller the object; but, so long as it is visible, it falls and always ends up at the bottom of the vessel. When at the bottom, as is well known, it does not tend again to rise, and this is one way of formulating Carnot's principle (the impossibility of perpetual motion of the second kind [an operation so perfect that no energy is lost]).
>
> These familiar ideas, however, only hold good when we examine middle-sized objects to which we are accustomed; it takes no more than a microscope to impress upon us new ideas that replace the old static conception of the fluid state with a kinetic one.
>
> Indeed, we could not examine preparations in a liquid medium for any length of time without observing that all the particles therein contained fall and ascend, not in regular movements, according to their density, but, on the contrary, are animated by movements that are perfectly irregular. They go and come, stop, start again, *mount*, descend, *remount again*, without in the least tending toward immobility. This is the *Brownian movement*, so named in memory of the naturalist [Robert] Brown, who described it in 1827 . . . then proved that the movement was not due to living animalculae, and recognised that the particles in suspension are agitated the more briskly the smaller they are.

In the more than three-quarters of a century before the work of Perrin and other "atomists," however, this zigzag dance of inanimate particles lacked a convincing mathematical or mechanical explanation. If not "living animalcules," what caused this apparent perpetual motion?

In the period 1905–1908, Einstein and others developed mathematical models for Brownian motion, and in 1909 Jean Perrin offered the first convincing experimental proof that molecular agitation was the cause. His two-part title, in French, "Mouvement Brownien et la réalité moléculaire," clearly defines the issues at stake: the cause of Brownian motion *and* the reality of molecules. The conjunction *et* in the title works in two ways: additive and coordinate. On the one hand, it links Brownian motion with molecules; on the other, it asserts the reality of molecules independent of Brownian motion. Some scientists still doubted the existence of molecules as late as the first decade of the twentieth century!

The body of this book-length article has four parts. Parts I through III apply to the title's *et* in its additive role, while Part IV covers *et* in its coordinate capacity. Part I traces the intellectual heritage of the concept that molecular agitation caused Brownian motion, and also explains Brownian motion by means of the existing kinetic theory. In Part II, Perrin focuses on his experiments with colloids—gelatinous substances containing very small particles that behave like large molecules in a gas. Part III reviews Einstein's mathematical explanation for Brownian motion and explains how Perrin's experimental results confirm that explanation, one that posits the discontinuity of matter. At this point Perrin reproduces the zig-zag figure on the next page.

As Perrin explains, this figure works somewhat like a Russian doll—the random patterns repeat as the scale get smaller and smaller:

[The graph] shows three drawings obtained by tracing the segments that join the consecutive positions of the same granules of mastic [a gum or resin] at intervals of 30 seconds. It is the half of the mean square of such segments that verifies Einstein's formula. One of these drawings shows [nearly 50] consecutive positions of the same granule. These positions give only a very feeble idea of the prodigiously complex character of the real trajectory. If the positions were indicated from second to second, each of these rectilinear segments would be replaced by a polygonal contour of 30 sides, relatively as complicated as the drawing here reproduced, an effect that would keep repeating itself as we plotted shorter and shorter periods of time. From such examples, one realizes how near the mathematicians are to the truth when, through logical instinct, they refuse to admit as experimental evidence for the continuity of matter the geometric truth that, for each point on a curve, there exists a tangent.

This last point may not be immediately clear. Perrin means that mathematics is not physics. At its limit, a polygon of infinite sides becomes a circle, that is, in

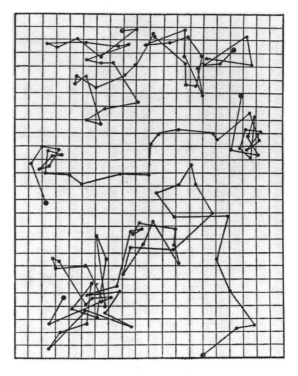

FIGURE 5-1 · Particle tracks in Brownian motion.
University of Minnesota Library.

....

mathematics, at the limit, the discontinuous becomes continuous. But it does not follow that the constitution of matter is such that the discontinuous will become continuous. It does not follow that, if you map molecular collisions at smaller and smaller intervals, points of collision will disappear and a smooth curve will supervene.

Part IV concentrates on proving molecular reality in other physical situations besides the Brownian one—the title's *et* now acting in its independent coordinate role. Central to this concluding part is a table comparing the "molecular reality" of Brownian motion with that of other natural phenomena, including the viscosity of gases, diffusion of dissolved substances, mobility of ions in water, the blueness of the sky, electric charges in microscopic dust, and radioactive decay. The column on the right displays values of N, Avogadro's constant (the number of molecules in a gram molecular weight, or mole, of any substance), calculated from the measurements in these diverse phenomena.

Perrin's underlying argument goes as follows. Einstein predicted that the extremely small suspended particles of colloids would behave just as gases

Phenomena studied.		$N.10^{-22}$.
Viscosity of gases taking into account	the volume of the liquid state ..	>45
	the dielectric power of the gas ..	<200
	the exact law of compressibility .	60
Brownian Movement.	Distribution of uniform emulsion	**70·5**
	Mean displacement in a given time	71·5
	Mean rotation in a given time	65
Diffusion of dissolved substances		40 to 90
Mobility of ions in water		60 to 150
Brightness of the blue of the sky		30 to 150
Direct measurement of the atomic charge.	Droplets condensed on the ions	60 to 90
	Ions attached to fine dust-particles .	64
Emission of α-projectiles.	Total charge radiated	62
	Period of change of radium	70·5
	Helium produced by radium	71
Energy of the infra-red spectrum.................		60 to 80

The most probable value always appears to me **70·5** $\times 10^{22}$.

FIGURE 5-2 · Table of Avogadro's numbers calculated by an array of methods. University of Minnesota Library.

do. If this is the case, then it should be possible to calculate Avogadro's constant from measurements of Brownian motion, a behavioral characteristic of colloids. Perrin does this and obtains the N values that the table notes opposite "Brownian motion" (about 70×10^{22}, not far from the present value of 60×10^{22}). The reasonable correspondence among all the calculated N values in the table means that the behavior of colloids and gases is analogous. From this analogy, we can infer that gases consist of discontinuous particles, that is, molecules and atoms. Concerning his table, Perrin writes: "I think it impossible that a mind, free from preconception, can reflect upon the extreme diversity of the phenomena that converge on the same result, without being very impressed. From now on, I think it will be difficult rationally to defend a hostile attitude to the molecular hypothesis."

A Table of Problematic Contents

David Weaver, Moema H. Reis, Christopher Albanese, Frank Constantini, David Baltimore, and Thereza Imanishi-Kari, 1986. "Altered repertoire of endogenous immunoglobulin gene expression in transgenic mice containing a rearranged mu heavy chain gene." *Cell*, Vol. 45, pp. 247–59.

On the surface, Table 2 in this paper seems a fairly ordinary example of the modern table of data. It concerns the central topic of the cited article: how immune-specific genes rearrange their DNA to form different antibodies, which destroy hostile invaders in the body. The table compares analysis of the spleen and lymph nodes in normal mice and "transgenic" mice. The transgenic mice differ from normal ones in that they have a gene from another mouse inserted into them when they are only a recently fertilized egg. This inserted gene (designated by the number 17.2.25 in the paper) then appears in every cell of the transgenic mice, making them an ideal research organism for immunologists.

As is typical of tables, the rows and columns in Table 2 permit the easy location of a particular datum: for example, the datum on the upper left shows that, from normal spleen, antibodies were produced by only 1 out of 144 hybridomas, or hybrid cells. The table also permits comparison in two dimensions of significance: column and row. By column, we can compare the production of antibodies in normal and transgentic mice; by row, we can compare the production of different types of antibodies (also called "immunoglobins") for a selected organ. In conveying the significance of data—so central to modern scientific communication—normal prose cannot achieve this clarity and communicative power.

The key finding appears in the first column of data: the transgenic mice produced many more antibodies than the normal mice (172 out of 340 versus 1 out of 244). That finding was totally unexpected. The next four columns quantify four main types of these antibodies in the mice spleens and lymph nodes.

Little did the authors know that their paper, and Table 2 in particular, would become the bane of their existence for the next decade. The trouble began when a post-doctoral fellow enlisted to build upon the original work (Margot O'Toole) questioned, for what would appear to have been legitimate reasons at the time, whether one of the authors (Thereza Imanishi-Kari) actually did the experiments and got the results claimed. Investigation of these charges escalated from the informal self-policing done by committees of scientists to the legalistic proceedings and grandstanding of a major congressional hearing. The name of a Nobel Prize–winning scientist on the list of authors (David Baltimore) ensured intense scrutiny by the scientific and popular press.

In general, points of contention in experimental papers revolve around five issues: Do the experimental procedures yield credible results? Are the results themselves trustworthy; that is, are they tainted because of human error, negligence, deliberate misrepresentation, or even fraud? Is the author's interpretation of the results correct? Is the underlying theory and assumptions behind this

Table 2. Frequency of 17.2.25 Idiotype-Producing Hybridomas in Normal and Transgenic Mice

Organ	17.2.25 Idiotype-Positive	17.2.25 Idiotype-Positive Plus:			
		Anti-NP (κ)	Anti-NP (λ)	μ^a	μ^b
Normal Spleen	1*/144 (<1%)	1†/144	2/144	0/144	0/144
Normal Lymph Nodes	0/100	0/100	0/100	0/100	0/100
Transgenic Spleen	43/150 (28%)	0/43	7‡/43	9§/43	1/43
Transgenic Lymph Nodes	129/190 (68%)	6/129	12/129	33/129	10/129

B cell hybridomas were isolated from spleen and lymph nodes of transgenic and normal C57BL/6 mice. Secreted Ig from hybridomas was assayed for binding to the anti-17.2.25 idiotypic antibody or for NP-binding with either λ or κ light chains. The portion of the hybridomas that contained the μ isotype was assayed by the anti-allotype antibodies (see Figure 1) to associate the anti-17.2.25 idiotype binding with either the μ^a or μ^b allotype.
* Weak reactivity of hybridoma to anti-17.2.25.
† Not 17.2.25 idiotype.
‡ 2/7 were observed to have μ^a allotype.
§ 2/9 were found to be λ-bearing Ig, and 7/9 were found to be κ-bearing Ig.

FIGURE 5-3 · Genetic characterization of the spleen and lymph nodes in normal mice and transgenic mice. Reprinted from *Cell*. Copyright (1986), with permission from Elsevier.

interpretation flawed? And finally, does the paper accurately communicate the author's experimental work?

In this decade-long controversy, all these issues came into play at various points, but the most serious of them was the second. A breach in reporting of results is grounds for a serious reprimand or even dismissal, that is, for most scientists a career death sentence. Here is where Table 2 enters the picture.

Margot O'Toole claimed that her supervisor at the time, Imanishi-Kari, could not have performed the necessary experiments to support the table's data—a charge later judged to be without sufficiently convincing evidence. However, two serious problems did emerge in the course of putting Table 2 under a microscope. What began as a contention over the second issue ended with the fifth issue, accurate communication.

First, Imanishi-Kari admitted that she did not do the additional testing as implied in the following discussion of the Table 2 data (italics for emphasis):

> Of the 172 idiotype-positive hybridomas [43 + 129 in first column of data], only 53 were IgM secretors [immunoglobulin mu type: 9 + 1 + 33 + 10 in final two columns and rows]. The remaining 119 clones produced other Ig heavy chain isotypes, the majority being γ2B (*data not shown*).

No such "data" for the gamma 2B type apparently existed. While this was a serious and embarrassing oversight, the author had gathered other data to support the basic assertion that the majority of clones were gamma type.

The second problem was terminological. Imanishi-Kari had used the words "hybridoma" in the table and "clone" in the text when she really meant tiny wells containing hybrid cell lines formed by cloning. So, strictly speaking, "119 clones" are not 119 isolated clones, but 119 wells containing one or possibly more clones. This imprecise wording called into question the transgenic mice percentages of 28% and 68%, given in parentheses in the first column of data. In response, the author argued that any respectable cellular immunologist would have known the intended meaning, and that aside, the wording choice here did not weaken her principal knowledge claims.

In the end, after ten years and multiple investigations (brilliantly documented by Daniel J. Kevles in *The Baltimore Case*), an appeals panel under the auspices of the National Institutes of Health exonerated Imanishi-Kari from having falsely fabricated or deliberately misrepresented her data.

PICTURES

Mapping the Earth's History

Alfred Wegener, 1912. "Die Entstehung der Kontinente" (The origin of the continents). *Petermanns Geographische Mittheilungen aus Justus Perthes' Geographischer Anstalt*, Vol. 58, pp. 185–95, 253–56, 305–9.

Every now and again, an explanation for the natural world comes along that appears, on the basis of everyday experience, ridiculous: the heliocentric universe, the evolution of Homo sapiens from unicellular organisms by natural selection, and the relativity of space and time, to name three of the more prominent examples. Alfred Wegener's continental drift theory also belongs in that august company.

While studying a map of the world in 1910, Wegener noted that the continents on either side of the Atlantic Ocean fit together like the pieces in a child's jigsaw puzzle. This observation was not really new. As long ago as 1596 a Dutch map maker, Abraham Ortelius, noted the close correspondence between the coasts of Africa and Europe, on the east side, and the Americas on the west. In the nineteenth century several others had proposed the separation of the three continents over time. No one, however, had amassed much evidence in support of the seemingly outlandish thesis that the continents were once joined and subsequently drifted apart.

Wegener's attempt at making a legitimate case for continental drift made its first substantive appearance in a pair of 1912 articles in German geologic journals, one in *Geographische Mitteilungen* (Geographical Reports) and the other in *Geologische Rundschau* (Geological Review). A few years later (1915) Wegener followed these preliminary publications with a book, *Die Entstehung der Kontinente und Ozeane* (The Origin of the Continents and the Oceans). In these publications, he argued that at one time the earth contained but a single supercontinent, near the north pole, which broke up some several hundred million years ago. The enormous fragments then slowly drifted through the ocean basins, like pilotless ships, occasionally colliding and separating, eventually to reach their present destinations. Wegener even proposed a mechanism for propelling such large land masses over thousands of miles — they float on a fluid material beneath the crust.

The first picture in Wegener's article in *Geographische Mitteilungen* makes his argument for the complementarity of opposing coastlines. The heavy lines trace the "continental masses" that Wegener refers to in his introduction:

This paper represents a first, crude effort to explain the large-scale features of the earth's surface, that is, the continental masses and the ocean basins. This task will be accomplished by means of a single overarching generating principle, namely, the horizontal movements of the continents. Although in the past we have posited old land-bridges sunk in the depths of the world ocean, now we will assume that the continents separated and drifted apart. This picture, which we derive from the nature of the earth's crust, is new and in many respects paradoxical, but is, as we shall show, not without a plausible physical cause. In the following examination, based only on general geological and geophysical principles we will reveal many surprising simplifications and interrelationships, so many that it seems to me that on these grounds alone, I am entitled, indeed I must, put aside the old hypothesis of sunken continents and replace it with the new, more efficacious working hypothesis. The shortcomings of the former are already evident; it flies in the face of the established doctrine of the permanent character of the oceans. Despite this broad foundation in fact and in theory, I designate this new principle as a working hypothesis, and should like to have it treated as such,

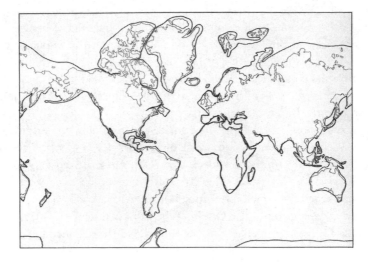

FIGURE 5-4 · Map of the continental land masses.
University of Minnesota Library.

until such time that, by means of exact astronomical measurements, we succeed in removing every doubt concerning the horizontal movement of the
continents in recent times. It is not beside the question to point out that this
is a first attempt at making a case. A thorough examination will probably
teach us that this hypothesis will have to be modified in many respects.

The next schematic is, as it were, the first one flipped on its side. It illustrates
Wegener's causal mechanism for continental drift, an incorrect theory that was
plausible at the time. Here, we are looking at a cross section of the great circle
through South America and Africa, drawn to scale. The vertical axis marks
distance in kilometers. The outer three spherical layers represent the Earth's
atmosphere. *Geokoronium* is a hypothetical gas that accounts for the green line
in the spectrum of the Northern Lights; *Wasserstoff* is a hydrogen gas, *Stickstoff*,
a nitrogen gas in the form of the air we breathe. (Wegener's main research
interest up to this time had been meteorology.) Below this gaseous zone is the
Earth's crust and underworld. *Sal* is sial, the upper silicon and aluminum-rich
zone of the continental crust of the earth; *Sima* is the basaltic rock that forms
the chief constituent of the ocean's floor; *Nife* is the iron-nickel core. Note
that as we move progressively outward from the *Nife*, we move also toward
increasing plasticity. It is on the more plastic *Sima* that, according to Wegener,
the continents (*Sal*) float, "like an iceberg floating in water."

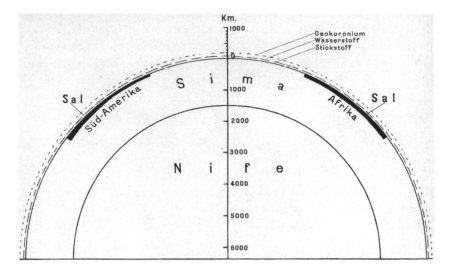

FIGURE 5-5 · Cross section of the great circle through South America and Africa.
University of Minnesota Library.

Wegener's theory as represented in the two figures had considerable explanatory power for the geological, geographical, and biological facts known at the time. It explained not only the complementarity of the coastlines between eastern South America and Western Africa, as well as between North America and Europe, but also the eruption of mountain ranges close to coastlines as well as inland near faults, the geologic evidence for dramatic climate changes in distant geological times, and the appearance of the same animal species and rock compositions on different continents and islands.

Wegener's supporting evidence notwithstanding, his theory remained "a beautiful dream" for decades for several reasons: it still seemed implausible that such large land masses could have migrated such great distances without the shore lines disintegrating, Wegener's proposed causal mechanism lacked evidence convincing to most geologists, and his theory threatened the status quo regarding the scientific practices, theories, and beliefs within the geological communities in America and Europe.

While attracting a few admirers and advocates, the many critics of continental drift dismissed it for several decades as a "German theory," "a fairy tale," "bizarre," and "utter damned rot." That pejorative attitude changed in the 1960s with the theory of plate tectonics and the accretion of scientific facts and explanations supporting it over the previous decades. Its basic premise is that convection currents involving the circular movement of molten rock far beneath the earth's surface move the tectonic plates above. Many nagging questions still remain about the mechanism behind plate tectonics. Nonetheless, Wegener's 1912 visualization of continental drift is now far more substantial than a beautiful dream or idle fancy, with the continental movements actually having been detected, as Wegener predicted they one day would be, by "exact astronomical measurements."

The Cloudmeister

C. T. R. Wilson, 1912. "On an expansion apparatus for making visible the tracks of ionizing particles in gases and some results obtained by its use." *Proceedings of the Royal Society of London*, Vol. A87, pp. 277–92.

Charles Thomson Rees Wilson began his research career by studying clouds and electrical phenomena in an observatory built on a damp and foggy Scottish mountain. This experience later led him to build a device for studying condensation and other weather phenomena in the comfort and controlled conditions of a scientific laboratory. He then had a bright idea: he would shoot different kinds of rays—alpha, beta, X—into his chamber, "making visible the track of ionizing particles . . . by condensing water upon the ions."

His original cloud chamber was a cylindrical glass chamber, slightly higher than a coffee mug but twice as wide. It rested on a hollow piston with a pool of water at the bottom. A valve linked this piston to a vacuum flask. When the valve was opened, the piston fell rapidly, thus reducing the pressure in the chamber, thus lowering the temperature, thus supersaturating the air within. As a consequence, water vapor condensed along the tracks of alpha, beta, and X-rays shot into the chamber. A strong magnetic or electrical field applied to the particles altered their paths. The particle tracks were illuminated by a strong light source and photographed for later analysis.

We include one of Wilson's first such photos. Without commentary, this photograph is meaningless. Is the phenomenon subatomic or celestial? Is the photograph a work of science or a work of abstract art? With Wilson's commentary appended, these questions are answered: a hitherto invisible world is revealed, a world for whose revelation Wilson won the Nobel Prize. In the figure legend, Wilson tells us that we are looking at "α and β rays from radium." In the main text he comments on the α ray at the bottom of the photograph, noticing its abrupt bend in the last two millimeters of its course (left side). He links this deviation to an observation of the great physicist Ernest Rutherford: "as Rutherford has contended, the scattering of a large amount is in the case

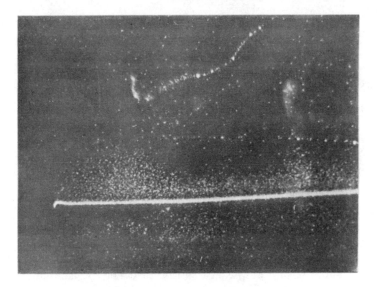

FIGURE 5-6 · Tracks of alpha and beta rays from radium source in cloud chamber. University of Minnesota Library.

of α-particles mainly ... the result of single deflections through considerable angles and not a cumulative effect due to a very large number of minute deviations." Next, Wilson notices and explains the bright scatter of star-like dots beside the track: "Some of the ions appear to have retained their mobility in the supersaturated atmosphere long enough to enable them to travel some distance under the action of the electric field before growing into drops. The effect is most marked above the main track, *i.e.*, on the side to which the negative ions would travel."

Finally, Wilson comments on and explains the behavior of the β-ray whose coiled up trail appears in the top half of the photograph: "It will be noted that the β-rays photographed do not show abrupt deflections like the α-rays but, except while the velocity remains very high, they show gradual bending resulting in large deviations. The scattering of the β-rays is thus mainly or entirely of the culmulative or 'compound' type, being due to a large number of successive deflections, each in itself inappreciable."

In total, Wilson's article has nineteen such photos, most of them with patterns reminiscent of lovely wisps of clouds, comets, and stars. Aesthetics aside, Wilson's cloud-chamber photographs constitute a landmark for scientific visualization. His cloud chamber began a trend in the practice of physics involving the creation of ever more elaborate (and expensive) machines for generating and interpreting images that track the behavior of subatomic particles.

The Quantum World Made Visible

R. P. Feynman, 1949. "Space-time approach to quantum electrodynamics." *Physical Review*, Vol. 76, pp. 769–89.

Few have surpassed the bongo-playing physicist Richard Feynman in the clear explanation of complex physics. His *Lectures on Physics* remain in print after forty years—astounding longevity for an undergraduate textbook. His discussion of quantum electrodynamics aimed at the general public—*QED: The Strange Theory of Light and Matter*—also remains in print and commonly available at well-stocked bookstores.

But Feynman did not only explain; he created. His principal achievement in the world of science was his mathematical welding of quantum mechanics to relativity theory. This new theory, called quantum electrodynamics or QED for short, makes use of the trajectory of subatomic particles along paths, an idea that at least at first glance would appear to be at odds with the random behavior central to quantum mechanics. To explain his new physics, Feynman devised a new kind of visual, the eponymous Feynman diagram. (He was so proud of his visual creation that he emblazoned his van with Feynman diagrams.) We reproduce here the first published one, showing the billiard ball–like paths of two interacting electrons in space and time.

To read this depiction of the subatomic world, you must leave your common sense, though not your thinking cap, at the door. You read the diagram from bottom to top. The straight lines with arrows represent the paths of electrons, the gammas (γ), along with their mus (μ), their four possible polarizations or orientations (think polarized sunglasses). These K_+'s are not the paths exactly, since the square of these "amplitudes" is not the event, but only the *probability* of the event. In Feynman's words: "It is not of interest to be able to say how the situation would look at each instant of time during a collision and how it progresses from instant to instant. Such ideas are only useful for events taking a long time and for which we can readily obtain information during the intervening period. For collisions it is much easier to treat the process as a whole."

The accompanying diagram begins with two imaginary electrons at points 1 and 2. They approach each other by paths 1 to 5 and 2 to 6. At point 5, they reach a close enough state that the first electron emits a photon, and changes directions away from the second electron, which absorbs the photon. Its amplitude is the wavy line from 5 to 6, and $\delta_+(S_{56}^2)$ is its "Dirac delta function," a measure of its

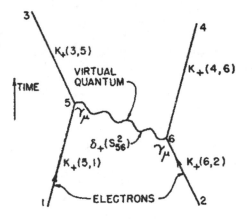

FIGURE 5-7 · Feynman graph for two
interacting electrons in space and time.
Reprinted from *Physical Review*. Copyright
(1949) by the American Physical Society.

trajectory important to physicists but not for a general understanding of what's
happening. Notice that the particle being exchanged actually appears to be going
backwards in time, since time moves ahead from bottom to top in the diagram.
The emitted photon having been absorbed, the second electron is deflected outward *before* the first, along path 6 to 4. Feynman uses the term "virtual quantum"
for this strange exchange process, one that cannot be observed, only inferred.

The various terms along each line segment, such as $K_+(5,1)$ and $K_+(6,2)$, figure
in Feynman's variation on the famous wave equation for the electron developed
by the mathematical physicist P. A. M. Dirac in 1928. So the diagram visualizes
electron interactions and functions as part of a mathematical explanation of one
of nature's laws. Overall, this first Feynman diagram appears remarkably simple.
It merely describes how two negative charges, the two electrons, approach each
other then repel. As the complexity of the situations grows, however, so too
the Feynman diagrams.

While the photographs taken in Wilson's cloud chamber actually depict the
subatomic world, Feynman diagrams do not: nothing in that world necessarily
corresponds to the paths in the diagram. Rather, Feynman diagrams are a useful
way of representing a three-dimensional electron interaction (two of space, one
of time) in the two dimensions of the printed page. Feynman diagrams depict,
not nature, but a theory of how nature works. Their communicative utility is
without question; an equivalent mathematical realization, by Julian Schwinger,
never caught on.

Structuring a Building Block of Life

···

Linus Pauling, Robert B. Corey, and H. R. Branson, 1951. "The structure of proteins: Two hydrogen-bonded helical configurations of the polypeptide chain." *Proceedings of the National Academy of Sciences*, Vol. 37, pp. 205–11.

Outside astronomy, few if any modern scientific disciplines have relied on pictures so critically and for so long as chemistry. We trace the emergence of this strong visual orientation to the mid-nineteenth century, when August Kekulé and others began drawing schematic chemical structures for hydrocarbons and other relatively small molecules. Over the past century and a half, several visual conventions have emerged: in particular, circles indicate atoms (or on occasion, a group of atoms); straight or dashed lines, bonds between atoms. One type of atom is differentiated from another (hydrogen versus carbon, say) in the same molecule by different shadings, colors, sizes, or markings. Working with those simple conventions, chemists draw two-dimensional structures on paper that other chemists instantly reconstruct as three-dimensional molecules. The visuals in this article are of this sort. Unlike Wilson's cloud chamber photographs, they are not reproductions of an event; unlike Feynman's diagrams, they are not pictures of a theory. Rather, they use theory to depict a simplified version of what a particular organic molecule looks like in three-dimensional space.

The chemical structure on display shows a protein molecule, the essential working part of all biological cells. It is from an article by chemists Linus Pauling and Robert Corey and, a genuine rarity at the time, an African-American physicist, Herman Branson. Before the 1940s, scientists viewed the protein molecule as a jumble of amino acids without discernible structure. Then in 1948, the indefatigable Pauling, unable to rest even when sick, concocted a single-helix structure while lying in bed with a bad cold and fiddling around with cardboard cutouts meant to stand in for planar amide groups. But a plausible theoretical model was not enough for publication. Pauling and his coworkers needed X-ray crystallographic data and calculations to construct a strong enough argument in support of the model. About three years later, in 1951, Pauling and his coworkers went to press with eight papers on protein structures that in the world of biochemistry are held in almost as high esteem and awe as Einstein's 1905 series of papers is in physics.

The main structure described in the cited article consists of the repeated sequence CH-CO-CN (where C is carbon; H, hydrogen; O, oxygen; and N, nitrogen) in a corkscrew arrangement forming a single helix. A crucial insight was that there are 3.7 residues within each turn of the helix, not a whole number, such as three or four, as previously believed. (In the article, this is referred to as the "integral number.") The "residue" (R in the chemical structure) is a side chain to the CH-CO-CN backbone and consists of as little as one hydrogen atom to as many as eighteen atoms. The authors also proposed another helix with 5.1 residues per turn, a structure yet to be found in a protein.

Suggesting a new structure for a molecule and presenting experimental data in support of it seldom constitute a sufficient argument for a modern scientific article. The authors must also explain how their structure differs from previous ones, and why those earlier ones are flawed. This is exactly what Pauling and his coworkers do in the following paragraph, leaving in their wake some of the leading biochemists of their time (italics added for emphasis):

> These helical structures have not previously been described. In addition to the extended polypeptide chain configuration, which for nearly thirty years has been assumed to be present in stretched hair and other proteins with the β-keratin structure, configurations for the polypeptide chain have been proposed by Astbury and Bell, and especially by Huggins and by Bragg, Kendrew, and Perutz. Huggins discussed a number of structures involving intramolecular hydrogen bonds, and Bragg, Kendrew, and Perutz extended the discussion to include additional structures, and investigated the compatibility of the structures with x-ray diffraction data for hemoglobin and myoglobin. None of these authors proposed either our 3.7-residue helix or our 5.1-residue helix. On the other hand, *we would eliminate, by our basic postulates, all of the structures proposed by them.* The reason for the difference in results obtained by other investigators and by us through essentially similar arguments is that both Bragg and his collaborators and Huggins discussed in detail only helical structures with an integral number of residues per turn, and moreover assumed only a rough approximation to the requirements about interatomic distances, bond angles, and planarity of the conjugated amide group, as given by our investigations of simpler substances. We contend that these stereochemical [three-dimensional] features must be very closely retained in stable configurations of polypeptide chains in proteins, and that there is no special stability associated with an integral number of residues per turn in the helical molecule.

The point here is that while others used "similar arguments" for their models, they incorrectly assumed "an integral number of residues" (standard practice at the time) and had at their disposal "only a rough approximation" of key protein properties. The phrase "only a rough approximation" highlights the central value of precise measurements in convincing other biochemists of the truth of the new protein model Pauling dreamed up while sick in bed.

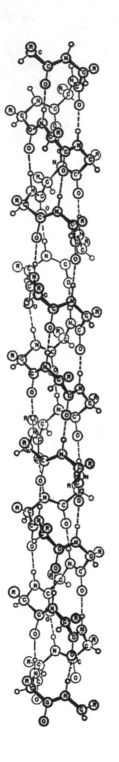

FIGURE 5-8 · Single-helix
structure of protein molecule.
Reprinted from *Proceedings of the
National Academic of Sciences*
(1951), with permission of Linus
Pauling estate.

Reconstructing an Enigmatic Animal from
500 Million Years in the Past

....

II. B. Whittington, 1975. "The enigmatic animal *Opabinia Regalis*, Middle Cambrian, Burgess Shale, British Columbia." *Philosophical Transactions of the Royal Society, London*, Vol. B271, pp. 1–43.

Harry Whittington begins his article with an anecdote about the public debut of his reconstruction of the *Opabinia regalis*, a marine creature that flourished over 500 million years ago: "When an earlier version of figure 82 was shown at a meeting of the Palaeontological Association in Oxford, it was greeted with loud laughter, presumably a tribute to the strangeness of this animal."

Stephen J. Gould hyperbolically refers to Whittington's article as "one of the great documents in the history of human knowledge," one that calls for a thorough revision of evolutionary history and theory. Whittington's explanation of the strangeness of this creature is far more conservative: "I suggest that *Opabinia* is neither an arthropod [lobsters are anthropods] nor an annelid [earthworms are annelids], but may be a representative of such an ancestral group of segmented animals." But the article is fascinating to us for a reason unrelated to evolutionary theory. The state of preservation of the fossils—dug up in the Burgess Shale, British Columbia—allows them to be dissected. As a consequence, *Opabinia* and its way of life can be reconstructed and pictured in startling detail.

According to Whittington, the animals were "trapped alive by a moving cloud of suspended sediment, and buried as ... [they] settled out [500 million years ago]." The sediments were compacted, compressing the animals but nonetheless preserving their outer covering, alimentary canal, and, possibly, their gut passages. Despite an eventual compression ratio "greater than 8:1," the fossils were not "crushed flat," but were preserved in a three-dimensional condition that permitted their dissection and study.

Separated from their surrounding rock, the fossils could be dissected with fine chisels and delicate drills. They could also be photographed in two different sort of light, revealing different features, as we can see by comparing Whittington's Figures 9 and 10. Still, it is difficult to compare the various specimens of fossils. They are chunks of rock, after all. Moreover, their parts and counterparts—the fossils themselves and the impressions they make in the surrounding rock—preserve different features. To overcome these difficulties, Whittington used a *camera lucida*, a contraption that projects an image that can be accurately sketched. With the aid of this device, he was able to sketch each specimen, combining part and counterpart, as in his Figure 12. The figure is oriented left (L) and right (R), and contains such details as the eyes (o and i),

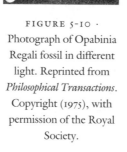

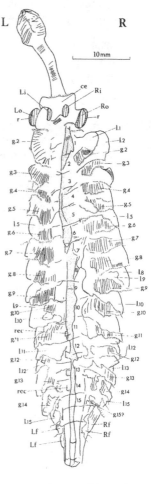

FIGURE 5-9 · Photograph of Opabinia Regali fossil. Reprinted from *Philosophical Transactions*. Copyright (1975), with permission of the Royal Society.

FIGURE 5-10 · Photograph of Opabinia Regali fossil in different light. Reprinted from *Philosophical Transactions*. Copyright (1975), with permission of the Royal Society.

FIGURE 5-11 · Schematic reconstruction of Opabinia Regali. Reprinted from *Philosophical Transactions*. Copyright (1975), with permission of the Royal Society.

the gills (g), the lobes (l), the fins (f), a reflective rectangular area (rec), the cephalon or frontal portion (ce), and the distal aspect or rear end (ds). Running down the center of this camera lucida sketch is the alimentary canal.

From these *camera lucida* sketches, Whittington was able to reconstruct the animal itself, the subject of the risible slide, the notorious Figure 82, showing views from above and one side. (Its five eyes might have been the punch line to this visual joke.)

From his meticulous visual reconstruction, Whittington is able to verbally recreate the "mode of life" of this early denizen of ocean floors, a creature only about four inches long:

The mode of burial of Burgess Shale fossils . . . implies that the animals were benthonic [sea-bottom inhabitants] in habit, otherwise they were unlikely to be trapped by suspended sediment moving along the sea bottom. The nature and position of the compound eyes of *Opabinia regalis* indicates that they were well adapted to detect changes in intensity of light, and thus movements, in waters beside or above a benthonic animal. The frontal process is interpreted as having been muscular, adapted to exploring the sediment for food, trapping it, and conveying it to the mouth. . . . No jaw structures are known, so the food was presumably soft. Digestion may have been aided by diverticula of the gut, which extended into the lateral lobes of the trunk. These lateral lobes must have been moderately rigid (wrinkling and variations in outline in the fossils being attributed to post-burial compression), able to support the body on the soft substrate, and an up and down movement would both aerate the gills and propel the animal. It may be assumed that such movements were slow, enabling *Opabinia* to plough shallowly in the surface of the bottom mud as the frontal process explored for food. The eyes and the tail fan would have been just above the surface of the mud, and the blades of the fan may have helped to create currents over the dorsal surface of the body, aiding in aeration of the gills. Flapping movements of the lateral lobes may have also enabled the animal to have moved over the substrate, swimming feebly. If such movements were in the form of a metachronal [sequential] wave along the body, the swimming powers would have been somewhat stronger, and the tail fan could have been used in steering. The fossils do not suggest more than a limited flexibility of the trunk, so that swimming by a body wave does not seem to have been possible.

Whittington's article is a textbook example of the synergy of words and pictures in bringing a long-extinct creature before our eyes.

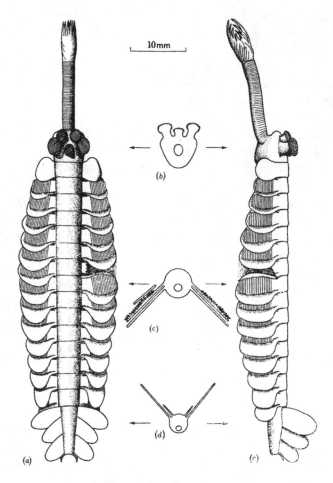

FIGURE 5-12 · Different orientations of reconstructed Opabinia Regali. Reprinted from *Philosophical Transactions*. Copyright (1975), with permission of the Royal Society.

On Photography in Science: Hedgehogs and Gooseberries

Christiane Nüsslein-Volhard and Eric Wieschaus, 1980. "Mutations affecting segment number and polarity in *Drosophila*." *Nature*, Vol. 287, pp. 795–801.

The realistic drawings we saw in our pre-twentieth century selections, like Hooke's flea, Perrault's chameleon, and Boyle's air pump, now have a major competitor in photographs, usually magnified in a way to make the very small more easily scrutinized. In general, these photographs are purposely selected, cropped, and arranged as visual support for the claim being made. Such is the case in the cited article, whose stated purpose is to better understand the processes that govern the embryonic development of *Drosophila* (fruit flies), one of the principal organisms of study for genetic experiments because of their mating habits and short life spans.

For their research, the German scientists Christiane Nüsslein-Volhard and Eric Wieschaus invented a method called "saturation mutagenesis." This involves creating lethal genes by means of mutation. These genes form patterned morphological anomalies fatal to the embryo, ones that can be observed and analyzed under a microscope.

The normal *Drosophila* larva is an elongated oval shape with three thoracic and eight abdominal segments (far left photograph, marked T1 to T3 and A1 to A8, respectively). One end eventually develops a head, the other, a tail. The authors' experiments revealed that, out of the many thousands of *Drosophila* genes, only about 15 mutants controlled embryonic development. In this work, they identify ten of these for the first time and give their fatal mutations whimsical names like "gooseberry," "hedgehog," and "patch."

Nüsslein-Volhard and Wieschaus divided these mutants into three groups: the first broadly affects the larva (named "gap" genes), the second works on small regions separated by two segments ("pair-rule" genes), and the third affects part of a segment ("segment-polarity" genes). In embryo development the first group of genes prepares a rough draft for the body plan, then the other two groups fill in the details. The following reproduces the article's short abstract summarizing these points in the authors' own words:

> In systematic searches for embryonic lethal mutants of *Drosophila melano-gaster* we have identified 15 loci which when mutated alter the segmental patterns of the larva. These loci probably represent the majority of the genes in *Drosophila*. The phenotypes of the mutant embryos indicate that the process of segmentation involves at least three levels of spatial organization: the entire egg as developmental unit, a repeat unit with the length of two segments, and the individual segment.

This research is not just about fruit flies; like Mendel's peas, fruit flies are a model organism, intended to represent all creatures that reproduce sexually. This research opens up a new field of scientific inquiry in biology because it demonstrates that the genes essential to an organism's development can be deduced by visually comparing the segment patterns of mutant embryos with

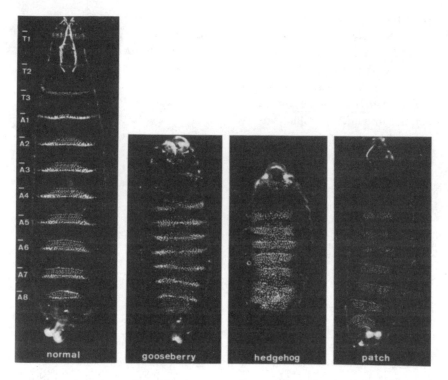

FIGURE 5-13 · Normal and mutant embryos of *Drosophila*. Reprinted by
permission from *Nature*. Copyright (1980) Macmillan Publishers Ltd.

a normal embryo. To facilitate their comparison, Nüsslein-Volhard and Wieschaus photograph significantly enlarged versions (140 times) of their embryos
and arrange them as though in a police lineup.

Gene Expression in Living Color

Michael B. Eisen, Paul T. Spellman, Patrick O. Brown, and David Botstein, 1998.
"Cluster analysis and display of genome-wide expression patterns." *Proceedings of the
National Academy of Sciences*, Vol. 95, pp. 14863–68.

Computers have changed scientific practices in innumerable ways. On the communicative side, they have changed the way in which the scientific manuscript
is prepared by authors, put through peer review, produced in final form, distributed to interested readers, and perused by those readers. To a great extent,
these communicative changes have simply automated what was once a more
labor-intensive process.

On the research side, computers now routinely assist in collecting and analyzing data from scientific instruments in the laboratory or field, and perform theoretical calculations at speeds and in quantities hard to imagine only a few decades ago. One byproduct of these technological advances has been mountains of data. As a consequence, some substantive changes have taken place in visual communication, as scientists try to cope with this deluge in order to better understand and communicate their results.

One data-driven research front of extraordinary vitality at the moment is microbiology—in particular, determining the DNA sequence and functions of different genes for organisms from humans to bacteria. Researchers have met with astounding success in this field largely because of the invention of computer-driven, automated techniques for sample analysis and data acquisition. One such technique goes by the name "DNA microarray"; it generates an immense amount of biological information on gene expression (hundreds of numbers for the behavior of each of several thousand genes). But displaying this massive amount of data in tables or conventional Cartesian plots overwhelms the eye's ability to discern patterns. To improve communication, computer-based techniques are being invented, such as the one described in the cited article by Michael Eisen and coworkers. Their method for mining microarray data transforms numbers into colors for more easily uncovering complex relationships. Here is how the authors explain their method in a passage reminiscent of the seventeenth-century selection by Dodart (see chapter 2), who also grappled with how to best visually represent his research:

Microarray-based genomic surveys and other high-throughput approaches (ranging from genomics to combinatorial chemistry) are becoming increasingly important in biology and chemistry. As a result, we need to develop our ability to "see" the information in the massive tables of quantitative measurements that these approaches produce. Our approach to this problem can be generalized as follows. First, we use a common-sense approach to organize the data, based on order inherent in the data. Next, recognizing that the rate-limiting step in exploring and searching large tables of numerical data is a trivial one: reading the numbers (human brains are not well adapted to assimilating quantitative data by reading digits), we represent the quantitative values in the table by using a naturalistic color scale rather than numbers. This alternative encoding preserves all the quantitative information, but transmits it to our brains by way of a much higher-bandwidth channel than the "number-reading" channel.

A natural way of viewing complex data sets is first to scan and survey the large-scale features and then to focus in on the interesting details. What we have found to be the most valuable feature of the approach described here is that it allows this natural and intuitive process to be applied to genomic data sets. The approach is a general one, with no inherent specificity to the particular method used to acquire data or even to gene-expression data. It is therefore likely that very similar approaches may be applied to many other kinds of very large data sets. In each case, it may be necessary to find alternative algorithms and computation methods to bring out inherent structures in the data, and, equally important, to find dense naturalistic visual representations that convey the quantitative information effectively. We recognize that the particular clustering algorithm we used is not the only, or even the best, method available. We have used and are actively exploring alternatives such as parametric ordering of genes and supervised clustering methods based on representative hand-picked or computer-generated expression profiles. The success of these very simple approaches has given us confidence to face the coming flood of functional genomic data.

You can view a sample color graph in the online version of the *Proceedings of the National Academy of Sciences* (www.pnas.org/cgi/reprint/95/25/14863.pdf). Look especially at their Figure 1, which shows cluster variation over time. The dendrogram on the left shows genetic relationships. By means of a visual argument by comparison, their Figure 3 shows readers that these patterns are not random, but meaningful.

Computers have made it possible to use not only color in more sophisticated ways, but also, for the first time ever, moving pictures and even sound tracks. For a very recent example in applied physics, see the online graph in which Seth Putterman's lab at UCLA reports achieving nuclear fusion in a simple tabletop device (www.nature.com/nature/journal/v434/n7037/extref/nature03575-s2 .mpg). Their multimedia graph registers detection of neutrons generated by the fusion reaction in real experimental time set to a Geiger-counter sound track. The viewer thus both sees and hears neutrons being created during a typical experimental run. This visualization makes for a very powerful argument that, after years of dubious reports on "cold fusion" and similar tabletop devices, this one actually works as advertised.

Plotting Constant Change Back to the Big Bang

J. K. Webb, M. T. Murphy, V. V. Flambaum, V. A. Dzuba, J. D. Barrow, C. W. Churchill, J. X. Prochaska, and A. M. Wolfe, 2001. "Further evidence for cosmological evolution of the fine structure constant." *Physical Review Letters*, Vol. 87, pp. 091301-1 to 091301-4.

We now turn to a very recent example of that most prominent of modern scientific visualizations, the Cartesian graph, independently invented in the late eighteenth century by Johann Lambert in Germany (see chapter 3), William Playfair in England, and James Watt in Scotland. This new technique moved data out of text and tables and into figures. Today, Cartesian graphs are the workhorse scientific visualization—few articles reporting experimental, calculational, or observational data can do without them.

The above-cited short article has a single graph with profound implications for the discipline of physics and cosmology if it survives critical challenge. The data indicate that slight changes in a fundamental constant have occurred over many billions of years. Fundamental constants such as atomic mass, Avogadro's number, Boltzmann's constant, elementary charge, proton mass, and speed of light in a vacuum frequently occupy the equations of physics; they are not supposed to change one iota throughout eternity. That's why they are called "constant."

Alpha (α), the first letter of the Greek alphabet, stands for one of the most peculiar and important physical constants in physics, the "fine structure constant." It is a composite constant, calculated from a simple equation made up of three other constants: the charge of an electron, Planck's constant, and the speed of light. That simple equation yields a fraction of about 1/137, representing the strength with which subatomic particles interact with one another and with light. Had it been much different from 1/137 our universe would not be here; electromagnetic forces hold atoms together in accord with this fraction, as by a law.

To track potential changes in this constant over time, the Webb group analyzed data obtained from a powerful telescope in Hawaii. The data measured the absorbance of light from clouds of gas and dust illuminated by different quasars. The astrophysicists chose these bright but distant celestial objects for their measurements because "observing quasars at a range of redshifts provides the substantial advantage of being able to probe α over most of the history of the Universe." From these astronomical measurements, the authors could calculate α over a vast cosmological time scale: 23 to 87 percent of the "look-back time" to the big bang.

The considerable challenge facing the authors was convincing other physicists that they could calculate alpha with enough accuracy to conclude whether it had changed ever so slightly over the last 13 billion years. Key to their visual argument is the reproduced Cartesian graph. The vertical axis plots $\Delta\alpha/\alpha$, a measure of the change in alpha between some time in the distant past and the present. The horizontal axis plots two variables: the redshift of the quasars at the top, determined from the wavelengths measured in the astronomical spectra,

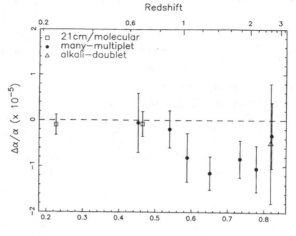

Fractional look–back time

FIG. 1. $\Delta\alpha/\alpha$ vs fractional look-back time to the big bang.
The conversion between redshift and look-back time assumes
$H_0 = 68 \ \mathrm{km\,s^{-1}\,Mpc^{-1}}$, $(\Omega_M, \Omega_\Lambda) = (0.3, 0.7)$, so that the age
of the universe is 13.9 Gyr. A total of 72 quasar absorption
systems contribute to this binned-data plot. The hollow squares
correspond to two HI 21 cm and molecular absorption systems
[16]. Those points assume no change in g_p, so should be interpreted with caution. The seven solid circles are binned results
for 49 quasar absorption systems. The lower redshift points (below $z \approx 1.6$) are based on (MgII/FeII) and the higher redshift
points on (ZnII, CrII, NiII, AlIII, AlII, SiII) [13]. Of these 49
systems, 28 correspond to the sample used in [4]. The hollow
triangle represents the average over 21 quasar SiIV absorption
doublets using the alkali-doublet method [14].

FIGURE 5-14 · Cartesian graph plotting fine structure
constant to Big Bang. Reprinted with permission of *Physical
Review Letters*. Copyright (2001) American Physical Society.

and the fraction of time elapsed since the universe's origin at the bottom, calculated from the redshift. A horizontal dashed line appears at $\Delta\alpha/\alpha = 0$; that
line signifies no change in the fine structure constant over time. The symbols of
square, circle, and triangle distinguish different measurement methods, defined
in the figure caption and the main text. The short vertical bars passing through
the data points indicate their statistical error range. They define the limits of the
authors' confidence in each data point.

You do not have to be a cosmologist to see that beyond a redshift of one
(fractional look-back time of about 0.54) the top of the error bars dip decidedly
below zero. That grouping of nonzero $\Delta\alpha/\alpha$ indicates a smaller alpha in the

distant past. The magnitude of the change is tiny, to say the least: about one part in one hundred thousand over the 13 billion years.

A calculation of such complexity hinges on convincing others that no errors have crept in, either by those making the measurements or those doing the calculation. Any one of a number of possible sources could have thrown the calculation off track. To counter any such criticism, the Webb group analyzed thirteen possible sources of error and found that none accounts for the changing α. Their conclusion from all this analysis is unequivocal: there is "no systematic effect [error] which can produce our results." As always in such statistical arguments, time will tell.

Exploring the Universe by Leaps of Ten

J. Richard Gott III, Mario Jurić, David Schlegel, Fiona Hoyle, Michael Vogeley, Max Tegmark, Neta Bahcall, and Jon Brinkmann, 2003. "A map of the universe." Preprint <arxiv.org/abs/astro-ph/0310571>.

Start with the number 1 and increase it tenfold, again and again: 1, 10, 100, 1,000, 10,000, 100,000, 1,000,000, . . . In the 1970s the husband-and-wife team of Charles and Ray Eames created a short film illustrating this numerical progression. It is one of the most inventive educational films ever made, and is still played in science museums on a regular basis.

We viewers, virtual space travelers, begin one meter above a male picnicker napping in a Chicago park on a sunny day. We take off at the rate of one power of ten every ten seconds as our field of vision also expands by a power of ten. With dizzying acceleration, we leave the earth behind, intersect our moon's orbit, fly by our sun, wave goodbye to our solar system, meet up with the nearest stars, then are spit out of our whirlpool of a galaxy, all within a matter of a few minutes. From there our space odyssey takes us past numerous other galaxies and galaxy clusters, right up to the darkness at the end of the visible universe—twenty-five powers of ten, in meters, from home. Next, we retrace our steps at an even more vertiginous rate: one power of ten every two seconds. In no time we land back with the picnicker in Chicago, where the journey continues as we, suddenly the size of a microbe, plunge into his hand and gradually shrink down to the subatomic level (10^{-18} meters). The whole journey from earth to the end of the universe and back again, down to a proton in a carbon atom within a DNA molecule from the picnicker, takes less than ten minutes.

In the cited article appearing in an e-print archive from Cornell University, Gott and his colleagues made a similarly inventive use of the powers of ten

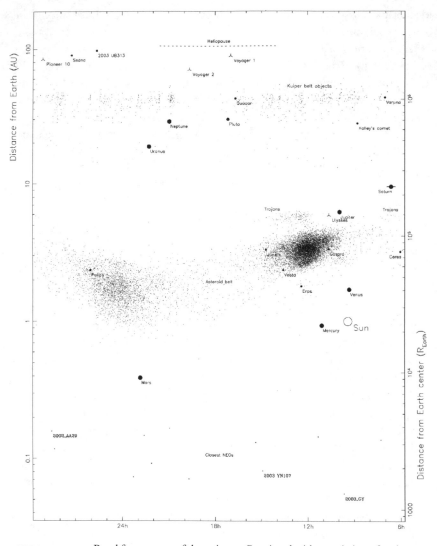

FIGURE 5-15 · Panel from a map of the universe. Reprinted with permission of authors.

for exploring the universe, but with a more scientific purpose. They drew a "conformal map" of the known universe—starting from the center of the earth at the very bottom and ascending to the edge of the universe over many powers of ten. The map is "conformal" in a similar sense to maps of the globe: it translates a sphere into a plane with fairly minimal distortion. Not only does this map communicate a feeling for astronomical distance from the earth, but it also emphasizes the outer space we know in greatest detail, our solar system. Because of this emphasis, the objects in the map become bigger as we get farther from the earth: advancing from sun and planets, to nearby stars, to galaxies, and to large celestial structures, the largest of which spans a billion light-years.

This earth-centered map is a direct descendant of similar maps that originated in the earliest days of astronomical research—the most famous being the sun-centered universe with circling planets drawn by Copernicus. The one included in this cited e-print is intended as a tool for testing of cosmological models and conducting further research into why the various objects in the heavens are distributed as they are. It is not meant as a representation of any new theory, as was the case in Copernicus's diagram, though it does reflect our present theoretical understanding of the beginning of the universe: the various clumps of celestial matter arise from the effect of big bang expansion combined with the effect of gravity as predicted by general relativity. No words could adequately convey this complex message. The complete map is also a work in progress: while a little more than one hundred thousand galaxies appear, the authors' goal is a million. Updating is no problem, since the map is on the World Wide Web.

The complete map spans twenty-four powers of ten in kilometers, the outer limits of the universe as far as we know. Because the full map is too large for display here, we reproduce on p. 186 only one of its six panels, a view of the solar system from the point of view of the earth. The logarithmic vertical scale, left-hand side, is in astronomical units (AUs), where one such unit equals the distance between our earth and sun (150 million kilometers); the logarithmic vertical scale on the right-hand side is in units of Earth radii (6,378 kilometers). So the top of the vertical scale equates with about 100 AUs or about ten powers of ten in kilometers. The linear horizontal scale relates to the 360 degrees of the celestial sphere (every 15 degrees equals one hour). The panel displayed here gives an earth-centered perspective of our solar system, including such occupants as Halley's comet, asteroid belts, and unmanned space vehicles at the outskirts of our solar system.

The Eames's educational film exploits the mathematical magic behind the powers of ten to traverse the universe. The selected map of the universe does that and more. It is both an original scientific image and a fitting tribute to past astronomical observations and theories, a great many of which date from the golden age of astronomy over the last fifty years.

6

ORGANIZING
SCIENTIFIC ARGUMENTS

T HE TYPICAL EARLY scientific article was either a straightforward description of the natural world, or a narrative centered around a series of loosely connected experiments or observations, often arranged chronologically. Over time, however, the article evolved a modular organization suited to authorship by a group of researchers and designed to present each article's arguments in a roughly similar way and level of detail. The chemist and Nobel Laureate Roald Hoffmann claims that the modular organization of the modern scientific article acquired its present "canonical or ritual form" during the 1830s and 1840s in the German chemical literature.

The "form" at that early stage included four fundamental components: introduction, methods, results and discussion, and a few footnotes referring to past research. In the nineteenth century, these components were not, in general, clearly differentiated and marked off from one another by headings or a different font size, as they are today. By the mid-twentieth century, however, Hoffmann's canonical form expanded into a fairly detailed set of reader expectations, typically:

- A Title that compactly conveys the gist of the main new knowledge claim or claims, along with list of authors and their affiliations.

- An Abstract, a brief digest of the important claims and methods.
- An Introduction that places readers in the scientific context in which its authors are working and defines a specific research problem.
- Methods and Materials that detail the procedures used or invented to solve the problem.
- Results that display the data generated by the methods and the intellectual context of their acquisition, often combined with a Discussion that interprets the data.
- A Conclusion that reiterates the central new claims in a single paragraph or two and addresses future research that would extend the present insights.
- References that identify literature sources the authors have relied on, contradicted, or extended.
- Acknowledgments that note personal or financial assistance provided in the article's research.

Combined with the use of headings, this organization permits scientists to read articles opportunistically rather than sequentially, scanning the various sections in search of useful bits of method, theory, fact, and references. It also facilitates the process of deciding whether a particular article is worth reading more attentively.

But one size does not fit all. The organization just outlined applies primarily to experimental articles, that is, those recounting the manipulations of natural and manufactured objects, usually in artificial settings such as research laboratories. Such articles, which provide an empirical basis for the continued conceptual evolution of science, dominate the pages of the twentieth and twenty-first century journal literature. But different types of articles demand variations on the typical organizing strategy. Let us look at three common ones: theoretical, observational, and review. Each has a typical organization tailored to suit its purpose.

Theoretical articles present new mathematical, physical, chemical, or biological explanations for the natural world by drawing upon previous experiments, theory, and observations. The typical theoretical article starts with an abstract that summarizes its main points and an introduction that presents a research problem, just as in the experimental type. The introduction also often states the primary assumptions made in attempting to solve the problem the article sets. Next comes the actual theory derived from those assumptions, along with additional factors like subordinate assumptions, definitions, boundary conditions, and so on. Some sort of a proof follows the theory statement. As a general rule, authors prove their theories through comparison with what is experimentally established or calculated. The conclusion recommends future work (experimental or theoretical) that would verify or extend the theory. The closing two elements are the same as for the experimental article: references and acknowledgments.

Observational articles describe natural objects, usually outside the laboratory. They involve such activities as describing a new hummingbird or measuring the spectrum of a new star; they do not involve manipulating natural objects under controlled conditions. Observational articles complement the experimental; they exist because part of the task of science will always involve describing the natural world outside the laboratory. In addition, as cosmology and paleontology demonstrate, some phenomena that interest science will always be closed to direct experimentation. When this type of article reports measurements based upon observations in the field or observatory, its basic structure mimics the experimental type. When it involves identifying a new plant, animal, or mineral, its organization centers on a description designed for specialists, one whose purpose is to establish its exact taxonomic place in the natural world.

Review articles describe and evaluate the recent literature in a field. Unlike the other types, however, these articles are written solely by means of the study of other texts. They are the product of trips to the library and office, rather than to the laboratory or field site. While their main purpose is to summarize and interpret past science, they serve an indispensable function—winnowing the fit from the unfit among past knowledge claims on a narrow topic.

Sometimes this process actually allows the authors to fashion brand-new knowledge claims from the old; most times it provides a synoptic view that will help other scientists conveniently acquaint themselves with unfamiliar territory. The utility of the literature review should not be underestimated, as evidenced by the high citation impacts of review articles. They have the power to define the principal workers at a new or developing research front, and even ignore others who believe they belong in that group.

We can typify the review-type organization only in the most general of terms: an introduction designed to secure attention and focus on a central research problem or issue, a discussion of the state of current and past research, a conclusion designed to shape future research, and a reference list that far surpasses in length the other types of article.

We divided each earlier chapter into topics related to the chapter title, then arranged the selected excerpts chronologically by topic. This chapter differs in that we organized it entirely by topic, giving examples of the different components of the modular organization in the canonical order just given. We end with a tour of an entire experimental-type paper, from its title to the acknowledgments.

BEGINNING

Front Matter: Title, Author, Abstract

··

Milan N. Stojanovic and Darko Stefanovic, 2003. "A deoxyribozyme-based molecular automaton." *Nature Biotechnology*, Vol. 21, pp. 1069–74.

Scientific articles normally begin with descriptive titles, followed by the names of authors and their affiliations, followed by a short abstract.

Even in something as serious as cutting-edge science an occasional joke or literary reference is permitted in the title. Recent examples include Gould and Lewontin's "The spandrels of San Marco and the Panglossian paradigm: a critique of the adaptionist programme" and Peebles and Silk's "A cosmic book of phenomena." (See the next chapter for more about these articles.) But modern scientific titles are usually far more sober, stripped of anything personal or openly literary. The title from our selection here, "A deoxyribozyme-based molecular automaton," is more typical. In contrast to their literary peers, scientific titles such as this one are full of technical terms. They seek to squeeze a maximum amount of information in a minimum amount of space. They are also dryly descriptive to minimize scientists' time in pinpointing the subject of the article. One could not imagine such a title headlining a newspaper or popular science account. For comparison, the title for the *Dallas Morning News* article reporting on this same discovery is the catchy full sentence "DNA computer gets in the game."

Next up is the byline. Its primary purpose is to identify the persons and institutions responsible for the article's contents. Should some facet prove wrong or fraudulent or poorly conceived, they must shoulder the blame. The authors' names and institutional affiliations can also contribute to the article's authority. It is only human nature that readers trust certain names more than others; the same goes for institutions. In this particular selection, the authors are not exactly household names, even within the scientific community. But their institutional affiliations do carry some weight: the first author works in the Division of Pharmacy and Experimental Therapeutics in the Department of Medicine at Columbia University; the second, in the Department of Computer Science in the Farris Engineering Center at the University of New Mexico.

Next comes the abstract. Abstracts of articles existed from the earliest days of the journal literature. As we mentioned in chapter 2, *Histoire de l'Académie Royale des Sciences*, first published in the early eighteenth century, was made up entirely of short summaries of the longer *mémoires*. But as far as we can determine, the heading abstract—a very brief digest of the article's central message inserted after the title and byline and before the introduction—first appeared sporadically in the journal literature of the 1920s. Not until the 1950s did it become routine practice. Along with the title and byline, the heading abstract enables readers to make a rapid, informed decision as to whether the remaining article is worth their time and effort.

Our sample abstract, quoted below, describes a new biotechnology: a DNA-based computer that plays tic-tac-toe against a human opponent and never loses. This invention sounds almost like science fiction, behaving "analogously to digital logic circuits used in electronic computers or to enzymes in metabolic circuits." In a short compass, this abstract summarizes what the authors invented, how it works, and what it does:

> We describe a molecular automaton, called MAYA, which encodes a version of the game of tic-tac-toe and interactively competes against a human opponent. The automaton is a Boolean network of deoxyribozymes that incorporates 23 molecular-scale logic gates and one constitutively active deoxyribozyme arrayed in nine wells (3×3) corresponding to the game board. To make a move, MAYA carries out an analysis of the input oligonucleotide keyed to a particular move by the human opponent and indicates a move by florescence signaling in a response well. The cycle of human player input and automaton response continues until there is a draw or a victory for the automaton. The automaton cannot be defeated because it implements a perfect strategy.

Like all good abstracts, the above captures the essence of the authors' accomplishment in a short paragraph. Only those curious about further technical details need read any further.

The specialized audience for this abstract is readily apparent from the highly technical vocabulary alone. A key piece of information essential for a wider audience is obviously missing: why this invention is important. That is, why invent a DNA computer that can play tic-tac-toe flawlessly? Isn't this merely an amazing trick, like teaching a horse to do simple arithmetic by tapping the answer out with its hoof?

What Stojanovic and Stefanovic do not tell us in their abstract is that they are part of a larger research community hoping to one day harness the extraordinary computing power of DNA to identify and treat a virus or cancer cells within patients. Their article is one short step in that direction. But a statement of the "big picture" seldom makes it into the few sentences allotted to a heading abstract for a typical scientific article. If it appears at all, that information is usually reserved for the introduction or conclusion or both, as is the case for the selected article. (And if the reader will forgive brief editorializing by us,

Stojanovic and Stefanovic aside, scientist-authors tend to do a poor job of expressing how their solving some small research problem fits within the big picture.)

Introduction: So What's the Problem?

M. K. Wu, J. R. Ashburn, C. J. Torng, P. H. Hor, R. L. Meng, L. Gao, Z. J. Huang, Y. Q. Wang, and C. W. Chu, 1987. "Superconductivity at 93 K in a new mixed-phase Y-Ba-Cu-O compound system at ambient pressure." *Physical Review Letters*, Vol. 58, pp. 908–10.

In a delightful parody of the scientific paper in theoretical physics ("The super G-string"), physicist Warren Siegel began with the following anti-introduction:

> Actually, this paper doesn't need an introduction, since anyone who's the least bit competent in the topic of the paper he's reading doesn't need to be introduced to it, and otherwise why's he reading it in the first place? Therefore, this section is for the referee.

This paragraph plays upon two conflicting issues in scientific introductions in general. On the one hand, some writers will assume that readers "the least bit competent in the topic" need the bare minimum of context-setting information, as in the mathematical paper that starts *in medias res*: "Let x stand for a prime number . . ." On the other hand, some readers do not possess the storehouse of knowledge assumed by such writers, and quickly become confused and frustrated. Their response to the minimalistic introduction might be "So what's the problem and why should I care?" Ideally, the introduction (in fact, the whole article) should reach all scientists who might be able to exploit the findings in their own research. These scientists should come away with a clear understanding of what problem has been addressed, and why solving it is important.

What should go into a typical introduction? The linguist John Swales has defined three core steps necessary to create a "research space," the essential task of introductions. In the first—establishing the historical context for the problem to come—the authors introduce the reader to a general technical topic. Research problems emerge out of past research, in some cases, a chain of problems and their solutions reaching back for centuries. So authors must incorporate a review, however truncated, of past literature. The next step—defining a specific problem or need—normally involves one of several actions: pointing out a contradiction or inconsistency in present knowledge, identifying a gap that needs filling, or posing a question no one had addressed before. In the last step of the introduction—summarizing the solution—the authors indicate what headway they have made on the stated problem and, sometimes, why solving the problem might be of interest to the readers. The solution might be complete or partial; the authors may even fail to solve the problem, though abject failure is seldom reported.

As an example of Swales's three moves, we selected one of the major technological breakthroughs in the late twentieth century—the discovery of a new material that becomes superconducting at "high temperatures," that is, a material that offers no resistance to the passage of electricity above a temperature that will sustain nitrogen in liquid form, a fairly inexpensive product widely available. This "high" temperature, of course, is still pretty cold by most other standards—77 degrees on the kelvin scale, which is equal to minus 196 centigrade or minus 321 Fahrenheit. For comparison, the coldest outdoor temperature ever recorded on Earth—in Antarctica—is only about minus 90 centigrade or minus 130 Fahrenheit.

The first paragraph begins by stating a problem or "challenge" (finding a high-temperature superconducting material), trumpeting its importance, and announcing the authors' solution. It covers Swales's second and third steps:

> The search for high-temperature superconductivity and novel superconducting mechanisms is one of the most challenging tasks of condensed-matter physicists and material scientists. To obtain a superconducting state reaching beyond the technological and psychological temperature barrier of 77 K, the liquid-nitrogen boiling point, will be one of the greatest triumphs of scientific endeavor of this kind. According to our studies, we would like to point out the possible attainment of a superconducting state with an onset temperature higher than 100 K, at ambient pressure, in compound systems generically represented by $(L_{1-x}M_x)_aA_bD_y$. In this Letter [really, short article], detailed results are presented on a specific new chemical compound system with L = Y, M = Ba, A = Cu, D = O, x = 0.4, a = 2, b = 1, and y $\leq$ 4 with a stable superconducting transition between 80 and 93 K. For the first time, a "zero-resistance" state ($\rho < 3 \times 10^{-8}\,\Omega$-cm, an upper limit only determined by the sensitivity of the apparatus) is achieved and maintained at ambient pressure in a simple liquid-nitrogen Dewar.

In their second paragraph, which we do not reproduce, Wu and company turn to Swales's first step by giving a short historical review on superconductivity research. The story they tell is of a once moribund research program suddenly energized by a new discovery a year earlier. About seventy-five years prior to the cited article, in 1911, a Dutch physicist (Heike Kamerlingh Onnes) had discovered that a metal (mercury) became superconducting at four degrees above absolute zero. In subsequent decades, by experimenting with other metals and metallic compounds, scientists nudged the superconducting temperature up by roughly four-tenths of a degree per year on average—eventually reaching about 25 degrees above absolute zero in 1973. After that, attempts to significantly boost the superconducting temperature by unconventional routes met with "gross failure." Then, in 1986, two IBM scientists, J. G. Bednorz and K. A. Müller, stumbled upon a complicated, brittle ceramic (composed of La-Ba-Cu-O) that increased the superconducting temperature by an astonishing 5 degrees. That discovery opened the floodgates. Within a year, the group led by Paul C. W. Chu at the University of Houston found other ceramics with much higher superconducting temperatures—well past the magic number of 77 kelvin. In short order, research groups from all over the world were searching for, and finding, new superconducting ceramics at ever higher temperatures.

This 1987 paper created a stir in the news media because of the possible practical applications of a high-temperature superconductor, including use for electric power transmission lines and high-speed digital electronics. Still, the authors introduce their problem as a scientific one (how to significantly increase the superconducting temperature over the prior record) rather than a technological or practical one (how to better transmit electricity or speed up processing in computers). In general, scientific articles limit themselves to the specific research problem at hand, while scientific journalism emphasizes the possible benefits to society down the road.

MIDDLE

Methods: How Was the Problem Solved?

Oliver H. Lowry, Nira J. Rosebrough, A. Lewis Farr, and Rose J. Randall, 1951. "Protein measurement with the Folin phenol reagent." *Journal of Biological Chemistry*, Vol. 193, pp. 265–75.

While the purpose of the introduction is to formulate a research niche, that of the methods section is to embody the author's strategy for filling that niche. The content of this section is as varied as the methods, materials, and theoretical principles employed within the numerous specialties that populate science. Nonetheless, the principal questions readers want answered are the same: What was done? How was it done? What materials were involved? Why were any nonstandard methods or materials chosen?

The genre of writing most closely akin to the methods section is the instruction manual or cookbook—a form of discourse that describes as opposed to argues. Despite its descriptive form, the methods section is an essential step in the article's overall argument: *these* methods produce *those* results, which have *this* significance.

In contrast to the instruction manual or cookbook, the details of the typical methods section prompt readers not so much to act as to imagine events. If critical readers judge those details as a plausible strategy for solving the problem stated in the introduction, then they will likely view the article as authentic science.

Despite its importance, the methods section in the typical experimental article is usually segregated from the rest of the text in such a way that it can be skipped. Indeed, it will be skipped or skimmed by all but a few interested readers unless a strong disbelief in the results prompts thoughts of replication, or else a strong belief in them inspires any new converts to adapt the original to their own research purposes. Replication solely for the sake of replication is rare in science because it is often a complex, time-consuming, and expensive process, with little reward whether the repeater fails or succeeds.

The selected passage comes from a much-cited methods article—a subtype of experimental article whose main purpose is to report a new method for others to follow. Such articles report new means of experimenting, observing, or calculating, and tend not to make new claims about the natural world. They are about creating new tools to do science. In this special case, the methods section is the center of attention. That is the case for the chosen text here. It reports a method for determining protein concentrations in various solutions such as serum, spinal fluid, and insulin, as well as after precipitation from such a solution.

The so-called Lowry method involves the traditional two main steps in modern analytical chemistry: first, preparation of a sample suitable for analysis, then the actual analysis with research equipment able to detect the desired substance at microscopic levels. In this method, the analytical equipment is a spectrophotometer—a device that shines light of a particular wavelength through a substance and calculates its absorbance, a measure of the capacity of a substance to absorb light of a specified wavelength. The authors did not invent the equipment; they were not even the first to recognize the possibility of measuring protein with the Folin phenol reagent. In fact, the latter had been done almost three decades before. What they did was devise a relatively simple, accurate method that overcame previous technical shortcomings. And that method proved immensely fruitful in research for all kinds of "biochemical purposes."

Because of this method's extraordinary utility, this article holds the record as the most highly cited in the *Science Citation Index*, with more than a quarter million citations. That many citations also suggest it is probably the most read twentieth-century scientific article ever. Quoted here are the first few sentences describing the method:

> Reagent A, 2 per cent Na_2CO_3 in 0.10 N NaOH. Reagent B, 0.5 per cent $CuSO_45 \cdot H_2O$ in 1 per cent sodium or potassium tartate. Reagent C, alkaline copper solution. Mix 50 ml. of Reagent A with 1 ml. of Reagent B. Discard after 1 day.

Let us stop here for a moment. As with the best-selling *Mastering the Art of French Cooking* by Julia Child, Louisette Bertholle, and Simone Beck, the authors assume considerable knowledge and expertise on the part of their readers who might attempt to reproduce their recipes. Such readers must be able to read between the lines and fill in the blanks. For example, Lowry and his collaborators write "Discard after 1 day." You might wonder, discard what? Presumably, that means discard the mixture of reagents A and B, mentioned in the previous sentence, after one day if you haven't actually used it. But discard them how and where? The readers hoping to execute this method must know how to act upon such minimalist statements, as well as others from later in the selection, like "Working standards may be prepared from human serum diluted 100- to 1000-fold (approximately 700 to 70 γ per ml.)," where gamma is a unit of magnetic field strength.

Let us continue with a much longer quotation:

> Reagent D, carbonate-copper solution, is the same as Reagent C except for the omission of NaOH. Reagent E, diluted Folin reagent. Titrate Folin-Ciocalteu phenol reagent (Eimer and Amend, Fisher Scientific Company, New York) with NaOH to a phenolphthalein end-point. On the basis of this titration dilute the Folin reagent (about 2-fold) to make it 1 N in acid. Working standards may be prepared from human serum diluted 100- to 1000-fold (approximately 700 to 70 γ per ml.). These in turn may be checked against a standard solution of crystalline bovine albumin (Armour and Company, Chicago); 1 γ is the equivalent of 0.97 γ of serum protein... Dilute solutions of bovine albumin have not proved satisfactory for working standards because of a marked tendency to undergo surface denaturation...
>
> To a sample of 5 to 100 γ of protein in 0.2 ml. or less in a 3 to 10 ml. test-tube, 1 ml. of Reagent C is added. Mix well and allow to stand for 10 minutes or longer at room temperature. 0.10 ml. of Reagent E is added very

rapidly and mixed within a second or two... After 30 minutes or longer, the sample is read in a colorimeter or spectrophotometer. For the range 5 to 25 γ of protein per ml. of final volume, it is desirable to make readings at or near λ [symbol for wavelength] = 750 mμ [millimicrons or nanometers]. To a sample of 5 to 100 γ of protein in 0.2 ml. or less in a 3 to 10 ml. test-tube, 1 ml. of Reagent C is added. Mix well and allow to stand for 10 minutes or longer at room temperature. 0.10 ml. of Reagent E is added very rapidly and mixed within a second or two... After 30 minutes or longer, the sample is read in a colorimeter or spectrophotometer. For the range 5 to 25 γ of protein per ml. of final volume, it is desirable to make readings at or near λ = 750 mμ, the absorption peak. For stronger solutions, the readings may be kept in a workable range by reading near λ = 500 mμ (Fig. 2). Calculate from a standard curve, and, if necessary, make appropriate correction for differences between the color value of the working standard and the particular proteins being measured....

For stronger solutions, the readings may be kept in a workable range by reading near λ = 500 mμ (Fig. 2). Calculate from a standard curve, and, if necessary, make appropriate correction for differences between the color value of the working standard and the particular proteins being measured....

It is unnecessary to bring all the samples and standards to the same volume before the addition of the alkaline copper reagent, provided corrections are made for small differences in final volume. The critical volumes are those of the alkaline copper and Folin reagents.

If the protein is present in an already very dilute solution (less than 25 γ per ml.), 0.5 ml. may be mixed with 0.5 ml. of an exactly double strength Reagent C and otherwise treated as above.

In this passage, we witness a dance of verb forms: indicatives, such as "it is," to state facts; optatives, such as "may be kept," to make suggestions and to indicate procedures; passives like "is added" to keep things impersonal; and imperatives like "calculate" to give instructions. Those who think that the phrase "scientific style" is an oxymoron should give close attention to this verbal dexterity. (For more on scientific style, see the next chapter.)

The methodological description such as Lowry's has an important difference from a cookbook. Just describing a new method in excruciating detail is seldom sufficient; the authors must also convince their readers that it actually works and performs better than, or at least has some advantage over, any alternatives. Accordingly, the second part of this particular article puts the new method through its paces for a variety of measurements in order to "sell" it to other researchers who might use it in their own experiments.

Results and Discussion: What Did the Method Yield and Why?

Oswald T. Avery, Colin M. MacLeod, and Macyln McCarty, 1944. "Studies on the chemical nature of the substance inducing transformation of pneumococcal types: Induction of transformation by a desoxyribonucleic acid fraction isolated from pneumococcus Type III." *Journal of Experimental Medicine*, Vol. 79, pp. 137–58.

In the results and discussion section, the authors argue that by means of the method outlined in the article, they solved the problem they defined in the introduction. Sometimes the results and discussion are combined in one; other times, they are separate. Whatever the arrangement, the crucial argument tends to be comparative, an argument often supported with statistical analysis. Such comparisons typically involve

- present results versus earlier published ones,
- experimental versus control group,
- baseline values versus different values under different conditions, and
- theoretical calculation versus experimental measurement.

This section is particularly critical because it is here that the authors hope to transform their findings into new knowledge that will arrest the attention of significant others. The stakes here are high—whether or not their hard work will receive any recognition. If the researchers are successful, readers will accept their knowledge claim, and in the best-case scenario, pass it along to other researchers in the form of citations—the professional capital in the intellectual marketplace of science. In the worst case, others will reject the claim and ultimately discredit it in the scientific literature. The vast majority of claims, though, suffer neither fate; they merely sink into the ocean of scientific publications without much notice.

The selected article belongs in the successful category—though recognition did not come immediately. The authors established that DNA is responsible for transmitting genetic traits, not protein or other macromolecules as earlier proposed by others. They demonstrated this claim in a pneumonia-inducing bacterium (Pneumococcus Type III) because it represented "the most striking example of inheritable and specific alterations in cell structure and function that can be experimentally induced and are reproducible under well defined and adequately controlled conditions." The following selection is a key passage with results that support the author's central claim by a comparison between experiment and theory:

> Four purified preparations [of the transforming substance] were analyzed for content of nitrogen, phosphorus, carbon, and hydrogen. The results are presented in Table I. The nitrogen-phosphorus ratios vary from 1.58 to 1.75 with an average value of 1.67 which is in close agreement with that calculated on the basis of the theoretical structure of sodium desoxyribonucleate (tetranucleotide). The analytical figures by themselves do not establish that the substance isolated is a pure chemical entity. However, on the basis of the nitrogen-phosphorus ratio, it would appear that little protein or other substances containing nitrogen or phosphorus are present as impurities since if they were this ratio would be considerably altered.

TABLE I. *Elementary chemical analysis of purified preparations of the transforming substance*

Preparation No.	Carbon per cent	Hydrogen per cent	Nitrogen Per cent	Phosphorus per cent	N/P ratio
37	34.27	3.89	14.21	8.57	1.66
38B	—	—	15.93	9.09	1.75
42	35.50	3.76	15.36	9.04	1.69
44	—	—	13.40	8.45	1.58
Theory for sodium desoxyribonucleate	34.20	3.21	15.32	9.05	1.69

In Table I, white spaces force the eye down each column to the borders of the rectangle formed by the experimental results from samples numbered 37, 38B, 42, and 44. At this point, the eye jumps the gap from this wide rectangle to the narrow rectangle of the satisfyingly close theoretical expectations for "sodium desoxyribonucleate."

Here is how Avery and coworkers expressed their central claim in the final sentence, derived from Table I and from many other experimental results: "The evidence presented supports the belief that a nucleic acid of the desoxyribose type is the fundamental unit of the transforming principle of Pneumococcus Type III." The authors chose their words very cautiously here. Since they only experimented with one substance, they could not claim that DNA is the transforming factor in all life forms. Their wording leaves open the door to different experiments that might find that protein or some other macromolecule could be storing the genetic information in bacteria other than Pneumococcus Type III. As is not infrequently the case, it took some time—and much additional research—before the greater significance of their results came to be recognized.

END

Conclusion: What's the Point Again?

Motoo Kimura, 1968. "Evolutionary rate at the molecular level." *Nature*, Vol. 217, pp. 624–26.

The following are elements for bringing closure to a scientific article's overall argument: reiterating claims or presenting them for the first time, indicating their wider significance, and recommending future work that will build on the original. By these means, authors hope to shape the research front to which they are contributing. Still, these three components are not nearly as firmly ingrained in scientific writing as the three we gave for the introduction. In fact, the third component is far from a given, in part, we suspect, because scientists do not want to reveal possible fruitful lines of inquiry to their rivals.

The two-step conclusion quoted below comes from a Japanese scientist who combined the then-new field of molecular biology and the long-established one of evolutionary biology. During the twentieth century, much research effort was spent on synthesizing Mendelian genetics and Darwinian natural selection. A contentious issue within this synthesis was whether *every* biological character in *every* species formed as a result of natural selection, or whether some characteristics formed by another process, random genetic drift being one possibility. By the early 1960s, according to the author, Motoo Kimura, the consensus greatly favored the former explanation. His article, based on molecular studies, as opposed to animals in the wild or the laboratory, argued otherwise.

Kimura arrived at a new theoretical explanation for evolution at the micro scale by analyzing data from past experimental studies of hemoglobin molecules. His short paper makes three major claims in order of ascending importance. First, the constituents of amino acids (the subunits of protein) substitute for one another at about the same rate in diverse mammalian lineages such as humans, rats, and pigs. Second, these substitutions occur randomly. Third, the rate of change at the DNA level for the entire genome of a species is much higher than ever thought possible. (Kimura calculated one nucleotide pair substituted every two years in a given population, as compared with the previous estimate of one every 300 generations.) From these three findings, he reasoned that at the DNA and amino acid level, much evolutionary change occurs because of random fixation of mutant genes that are selectively equivalent ("neutral" from an evolutionary perspective). That is his main conclusion.

Kimura wrapped up his article by reiterating that conclusion, extrapolating it to major genetic changes in biological populations, and ending with a picturesque analogy comparing biological and geological change:

> Finally, if my chief conclusion is correct, and if the neutral or nearly neutral mutation is being produced in each generation at a much higher rate than has been considered before, then we must recognize the great importance of random genetic drift due to finite population number in forming the genetic structure of biological populations. The significance of random genetic drift has been deprecated during the past decade. This attribute has been influenced by the opinion that almost no mutations are neutral, and also that the number of individuals forming a species is usually so large that random sampling of gametes should be negligible in determining the course of evolution, except possibly through the "founder principle." To emphasize the founder principle but deny the importance of random genetic drift due to finite population number is, in my opinion, rather similar to assuming a great flood to explain the formation of deep valleys but rejecting a gradual but long lasting process of erosion by water as insufficient to produce such a result.

The "founder principle" that Kimura mentions is the reduction of genetic variability because an entire population was started by a single pair or a single fertilized offspring. Kimura's argument is that his controversial conclusion is no more far-fetched than the well-accepted founder principle.

References: Establishing the Citation Web

M. Gell-Mann, 1964. "A schematic model of baryons and mesons." *Physics Letters*, Vol. 8, pp. 214–15.

Over the centuries, references have gradually increased in importance. One finds very few of them in the early scientific articles, the bibliographic information usually being inserted haphazardly into the text. By the nineteenth century, references had increased in prevalence so that most articles had at least a few. More often than not, these appeared as footnotes bunched together in the introductory pages. Today, the typical scientific article has about two dozen citations dispersed throughout the main text, their bibliographic information gathered together in a concluding list. (For review-type articles, that average is of course many times more.)

The original texts for the cited sources tell a story: they form the historical context out of which the present research emerged. We reproduce below the reference list from a much-cited theoretical article by physicist Murray Gell-Mann:

1) M. Gell-Mann, California Institute of Technology Synchrotron Laboratory Report CTSL-20 (1961).

2) Y. Ne'eman, Nuclear Phys. 26 (1961) 222.

3) M. Gell-Mann. Phys. Rev. 125 (1962) 1067.

4) E.g.: R. H. Capps, Phys. Rev. Letters 10 (1963) 312; R. E. Cutkosky, J. Kalckar and P. Tarjanne, Physics Letters 1 (1962) 93; E. Abers, F. Zachariasen and A. C. Zemach, Phys. Rev. 132 (1963) 1831; S. Glashow, Phys. Rev. 130 (1963) 2132; R. E. Cutkosky and P. Tarjanne, Phys. Rev. 132 (1963) 1354.

5) P. Tarjanne and V. L. Teplitz, Phys. Rev. Letters 11 (1963) 447.

6) James Joyce, Finnegan's [sic] Wake (Viking Press, New York, 1939) p. 383.

7) M. Gell-Mann and M. Lévy, Nuovo Cimento 16 (1960) 705.

8) N. Cabibbo, Phys. Rev. Letters 10 (1963) 531.

The reference lists in most other twentieth-century articles follow similar conventions of typeface, abbreviation, and punctuation, though these vary somewhat from journal to journal, and within a journal over time.

This particular reference list has citations to twelve documents while the main text is less than two pages long. This thicket of citations is typical of contemporary science. It is typical of rapidly advancing scientific fields that only one reference is more than five years old. But there is one oddity in the quoted list in date of publication and subject matter, the sixth reference. Gell-Mann had appropriated the word "quark" from James Joyce's opaque experimental novel to name what he postulated as the fundamental constituent of matter. Gell-Mann's "quark" seems to have spawned a whole generation of whimsical technical terms in particle physics and other fields as well.

The relative density of citations within scientific texts in general, combined with the uniformity of bibliographic presentation, has permitted the creation of a database that allows computerized tracking of citations made to a selected author, research institution or group, subdiscipline, and so forth. Without this powerful database (Eugene Garfield's *Science Citation Index*), science would operate less efficiently. We base this assertion on an intriguing theory recently published by Georg Franck, a theory about how science works in which citations play a central role. Franck views a publication as a form of intellectual property, and citation as a form of payment for using someone else's property. In other words, scientists pay other scientists for using their published research by means of citation; and the more citations scientists receive, the greater their accumulated wealth, that is, the greater their reputation. This system thus concentrates the scientific community's attention on published research that helps make others productive. So the modern reference list has taken on importance far beyond simply dry bibliographic information.

The crudity of this measure must be emphasized. Articles are cited for a variety of reasons, a variety that may affect any index of citations as a measure of reputation. Articles may be much cited simply because they add a new method to the repertory. Articles are also cited only to take issue with their authors, a process that, while it does not necessarily have a negative impact on a reputation, must affect reputation differently from citation with approval. Similarly, can we take seriously those blanket citations by means of which the authors refer to virtually every article on the subject? Citation counts can also be inflated by citation cartels: scientists who tend to cite each other's work, regardless of strict relevance. Presumably, this last process can be inverted. Scientists can purposely avoid citing certain scientists, despite the relevance of their work. Finally, some scientists become so famous that their work becomes part of textbook science and is not cited. The early DNA work of Watson and Crick is an example. Nevertheless, with all their flaws, citations are the best quantitative indicator we currently have of the degree of attention a scientific contribution gets over time.

Acknowledgments: Crediting Assistance

Percy L. Julian and Josef Pikl, 1935. "Studies in the indole series. V. The complete synthesis of physostigmine (eserine)." *Journal of the American Chemical Society*, Vol. 57, pp. 755–57.

The acknowledgment supplements both the list of authors and the citations, giving credit to those people or organizations who provided support but were not directly involved as scientists in creating new science. This list typically includes funding agencies, technicians who helped carry out experiments, and other scientists who donated materials, were consulted during the course of the research, or critically reviewed a draft of the manuscript.

Inclusion of an acknowledgments paragraph or section is rare before the twentieth century. We suspect that it arose in that century because of the growing complexity of work arrangements. The rise of the acknowledgment section may also reflect the democratization of scientific culture. Just as in movie credits, even technical advisors, student interns, secretaries, and electricians may get an acknowledgment.

The quoted acknowledgment has special significance because it comes from a key paper by one of the first African-American scientists to rise into the highest ranks of the scientific community. In collaboration with his assistant Josef Pikl, Percy Julian synthesized the anti-glaucoma drug, physostigmine, a derivative of the Calabar bean. First isolated in 1864 by Jobst and Hesse, this compound had resisted repeated attempts at its synthesis in the laboratory. The Julian-Pikl method led to its cheap production in large quantities for medicinal purposes.

Julian's acknowledgment of financial assistance from a Jewish patron is one of those rare instances where the scientific and sociopolitical openly intersect in a scientific text:

> In acknowledging a generous grant from the Rosenwald Fund, the senior author respectfully dedicates this finished project to the memory of Julius Rosenwald, who has made possible innumerable cultural contributions on the part of young negroes to his country's civilization. And he is none the less grateful to Dean W. M. Blanchard, Senior Professor of Chemistry [DePauw University], without whose courageous support this work would have been impossible.

Acknowledgments are not always a matter of decorum alone. They provide opportunity for crediting important scientific contributions. Maria Goeppert Mayer, in the final sentence to her classical article proposing the shell model for the atomic nucleus (*Physical Review*, 1949), notes that "Thanks are due to Enrico Fermi for the remark, 'Is there any indication of spin-orbit coupling?' which was the origin of this paper."

FROM START TO FINISH

Experimental Type: Reducing the Speed of Light to Zero

Chien Liu, Zachary Dutton, Cyrus H. Behroozi, and Lene Vestergaard Hau, 2001. "Observation of coherent optical information storage in an atomic medium using halted light pulses." *Nature*, Vol. 409, pp. 490–93.

To better illustrate how the various components in a typical scientific paper work together, we depart from our expository approach of only quoting and commenting upon a short extract or two per selection and take you on a guided tour through a complete scientific paper, from title to acknowledgments. For that purpose, we selected a recent experimental "letter" published in *Nature*. Such letters are not really "letters" in the normal sense of that word, but short papers—a popular way of sharing important results quickly. The famous letter of Watson and Crick reporting the DNA structure appeared in the same journal about fifty years earlier. We chose the Liu and coworkers letter because the complete text, while taxing reading for the uninitiated even with our commentary on it, is relatively brief and has much to admire: not only the science but the composition as well. You can view the full text *in situ* by going to www.nature.com.

The letter reports a remarkable experimental result in applied quantum physics: the complete arrest of light in the midst of a cloud of liquid sodium cooled almost to absolute zero and suspended by electromagnetic forces. In a vacuum, light always travels at 700 million miles per hour. By means of an ingenious experimental arrangement in their lab, the research team slowed a laser beam all the way to zero velocity in their sodium cloud, stored it very briefly, then started it on its way again. They think that their research program may make a contribution to quantum computing—the manipulation of photons and quantum states for storing and processing information.

OVERALL ORGANIZATION.   As we mentioned in the introduction to this chapter, modern experimental papers generally have the following components:

> Front Matter (title, byline, abstract)
> Introduction
> Methods
> Results and Discussion (separate or combined)
> Conclusion
> Back Matter (acknowledgments, references)

This is only a template, not a straightjacket. It is a guide *to* writing, not a prescription *for* writing. The letter that is our example deviates from this order in several respects. It has no heading abstract independent of the main text, but short papers like this do not need such abstracts. It also has a theoretical paragraph before the methods section (for reasons we will discuss later). Moreover, some elements of method and the conclusion are placed elsewhere than in their respective sections. Letters are also generally too short for headings to separate the sections, as typically done in longer papers. The above sections are still there, generally in the order we give, but they are not marked by headings as discrete units. These deviations are not errors. They are examples of author compositional choices and the journal style—a style designed to maximize the information content within the limited space allotted each letter and to maintain some consistency from one letter to the next.

FRONT MATTER/INTRODUCTION. What we call "front matter" consists of the title, the byline, and the abstract. Almost anyone reading through the table of contents or scanning through a journal with the slightest interest in a scientific paper will at least examine the title and byline. Prospective readers will go no further than the title if it does not either grab their attention because of its novelty or hint that the findings may help them solve their own research problems, or create a new problem to solve.

The title in our selected experimental letter,

Observation of *coherent optical information storage* in an *atomic medium* using *halted light pulses*

is fairly typical of scientific titles in general: they are sentence fragments filled with technical terminology. In contrast, the title for a semi-popular science magazine article reporting on this discovery is the poetic "Frozen Light," while a newspaper headline opts for the more prosaic "Scientists bring light to full stop, hold it, then send it on its way."

The real scientific title is not short. It is not catchy. It is not openly soliciting our attention. It is a collection of three key phrases composed of technical words (we added the italics for emphasis) linked by a preposition and a participle. It is fairly typical for the highly specialized journal in which it appears. The title tells us that at the center of the letter is a series of observations related to the storing of information by bringing light pulses to a stop in a specially prepared atomic medium.

After the title the readers' eyes normally jump to the byline:

Chien Liu[1,2], **Zachary Dutton**[1,3], **Cyrus Behroozi**[1,2] & **Lene Vestergaard Hau**[1,2,3]

[1]*Rowland Institute for Science, 100 Edwin H. Land Boulevard, Cambridge, Massachusetts 02142 USA,* [2]*Division of Engineering and Applied Sciences,* [3]*Department of Physics, Harvard University, Cambridge, Massachusetts 02138, USA*

In general, the typical scientific writing style leaves the impression of some impersonal agent for the actions being reported. The byline tells us right up front who did the work, and what institution they are from should any reader wish to contact any one of them. (Also given in our selected paper is the email address of which author to contact for general inquiries.) The byline distributes both credit and responsibility. Both the persons and institutions involved benefit from success, and suffer from failure or fraudulent reporting.

In this particular paper, the authors' names are differentiated from their affiliations by typeface: boldface for names, italic for affiliations. The two categories are linked by a series of superscripts. Of course, the style for typeface varies from journal to journal. It is only the differentiation by typographical conventions that will hold steady.

There is always a reason behind the arrangement of author names. On occasion, it is strictly alphabetical. Usually, it has to do with the roles of each author. In this case, the first author is the one with primary responsibility for this series of experiments; the last is the leader of the team. We assume that the second and third authors are listed in order of their contribution to the project.

Readers intrigued by the title and reassured by the credentials of the authors next turn to the paper itself. Let's look at the introductory paragraph that serves double duty as the abstract and the introduction. This convergence is a function of the need to conserve space in this journal. As part of the journal style, the introduction is boldfaced to indicate that it is also an abstract; we number each sentence to facilitate discussion:

1) Electromagnetically induced transparency is a quantum interference effect that permits the propagation of light through an otherwise opaque atomic medium; a 'coupling' laser is used to create the interference necessary to allow the transmission of resonant pulses from a 'probe' laser. 2) This technique has been used to slow and spatially compress light pulses by seven orders of magnitude, resulting in their complete localization and containment within an atomic cloud. 3) Here we use electromagnetically induced transparency to bring laser pulses to a complete stop in a magnetically trapped, cold cloud of sodium atoms. 4) Within the spatially localized pulse region, the atoms are in a superposition state determined by the amplitudes and phases of the coupling and probe laser fields. 5) Upon sudden turn-off of the coupling laser, the compressed probe pulse is effectively stopped; coherent information initially contained in the laser fields is 'frozen' in the atomic medium for up to 1 ms. 6) The coupling laser is turned back on at a later time and the probe pulse is regenerated: the stored coherence is read out and transferred back into the radiation field. 7) We present a theoretical model that reveals that the system is self-adjusting to minimize dissipative loss during the 'read' and 'write' operations.

8) **We anticipate applications of this phenomenon for quantum information processing.**

In its role as abstract and introduction combined, the paragraph in boldface summarizes the letter's argument: the claim it makes, the reasons in support of the claim, and the significance of the claim. In the first two sentences, the key term, "electromagnetically induced transparency," is defined, along with its history of use. The third sentence states the authors' main claim—that is, their having gone beyond previous research with this advanced technique and brought a laser beam to a "complete stop." Sentences four through six chronicle their results. Sentence seven is devoted to a theoretical model of the experimental system. The last sentence concerns a possible application of the experiment.

THEORY AND METHOD. The second paragraph in our experimental letter provides a key part of the theoretical model promised at the end of the opening paragraph. The text is in descriptive style: the authors limn a physical state, then model it with an equation. One might ask, why not save the model for the discussion section and go directly to method? That would be consistent with our structure for the experimental paper given earlier. The expository problem is this: without the theoretical model, the authors cannot describe what is happening in their experiment at the quantum level. The experiment hinges on the researchers' control of the quantum state of the atomic cloud with their coupling and probe lasers. Theory thus sets the context for the method to follow.

The authors also represent both the theory and the method in their complex and interesting first figure (see p. 211). This figure has three parts: two visual, one verbal.

The top part (labeled a) visually represents the theoretical model described in the letter's second paragraph. The authors assume interested readers will have seen this type of schematic before and can decipher its various elements. For example, $|1>, |2>, |3>$ and $|4>$ stand for different quantum states (marked by horizontal lines) in the supercooled cloud of sodium (Na) atoms and lasers. Key here is the relationship depicted among three of the four states and the coupling and laser probes (arrows). This schematic portrays sentence 4 from the introductory paragraph: "Within the spatially localized pulse region, the atoms are in a superposition state determined by the amplitudes and phases of the coupling and probe laser fields."

The middle part (labeled b) is a schematic representing the experimental setup that creates the physical state theoretically portrayed in the uppermost figure. It has three main regions reading from left to right: input to the main experimental site, the site itself, and output. On the left are the probe and coupling lasers. The center shows the sodium cloud and associated equipment. The right

Observation of coherent optical information storage in an atomic medium using halted light pulses

Chien Liu*†, Zachary Dutton*‡, Cyrus H. Behroozi*†
& Lene Vestergaard Hau*†‡

* Rowland Institute for Science, 100 Edwin H. Land Boulevard, Cambridge, Massachusetts 02142, USA
† Division of Engineering and Applied Sciences, ‡ Department of Physics, Harvard University, Cambridge, Massachusetts 02138, USA

Electromagnetically induced transparency[1-3] is a quantum inter-ference effect that permits the propagation of light through an otherwise opaque atomic medium; a 'coupling' laser is used to create the interference necessary to allow the transmission of resonant pulses from a 'probe' laser. This technique has been used[4-6] to slow and spatially compress light pulses by seven orders of magnitude, resulting in their complete localization and con-tainment within an atomic cloud[4]. Here we use electromagneti-cally induced transparency to bring laser pulses to a complete stop in a magnetically trapped, cold cloud of sodium atoms. Within the spatially localized pulse region, the atoms are in a superposition state determined by the amplitudes and phases of the coupling and probe laser fields. Upon sudden turn-off of the coupling laser, the compressed probe pulse is effectively stopped; coherent information initially contained in the laser fields is 'frozen' in the atomic medium for up to 1 ms. The coupling laser is turned back on at a later time and the probe pulse is regenerated: the stored coherence is read out and transferred back into the radiation field. We present a theoretical model that reveals that the system is self-adjusting to minimize dissipative loss during the 'read' and 'write' operations. We anticipate applications of this phenomenon for quantum information processing.

With the coupling and probe lasers used in the experiment, the atoms are accurately modelled as three-level atoms interacting with the two laser fields (Fig. 1a). Under perfect electromagnetically-induced transparency (EIT) conditions (two-photon resonance), a stationary eigenstate exists for the system of a three-level atom and resonant laser fields, where the atom is in a 'dark', coherent super-position of states $|1\rangle$ and $|2\rangle$:

$$|D\rangle = \frac{\Omega_c|1\rangle - \Omega_p|2\rangle\exp[i(\mathbf{k}_p - \mathbf{k}_c)\cdot\mathbf{r} - i(\omega_p - \omega_c)t]}{\sqrt{\Omega_c^2 + \Omega_p^2}} \quad (1)$$

Here Ω_p and Ω_c are the Rabi frequencies, $\mathbf{k}_p$ and $\mathbf{k}_c$ the wavevectors, and ω_p and ω_c the optical angular frequencies of the probe and coupling lasers, respectively. The Rabi frequencies are defined as $\Omega_{p,c} \equiv e\,\mathbf{E}_{p,c}\cdot\mathbf{r}_{13,23}/\hbar$, where e is the electron charge, $\mathbf{E}_{p,c}$ are the slowly varying envelopes of probe and coupling field amplitudes, and $e\,\mathbf{r}_{13,23}$ are the electric dipole moments of the atomic transitions. The dark state does not couple to the radiatively decaying state $|3\rangle$, which eliminates absorption of the laser fields[1-3].

FIGURE 6-1 · First page of paper on the stopping of laser pulse, presenting byline, title, abstract/introduction, and theory. Reprinted with permission of *Nature*. Copyright (2001) Macmillan Publishers Ltd.

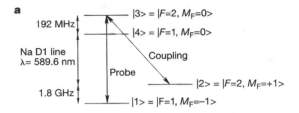

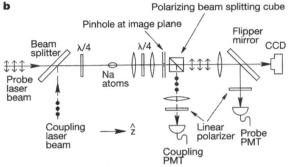

Figure 1 Experimental set-up and procedure. **a**, States |1⟩, |2⟩ and |3⟩ form the three-level EIT system. The cooled atoms are initially magnetically trapped in state |1⟩ = |3S, $F = 1, M_F = -1$⟩. Stimulated photon exchanges between the probe and coupling laser fields create a 'dark' superposition of states |1⟩ and |2⟩, which renders the medium transparent for the resonant probe pulses. **b**, We apply a 2.2-mm diameter, σ^--polarized coupling laser, resonant with the |3S, $F = 2, M_F = +1$⟩ → |3P, $F = 2$, $M_F = 0$⟩ transition, and a co-propagating, 1.2-mm diameter σ^+-polarized probe pulse tuned to the |3S, $F = 1, M_F = -1$⟩ → |3P, $F = 2, M_F = 0$⟩ transition. The two laser beams start out with orthogonal linear polarizations (two-headed arrows and filled circles show the directions of linear polarization of the probe and coupling lasers, respectively). They are combined with a beam splitter, circularly polarized with a quarter-wave plate ($\lambda/4$), and then injected into the atom cloud. After leaving the cloud, the laser beams pass a second quarter-wave plate and regain their original linear polarizations before being separated with a polarizing beam-splitting cube. The atom cloud is imaged first onto an external image plane and then onto a CCD (charge-coupled device) camera. A pinhole is placed in the external image plane and positioned at the centre of the cloud image. With the pinhole and flipper mirror in place, only those portions of the probe and coupling laser beams that have passed through the central region of the cloud are selected and monitored simultaneously by two photomultiplier tubes (PMTs). States |1⟩ and |2⟩ have identical first-order Zeeman shifts so the two-photon resonance is maintained across the trapped atom clouds. Cold atoms and co-propagating lasers eliminate Doppler effects. However, off-resonance transitions to state |4⟩ prevent perfect transmission of the light pulses in this case.

FIGURE 6-2 · Picture showing experimental arrangement and theory behind experiment to trap laser pulse in supercold sodium cloud. Reprinted with permission of *Nature*. Copyright (2001) Macmillan Publishers Ltd.

has various devices for detecting the beam and measuring its properties. The schematic thus depicts the relative positions of each segment of the experimental apparatus, their orientation in space, and their interactions with the laser beams that produce the experimental effect. Prose cannot perform these functions, or at least not nearly as effectively.

The figure has a very long caption describing both the contents of the two images and their interrelationship. This text takes advantage of the power of words to narrate, describe, and argue in a way not possible with visuals alone. Unlike purely artistic images, scientific ones seldom stand by themselves without commentary. The format for this particular journal is to pack that sort of information in the figure caption, not in the main text as done more typically in other journals. This format represents an interesting variation in the evolution of integrating text and image in the scientific article.

The methods section in the main text follows a narrative style. It links quantum theory and experiment in a narrative that explains what happens at the experimental site:

> Atoms are prepared (magnetically trapped) in a particular internal quantum state $|1 >$ (Figure 1a). The atom cloud is first illuminated by a coupling laser, resonant with the $|2-|3 >$ transition. With only the coupling laser on and all atoms in $|1 >$, the system is in a dark state (equation (1) with $\Omega_p = 0$). A probe laser pulse, tuned to the$|1 - |3 >$ transition and co-propagating with the coupling laser, is subsequently sent through the atomic medium. Atoms within the pulse region are driven into the dark-state superposition of states $|1 >$ and $|2 >$.... The presence of the coupling laser field creates transparency...

This narrative expands upon Fig. 1a. Next, the authors turn to Fig. 1b in a narrative that describes the experimental site in physical and quantitative terms:

> A typical cloud of 11 million sodium atoms is cooled to 0.9 μK, which is just above the critical temperature for Bose-Einstein condensation. The cloud has a length of 339 μm in the z direction, a width of 55 μm in the transverse directions, and a peak density of 11 μm^{-3}. Those portions of the co-propagating probe and coupling laser beams that have passed through the 15-μm-diameter centre region of the cloud are selected and monitored simultaneously by two photomultiplier tubes (PMTs).

Let us look more closely at the style of these sentences. They are uniform in that they always have physical objects or states as their subjects. Sometimes this priority leads to sentences having verbs in the passive voice (some form of verb *to be* plus past participle, discussed further in chapter 7). There are good reasons to use the passive voice in a methods narration. Let us look at one of these reasons, using the sentences we have just quoted: "The atom cloud is first illuminated by a coupling laser, resonant with the |2 - |3 > transition." The verb "to illuminate" is in the passive voice. You can easily turn it into the active voice by making the cause of the illumination the subject of the sentence: "A coupling laser, resonant with the |2 - |3 > transition, illuminates the atom cloud." But this would make the paragraph harder to follow because the preceding sentence concerns the sodium atom cloud, not the coupling laser.

Here is another sentence from the methods section, also in the passive voice: "A typical cloud of 11 million sodium atoms is cooled to 0.9 μK." We can easily change it to the active: "We cooled a typical cloud of 11 million sodium atoms to 0.9 μK." But this would make the paragraph about the researchers and what they did in the lab, not about the atomic clouds.

As we said, the expository style of the methods section is impersonal narrative, not argument. But in the context of the paper as a whole, this narrative functions as a major part of an argument. Unless the results of the experiment were achieved by appropriate methods, the inferences made from these results would not themselves be appropriate. In short, they would not be good science.

RESULTS/DISCUSSION. A combined results/discussion follows the methods in this letter. In general, the purpose of the results/discussion section is to convince readers that the data obtained from having applied the methods justify the claims in the introductory paragraph and conclusion. In this section, scientists normally present their data in both visual and verbal form. They also comment, interpret, and argue about these data. In making assertions about the data, they must not go beyond what is warranted by their methods. Sometimes there is a very fine line between legitimate inference and illegitimate speculation. Part of the job of journal referees is to make that call before publication.

Now let us consider this particular results/discussion section. Because results report states, the dominant style is quantitative description:

At time t = 6.3 μs, indicated by the arrow in Fig. 2a, the probe pulse is spatially compressed and contained completely within the atomic cloud. The probe pulse in free space is 3.4 km long and contains 27,000 photons within a 15-μm diameter at its centre. It is compressed in the atomic medium to match the size of the cloud (339 μm), and the remaining optical energy in the probe field is only 1/400 of a free-space photon.

This passage describes the mind-boggling compression of the light pulse from the probe laser mentioned in the introductory paragraph (a factor of 10^7, a reduction in pulse length from 3.4 kilometers in free space to 339 micrometers at the experimental site) and obtained by the method just described. Although the primary expository style is descriptive, the descriptions are subordinate to the argument of the whole letter: these data are the evidence for the major claims of the paper.

Here is a typical paragraph of discussion (italics added for emphasis):

> When the probe pulse is contained within the medium, the coherence of the laser fields is already imprinted on the atoms. As the coupling laser is turned off, the probe field is depleted to maintain the dark state (equation (1)) and the (negligible) atomic amplitude is transferred from state $|1>$ to state $|2>$ *through stimulated photon exchange between two light fields. Because of the extremely low energy remaining in the compressed pulse,* as noted above, it is completely depleted before the atomic population amplitudes have changed by an appreciable amount.

This paragraph is in a more argumentative style: we learn not only what events happened at the experimental site, but why and how. The first sentence in the above quote describes a state, while the next two sentences offer causal explanations for its appearance, explanations contained in the two italicized phrases.

This results/discussion section argues a third way: by means of equations. These form a separate subsystem set apart by white space and distinct numbering. Like the figures, they are part of the larger communicative system that constitutes the letter as a whole. We know this because they are incorporated into the text in such sentences as "As seen from equation (1), the difference between the wavevectors of the two laser fields determines the wavelength of the periodic phase pattern imprinted in the medium, which is 10^5 times larger than the individual laser wavelength."

Arguments by comparison like this dominate results/discussion. In our example, we find three general kinds: 1) between the experimental data and calculated results from the equation, 2) between the data and the theoretical picture underlying the experimental results, and 3) between the data and the needs of quantum information processing. For example, the following passage demonstrates the predictive validity of equation (1). In it, the authors compare experimental data extracted from a figure and calculations from the equation:

> The delay of this probe pulse, relative of the reference pulse, is 11.8 µs corresponding to a group velocity of 28 ms^{-1}, a reduction by a factor of 10^7 from its vacuum value. [*Our comment: a use of experimental data from Figure 2a.*]

The measured delay agrees with the theoretical prediction of 12.2 μs based on a measured coupling Rabi frequency Ω_c of 2.57 MHz × 2 π and an observed atomic column density of 3,670 μm^{-2}. [*Our comment: a prediction from equation (1).*]

These two sentences also illustrate the mutual reinforcement of text, figures, and equations.

As an example of the interaction of figures and text, let's look at another exemplary visual from this letter, Figure 2. It presents four Cartesian graphs stacked vertically. It tells a story in four parts, each more remarkable than the last. Here is our plot summary:

a: Working with a coupling laser and probe laser, experimenters from Harvard University bring probe pulse to a dead stop at 6.3 μs or millionths of a second (marked by vertical arrow).

b: Experimenters turn off coupling laser at 6.3 μs, then turn it back on at 44.3 μs. The halted probe pulse reappears with the stored optical information basically intact (signified by bump in solid curve with filled circles at just beyond 40 μs).

c: Authors repeat previous experiment except coupling laser not switched on until 839.3 μs.

d (coda): The stored optical information remains "coherent" for 900 μs (about 1 thousandth of a second). This coherence is evident from the virtual coincidence of the data (solid circles) and the exponential decay curve (solid line).

Telling this story simply cannot be done by visual images alone: it also requires verbal description, explanation, and commentary. Too often, scientist-writers do not provide enough verbal accompaniment for the reader to follow the visual story. Symbols and curves are not defined. Units of measure are not reported. More important, the data are not interpreted, or barely so. That is not the case here. The verbal text accompanying the visual has three components:

1) *Caption*. In this case, the main caption is a very short descriptive phrase: "Measurements of delayed and revived probe pulses." Following this caption is a discussion of each graph. It covers what each graph measures: for Figure 2a, "Probe pulse delayed by 11.8 μs." It also provides necessary keys for interpreting the figures: "Dashed curves and filled circles (fitted to solid gaussian curves) show simultaneously measured intensities of coupling and

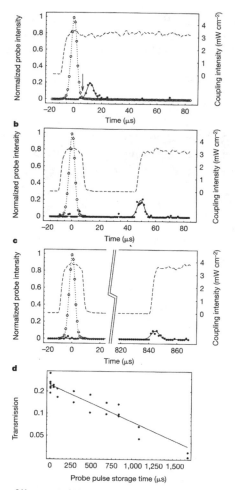

Figure 2 Measurements of delayed and revived probe pulses. Open circles (fitted to the dotted gaussian curves) show reference pulses obtained as the average of 100 probe pulses recorded in the absence of atoms. Dashed curves and filled circles (fitted to the solid gaussian curves) show simultaneously measured intensities of coupling and probe pulses that have propagated under EIT conditions through a 339-μm-long atom cloud cooled to 0.9 μK. The measured probe intensities are normalized to the peak intensity of the reference pulses (typically, $\Omega_p/\Omega_c = 0.3$ at the peak). **a**, Probe pulse delayed by 11.8 μs. The arrow at 6.3 μs indicates the time when the probe pulse is spatially compressed and contained completely within the atomic cloud. (The intersection of the back edge of the reference pulse and the front edge of the delayed pulse defines a moment when the tail of the probe pulse has just entered the cloud and the leading edge is just about to exit.) **b**, **c**, Revival of a probe pulse after the coupling field is turned off at $t = 6.3$ μs and turned back on at $t = 44.3$ μs and $t = 839.3$ μs, respectively. During the time interval when the coupling laser is off, coherent information imprinted by the probe pulse, is stored in the atomic medium. Upon subsequent turn-on of the coupling field, the probe pulse is regenerated through coherent stimulation. The time constants for the probe and coupling PMT amplifiers are 0.3 μs and 3 μs, respectively. The actual turn on/off time for the coupling field is 1 μs, as measured with a fast photodiode. **d**, Measured transmission of the probe pulse energy versus storage time. The solid line is a fit to the data, which gives a 1/e decay time of 0.9 ms for the atomic coherence.

FIGURE 6-3 · Picture plotting results from experiment to trap laser pulse in supercold sodium cloud. Reprinted with permission of *Nature*. Copyright (2001) Macmillan Publishers Ltd.

probe pulses that have propagated through a 339 μm-long atomic cloud cooled to 0.9 μK." The lengthy caption (about 300 words) is in keeping with the journal style. Some journals would have placed much of this discussion in the main text.

2) *Descriptive labels.* In the case of Figure 2a, for example, the left *y* axis is labeled "Normalized probe intensity," the right *y* axis, "Coupling intensity (mW cm^{-2})," the *x* axis, "Time (μs)." Also note the streak of lightning through the middle of Figure 2c. That signifies the excised time period from about 30 to 800 μs. As a consequence, readers can more easily compare graphs a, b, and c.

3) *Commentary.* The figure data and curves must be commented on so that the relationships between lines and points are understood. For example, Figure 2d is a scatter plot of data points intersected by a solid line sloping downward. What does the solid line mean? We are told in the figure legend that "The solid line is a fit to the data, which gives a 1/e decay time of 0.9 ms for the atomic coherence." Much more extensive interpretation of the four graphs appears in the main text.

Taken together, these four graphs and the relevant text make a strong visual and verbal argument that the authors accomplished what they claimed in their title and opening paragraph.

Comparisons and causal explanations are not the only kinds of argument critical to a persuasive results/discussion. Another kind is anticipating possible objections by others and responding in some way. In this particular letter, we presume that is the motivation behind such a statement as "We have verified experimentally that the probe pulse is regenerated through stimulated rather than spontaneous emission." In other words, if a critical reader questioned whether or not turning on the coupling laser really did rejuvenate the probe pulse, as opposed to some other explanation, the authors have a reply based on additional experiments not covered in the methods. The better authors are at anticipating such objections, the more powerful the overall argument.

CONCLUSION. In the conclusion of our sample paper, the scientists cover the three topics standard for conclusions:

1) Reiterate main points
"We have demonstrated experimentally that coherent optical information can be stored in an atomic medium and subsequently read out by using the effect of EIT in a magnetically trapped, cooled atom cloud . . ."
2) Point to wider significance

"We believe that this system could be used for quantum information transfer; for example, to inter-convert stationary and flying qubits."

3) Suggest future research

"By injection of multiple probe pulses into a Bose-Einstein condensate— where we expect that most atomic collisions are coherence preserving—and with the use of controlled atom-to-atom interactions, quantum information processing may be possible during storage time."

The third topic standard for conclusions, further research, also appears earlier in a paragraph concerning experimental outcomes demonstrating decoherence: "Further studies of the decoherence mechanisms are planned but are beyond the scope of this Letter." As we noted earlier, authors seldom rigorously adhere to the structure we outlined at the chapter's beginning.

Note that the closing sentences are cast in a tad more personal style than the method or results/discussion. They are primarily about the researchers, not the research. The authors tell us what *they* have discovered, and what *they* believe is possible with continued research.

BACK MATTER.   As we have said, what we refer to as back matter serves a similar function to the rolling of credits at the end of a movie. This particular one has three distinct parts: a list of the references cited in the main text, a section acknowledging help, and lastly, a short statement informing readers which one of the four authors to contact with questions or criticisms. The first deals with past researches by others, the second with the present research, and the third with future inquiries regarding the present research.

1. Harris, S. E. Electromagnetically induced transparency. *Phys. Today* **50**, 36–42 (1997).
2. Scully, M. O. & Zubairy, M. S. *Quantum Optics* (Cambridge Univ. Press, Cambridge, 1997).
3. Arimondo, E. in *Progress in Optics* (ed. Wolf, E.) 257–354 (Elsevier Science, Amsterdam, 1996).
4. Hau, L. V., Harris, S. E., Dutton, Z. & Behroozi, C. H. Light speed reduction to 17 metres per second in an ultracold atomic gas. *Nature* **397**, 594–598 (1999).
5. Kash, M. M. *et al.* Ultraslow group velocity and enhanced nonlinear optical effects in a coherently driven hot atomic gas. *Phys. Rev. Lett.* **82**, 5229–5232 (1999).
6. Budker, D., Kimball, D. F., Rochester, S. M. & Yashchuk, V. V. Nonlinear magneto-optics and reduced group velocity of light in atomic vapor with slow ground state relaxation. *Phys. Rev. Lett.* **83**, 1767–1770 (1999).
7. Harris, S. E., Field, J. E. & Kasapi, A. Dispersive properties of electromagnetically induced transparency. *Phys. Rev. A* **46**, R29–R32 (1992).
8. Grobe, R., Hioe, F. T. & Eberly, J. H. Formation of shape-preserving pulses in a nonlinear adiabatically integrable system. *Phys. Rev. Lett.* **73**, 3183–3186 (1994).
9. Xiao, M., Li, Y.-Q., Jin, S.-Z. & Gea-Banacloche, J. Measurement of dispersive properties of electromagnetically induced transparency in rubidium atoms. *Phys. Rev. Lett.* **74**, 666 669 (1995).
10. Kasapi, A., Jain, M., Yin, G. Y. & Harris, S. E. Electromagnetically induced transparency: propagation dynamics. *Phys. Rev. Lett.* **74**, 2447–2450 (1995).
11. Hau, L. V. *et al.* Near-resonant spatial images of confined Bose-Einstein condensates in a 4-Dee magnetic bottle. *Phys. Rev. A* **58**, R54–R57 (1998).
12. Fleischhauer, M. & Lukin, M. D. Dark-state polaritons in electromagnetically induced transparency. *Phys. Rev. Lett.* **84**, 5094–5097 (2000).
13. Harris, S. E. Normal modes for electromagnetically induced transparency. *Phys. Rev. Lett.* **72**, 52–55 (1994).
14. Fleischhauer, M. & Manak, A. S. Propagation of laser pulses and coherent population transfer in dissipative three-level systems: An adiabatic dressed-state picture. *Phys. Rev. A* **54**, 794–803 (1996).
15. DiVincenzo, D. P. The physical implementation of quantum computation. Preprint quant-ph/0002077 at ⟨http://xxx.lanl.gov⟩ (2000).

Acknowledgements

We thank J. Golovchenko for discussions during which the idea of the rapid turn off and on of the coupling laser first emerged. We also thank M. Burns for critical reading of the manuscript. This work was supported by the Rowland Institute for Science, the Defense Advanced Research Projects Agency, the US Airforce Office of Scientific Research, and the US Army Research Office OSD Multidisciplinary University Research Initiative Program.

Correspondence and requests for materials should be addressed to C.L. (e-mail: chien@deas.harvard.edu).

FIGURE 6-4 · References and acknowledgments from paper on stopping of laser pulse. Reprinted with permission of *Nature*. Copyright (2001) Macmillan Publishers Ltd.

7

SCIENTIFIC WRITING STYLE: NORMS *and* PERTURBATIONS

R ETURN BRIEFLY, if you will, to the botanical passage by Martin Lister from the seventeenth-century *Philosophical Transactions* in chapter 1. It possesses all the elements of good storytelling: strong verbs in the active voice, first person narrative, figurative language, rhetorical flourishes, circumstantial details, no esoteric words other than the Latin names for plants, and few precise quantitative expressions or grammatical constructions departing from common practice. The typical modern scientific article is the exact opposite.

Since the days of Lister, the scientific article has evolved a specialized language and prose style adapted for efficient communication to other professionals. The deliberately impersonal style is designed to focus the reader's mind on the things and processes of the laboratory and the natural world. Technical nouns and complex noun phrases do most of the work. Phrases like "the results of RNA analyses" and "the specific activity of this RNA" appear in sentences from which words without scientific meaning have been, as it were, wrung out. These noun phrases contain the action. Their verbs pale by comparison: "is," "show," and "indicate" are typical. One seldom finds emotional outbursts, exaggeration, or openly literary language. The emphasis is on detailed methodological descriptions, quantitative comparisons, and precise discrimination.

In the contemporary scientific article, so great an emphasis is placed on exactness of expression that even the mildest literary inventiveness is viewed with suspicion by editors, referees, and many scientist-readers. In a 1962 paper in *Physical Review* on particle physics, Steven Weinberg wanted to include an epigraph from *King Lear*: "Nothing will come of nothing: speak again." According to Weinberg, the editors "protected the purity of the physics literature, and removed the quote."

The norms of modern scientific writing constitute the subject of the first section in this chapter; our extracts are meant to be typical of current practices in English. It is worth noting that the most significant journals of science, no matter their nationality, now publish in what linguist M. A. K. Halliday has called "scientific English," which involves not only a specific language but also a suite of stylistic features. At no time in the long history of the scientific article has a single mode of expression so dominated the literature.

The rest of this chapter presents deviations from the norms. Some are included solely to reveal a human side of science. Others are meant to shed light on the norms, much like Marchant's "monsters" mentioned in chapter 2. What we are privy to in these isolated examples is not in any way science as usual, but science on holiday. Herein you will find scientists at play. You will find elaborate puns and startling metaphors, humor used to good effect, a lonely fisherman hidden in a technical illustration, a whole article written in verse, and even a chemical article in the form of a song. In addition to a childlike playfulness, you will find real belligerence: scientists giving vent to their emotions, getting their dander up, exhibiting genuine anger in print. Finally, you will find what you might least expect to find: the gracefulness of expression we expect mainly of poets and novelists.

NORMS

Cautious Daring

W. Baade and F. Zwicky, 1934. "Supernovae and cosmic rays." *Physical Review*, Vol. 45, p. 138.

Frequently, scientist-authors must resort to assertions about which they are not 100 percent certain. Stylistically, they signal their degree of uncertainty by the use of hedges as in "we wish *to suggest the possibility* that . . ." Of course, hedging does not mean any statement goes as long as it is qualified in some way. Should some important statement prove grossly in error, no matter how much it is hedged, authors know their reputations will suffer. Hedging does serve the readers, however, by helping them gauge the authors' epistemic commitment.

Its function is to communicate doubt within the boundaries of a prose where the absence of the personal conveys the impression of authority and neutrality.

Within the confines of this short article by Walter Baade and Fritz Zwicky, quoted in full, we witness scientific reasoning at its most daring and speculative. For that reason, the authors hedge throughout (indicated by boldface type; italics in original):

> Supernovae flare up in every stellar system (nebula) once in several centuries. The lifetime of a supernova is about twenty days and its absolute brightness at maximum **may be** as high as Mvis = -14^M. The visible radiation L_v of a supernova is about 10^8 times the radiation of our sun, that is, $L_v = 3.78 \times 10^{41}$ ergs/sec. Calculations indicate that the total radiation, visible and invisible, is of the order $L_\tau = 10^7\ L_v = 3.78 \times 10^{48}$ ergs/sec. The supernova therefore emits during its life a total energy $E_\tau \geq 10^5\ L_v = 3.78 \times 10^{53}$ ergs. If supernovae initially are quite ordinary stars of mass $M < 10^{34}$ g, E_τ/c^2 is of the same order as M itself. In the *supernova* process *mass in bulk is annihilated*. In addition **the hypothesis suggests** itself *that cosmic rays are produced by supernovae.* Assuming that in every nebula one supernova occurs ever thousand years, the intensity of the cosmic rays to be observed on earth **should be** of the order $\sigma = 2 \times 10^{-3}$ erg/cm^{2}sec. The observational values are about $\sigma = 3 \times 10^{-3}$ erg/cm^{2}sec. (Millikan, Regener). **With all reserve we advance the view** that supernovae represent the transitions from ordinary stars into *neutron stars*, which in their final stages consist of extremely closely packed neutrons.

The first sentence is a declarative statement without hedging: "Supernovae [giant star that has imploded] flare up in every stellar system (nebula) once in several centuries." Behind that assertion lies the authors' estimate based on the fifteen supernovae reported in the astronomical literature, the first of which was from Tycho Brahe in 1572. The data in the next two sentences represent estimates for the properties of a "typical" supernova based on reasonable assumptions from the limited data at hand.

Baade and Zwicky then calculate the total energy emitted by a supernova during its estimated insect-like lifetime of only twenty days. This calculation leads them to their "hypothesis," an informed guess: because supernovae are large stars that have collapsed into dense masses composed of neutrons, they should produce cosmic rays detectable on earth. In support of this hypothesis, the authors make a comparison. They calculate the intensity of the cosmic rays based upon their various assumptions, then compare that value with the previously observed value by Millikan and Regener. The two numbers agree remarkably well.

In the final sentence, acknowledging the speculative nature of their reasoning, the authors make a carefully hedged assertion: "With all reserve we advance the view that supernovae represent the transitions from ordinary stars to neutron stars." Even though heavily hedged, that statement would eventually win the authors entry into the history books. We say "eventually" because Baade and Zwicky's accomplishment was not widely recognized until three decades after the initial publication. That is when a Cambridge University graduate student in astronomy, Jocelyn Bell, detected a pulsing celestial body that turned out to be a neutron star, providing the first observational evidence for Baade and Zwicky's imaginative theoretical calculations.

Keeping It Impersonal

E. G. Bligh and W. J. Dyer, 1959. "A rapid method of total lipid extraction and purification." *Canadian Journal of Biochemistry and Physiology*, Vol. 8, pp. 911–17.

Scientist-writers attain the stylistic impersonality they seek mainly by what they leave out or minimize: openly literary language, emotional responses to the events described, and the mention of the individuals responsible for the actions. This last is managed in scientific prose by a preference for the passive over the active voice of the verb. The passive voice is a way that English allows writers to remove explicit mention of the human agent from a sentence. An object, process, or concept then appears in the subject position. Modern scientific prose relies heavily upon verbs in the passive with the human actor omitted:

Another experiment *was done* to verify the earlier result.

Revising the above example so that it has an active verb, we get

To verify the earlier result, we *did* another experiment.

In the selection below, nearly all the verbs are passive voice (indicated by bold type).

The authors are describing an important new method for extracting and purifying lipids from animals for biomedical research purposes. Other methods already existed for the same purpose, but this one was faster and easier and, therefore, became the method of choice in many biochemistry labs throughout the world. In this typical methods presentation, we have cookbook actions—mostly combining carefully measured proportions, blending, and separating—but not cookbook style, a sequence of commands in the active voice. Instead, we have a series of sentences that differ from the ordinary in that their grammatical subjects are consistently things or processes rather than people:

> The following procedure applies to tissues like cod muscle that contain 80 ± 1% water and about 1% lipid. Each *100-g sample of the fresh or frozen tissue* **is homogenized** in a Waring Blendor for 2 minutes with a mixture of 100 ml chloroform and 200 ml methanol. To the mixture **is then added** *100 ml chloroform* and after blending for 30 seconds, *100 ml distilled water* **is added** and blending continued for another 30 seconds. The *homogenate* **is filtered** through Whatman No. 1 filter paper on a Coors No. 3 Büchner funnel with slight suction. Filtration is normally quite rapid and when the residue becomes dry, pressure **is applied** with the bottom of a beaker to ensure maximum recovery of solvent. The *filtrate* **is transferred** to a 500-ml graduated cylinder, and, after allowing a few minutes for complete separation and clarification, the *volume of the chloroform layer* (at least 150 ml) **is recorded** and the *alcoholic layer* [is] **removed** by aspiration. A *small volume of the chloroform* **is also removed** to ensure complete removal of the top layer. The chloroform layer contains the purified lipid.

In this passage unmentioned actors perform various actions (boldface) on materials in the laboratory (italics). No specific person is anywhere to be found.

Our own research has shown that scientists used passive voice, on average, about once per hundred words in the seventeenth century. That rate steadily increased over time until it had doubled by the early twentieth century. Thereafter, it remained flat. Still, at over two uses per hundred words, clearly, modern scientists do not practice the advice preached by some writing gurus who vilify this practice. But the scientists are right; those gurus, wrong: science is typically about processes and things, not people. (It is equally wrongheaded to consider the first person pronouns *I* and *we* strictly persona non grata in the scientific article, as some scientists apparently do.)

Keeping It Simple

F. Sanger, G. M. Air, B. G. Barrell, N. L. Brown, A. R. Coulson, J. C. Fiddes, C. A. Hutchinson III, P. M. Slocombe, and M. Smith, 1977. "Nucleotide sequence of bacteriophage ΦX_{174} DNA." *Nature*, Vol. 265, pp. 687–95.

Modern scientific English differs in sentence structure from the standard English you will find in newspapers, magazines, and books for a popular audience. The basic sentence in standard English has three grammatical components:

noun or pronoun → verb → noun or pronoun

Here are two examples:

We [*pronoun*] performed [*verb*] an experiment [*noun*].
An experiment [*noun*] was performed [*verb*] by us [*pronoun*].

Such short and simple sentences seldom appear in modern scientific writing. Instead, one typically finds something more complicated:

complex noun phrase → verb → complex noun phrase

By no means do all scientific sentences have this basic structure, but the majority do. To illustrate, we reproduce an abstract to a paper by Fred Sanger and colleagues reporting the first ever sequencing of an entire genome for a bacteriophage, a virus that infects bacteria:

A **DNA sequence for the genome of bacteriophage** ΦX174 **of approximately 5,375 nucleotides** has been determined using **the rapid and simple 'plus and minus' method**. The sequence identifies **many of the features responsible for the production of the proteins of the nine known genes of the organism, including initiation and termination sites for the proteins and RNAs. Two pairs of genes** are coded by the **same region of DNA** using **different reading frames**.

You can readily see that this passage is composed mostly of complex noun phrases (in bold type). The only exception is the subject to the second sentence, "the sequence," which refers back to the complex noun phrase fronting the first sentence.

Let's examine more closely the first sentence in the example passage:

A DNA sequence for the genome of bacteriophage ΦX$_{174}$ of approximately 5,375 nucleotides [*complex noun phrase*] has been determined [*verb*] using the rapid and simple 'plus and minus' method [*complex noun phrase*].

The authors start with a general phrase that describes their object of study: "a DNA sequence," meaning a sequence of nucleotides derived from a specimen of deoxyribonucleic acid. But there are an infinite number of DNA sequences, so they have to specify which one: "a DNA sequence for the genome of bacteriophage ΦX_{174}." The authors assume their readership of molecular biochemists knows the definitions of these terms already. However, even experts in the field might not know anything much about ΦX_{174}. So they have to be even more specific. To that end, they add its size: "a DNA sequence for the genome of bacteriophage ΦX_{174} of approximately 5,375 nucleotides." In other words, to those in the know, a fairly small and simple genome: a sensible choice for the first ever sequencing of a complete genome. This long noun phrase with multiple modifiers is the subject of its sentence; it has the advantage of also being an extended version of the paper's title ("Nucleotide sequence of bacteriophage ΦX_{174} DNA"). So it serves double duty: not only the subject of its sentence, it is the subject of the entire paper.

The step-by-step progress of science—from one experiment to the next, from theory to experiment, and from experiment to theory—demands a grammatical form equal to tracking this progress with perspicacity. In English, the best way to accomplish this task is to construct two noun phrases with a verb in between to tell us how the second follows from the first, as illustrated by the Sanger et al. passage. "Is this form of language more complex?" linguist M. A. K. Halliday asks. "Not necessarily; it depends how we define complexity. If we take lexical density (the number of words per clause), and the structure of nominal elements (nominal groups and nominalizations), it undoubtedly is more complex. On the other hand, if we consider the intricacy of the sentence structure (the number of clauses in the sentence, and their interdependencies), then it will appear as simpler: mainly one-clause sentences; and likewise with the clausal structure—usually only two or three elements in the clause."

Complexities of Scientific English

J. Guillermo Paez, Psai A. Jänne, Jeffrey C. Lee, Sean Tracy, Heidi Greulich, Stacey Gabriel, Paula Herman, Frederic J. Kaye, Neal Lindeman, Titus J. Boggon, Katsuhiko Naoki, Hidefumi Sasaki, Yoshitaka Fujii, Michael J. Eck, William R. Sellers, Bruce E. Johnson, and Matthew Meyerson, 2004. "EGFR mutations in lung cancer: Correlation with clinical response to gefitinib therapy." *Science*, Vol. 304, pp. 1497–1500.

At the same time that scientific English has taken over as the international language of science, it has evolved a vast terminology efficiently to communicate new knowledge and research practices among specialists. In fact, it is not hard to find passages in which everyday English words are banished with the exception of verbs and connecting words like "prior to" and "because." In such passages, most of the nouns and their modifiers are of a highly technical nature; even everyday words are enlisted in the service of science, words such as "cell line" and "wild type." But the rise of a specialized lexicon is not the only contributor to this complexity. In our view, three other prominent contributors are quantifications, abbreviations, and noun strings. The first confronts the reader with a sea of numbers; the other two make an already information-rich text even more compact.

Quantifications permeate modern scientific prose. Our own analysis showed them to triple in density in sentences between the seventeenth and twentieth centuries. Even that is not the whole story because much data has migrated from the text into tables and graphs. Similarly, technical abbreviations, such as "DNA" for *deoxyribonucleic acid* and "SEM" for *scanning electron microscopy*, have grown over time. Over the last hundred years, their density in sentences from all scientific disciplines has increased sixfold. A third common feature of scientific writing in English is the noun string. Noun strings are possible because English allows you to turn modifying phrases and clauses into sequences of nouns without any connecting elements. As a very simple example, *plasma temperature* is a two-noun string meaning *temperature of the plasma*. Noun strings act as a kind of shorthand, much like technical abbreviations. They are extremely useful in scientific writing and have increased by a factor of about three over the past hundred years. They can also be maddeningly perplexing to the uninitiated, as in an expression like *15-day-old female mouse embryo dorsal root ganglia*. (Translation: the dorsal root of the ganglia of an embryo from a female mouse that is fifteen days old.)

The passage we select to illustrate cognitive and semantic complexity comes from an article in clinical medicine. In this article, we read that Japanese and American patients with lung cancer who also tested positive for a gene called "epidermal growth factor receptor," or "EGFR" for short, responded very favorably to a new designer drug called "gefitnib." And for reasons unknown, Japanese patients responded significantly better than Americans.

We quote a passage where the authors are trying to make sense of their data. Our version defines all undefined abbreviations within brackets, underlines all quantifications, and emphasizes the noun strings with boldface type. We think these modifications dramatize the density of these features in the selection:

> To determine whether mutations in EGFR [epidermal growth factor receptor] confer **gefitinib sensitivity** in vitro, the **mutation status** and response to gefitinib were determined in **four lung adenocarcinoma and bronchioloalveolar carcinoma cell lines**. The H3225 line [defined in an accompanying figure] was originally derived from a malignant pleural effusion from a **Caucasian female nonsmoker** with **lung adenocarcinoma**. This **cell line** was 50 times as sensitive to gefitinib as the other lines, with an IC50

[undefined in the article, meaning "inhibitory concentration 50%"] of 40 nM [nanomolar] for **cell survival** in a <u>72</u>-**hour assay** (Fig. 3A).

Treatment with <u>**100 nM gefitinib**</u> completely inhibited **EGFR autophosphorylation** in <u>H3255</u> (Fig. 3B). Such treatment also inhibited the phosphorylation of **known downstream targets** of EGFR such as the **extracellular signal regulated kinase** <u>1/2</u> (ERK <u>1/2</u>) and **the v-akt murine thymoma viral oncogene homolog** (AKT kinase) (Fig. 3B), a correlation that has been noted by others. In contrast, the **other** <u>three</u> **cell lines** showed comparable levels of inhibition of **target phosphorylation** only when gefitinib was present at concentrations roughly <u>100</u> times as high (Fig. 3B).

The **sequence analysis** of **EGFR cDNA** [complementary deoxyribonucleic acid] in **these** <u>four</u> **cell lines** showed the <u>**L858R mutations**</u> (table S3), whereas the **other** <u>three</u> **cell lines** did not contain **EGFR mutations**. We also confirmed the presence of the <u>**L858R mutation**</u> in the primary tumor from which <u>H3255</u> was derived (table S3, IRG [found in online supplementary material]), although no matched normal tissue was available. The results suggest that <u>**L858R mutant EGFR**</u> is particularly sensitive to inhibition by gefitinib compared with the **wild-type enzyme** and that this likely accounts for the **extraordinary drug sensitivity** of the <u>H3225</u> **cell line**.

As these sample paragraphs illustrate, the authors add to the thick descriptions typical of medical articles in earlier centuries, the quantifications, arcane technical terminology, noun strings, and standard abbreviations prevalent in all types of contemporary scientific article. The undefined abbreviation in the last sentence of the first paragraph above, "IC50," is emblematic. It stands for "the median concentration of a biological substance that causes 50% inhibition." The authors and journal editors no doubt decided this abbreviation did not need to be defined because it had been used so often in the past scientific literature and conversations as to virtually overshadow the actual words it represents, as similarly happened with "DNA" and "RNA." While such terms and grammatical structures illustrated by this selected passage make for more efficient communication, they also obviously make for a prose that is not only intellectually but also verbally dense.

PERTURBATIONS: PLAYFULNESS

The ABCs of the Big Bang

R. A. Alpher, H. Bethe, and G. Gamow, 1948. "The origin of chemical elements." *Physical Review*, Vol. 73, pp. 803–4.

The Origin of Chemical Elements

R. A. ALPHER*

*Applied Physics Laboratory, The Johns Hopkins University,
Silver Spring, Maryland*

AND

H. BETHE

Cornell University, Ithaca, New York

AND

G. GAMOW

The George Washington University, Washington, D. C.

February 18, 1948

FIGURE 7-1 · List of authors in Alpher-Bethe-Gamow
paper. Reprinted with permission of *Physical Review*.
Copyright (1948) American Physical Society.

Before the twentieth century, multiple-authored papers were unusual; now the reverse is true. Because winning credit for one's research is so important to career advancement, standard practice is for the authors' names in the byline to be listed in order of importance to the research project. Any contributor of marginal importance goes in the acknowledgments. The cited article is an amusing exception. It is an early theoretical paper related to the big bang theory, published by coincidence on April Fool's Day of 1948.

The byline lists the authors' names alphabetically, with George Gamow being the senior author, and Ralph Alpher his graduate student. According to Gamow, it seemed an affront to the Greek alphabet to include only him and Alpher as the authors, and thus the well-known theoretical physicist Hans Bethe was listed as the second contributor, although he contributed nothing but his name. This combination of last names allowed this theoretical explanation of the origin of the chemical elements to be called the "$\alpha\beta\gamma$ [alpha-beta-gamma] theory," the first three letters of the Greek alphabet standing in for the beginning of the universe. In the manuscript originally submitted for publication, Gamow wrote "(*in absentia*)" after Bethe's name for the sake of accuracy. Confronted by this peculiar notation, the editor for *Physical Review* sent the manuscript to Bethe for his approval. Reportedly, Bethe read the article, approved its contents, struck out the Latin qualification after his name, and recommended publication. Gamow later quipped that, when a rumor circulated that the $\alpha\beta\gamma$ theory might be cast aside in favor of a different theoretical explanation, Bethe thought about "changing his name to Zacharias."

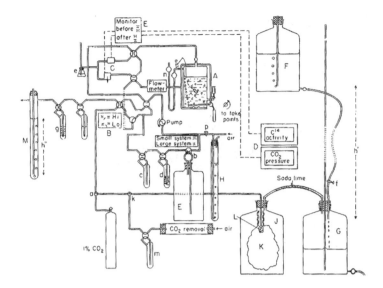

FIGURE 7-2 · Arrangement of apparatus for experiments on
photosynthesis cycle. Reprinted with permission of *Journal of American
Chemical Society.* Copyright (1955) American Chemical Society.

Scientists Gone Fishing

A. T. Wilson and M. Calvin, 1955. "The photosynthetic cycle: CO_2 dependent
transients." *Journal of the American Chemical Society*, Vol. 77, pp. 5948–57.

The experimental configuration shown in the figure above measures changes
that occur in the photosynthetic products from algae suspensions as a function
of carbon dioxide pressure, light, temperature, and other variables. Relying
on their data and on earlier studies, the authors described the complete cycle
of reactions involved in photosynthesis, the process by which plants assimi-
late life-sustaining carbon dioxide using the energy of sunlight. Thus it is no
exaggeration when the authors boast: "Perhaps the most important result of
the work is the insight it gives into the complicated, finely balanced system of
interrelated chemical reactions we call life."

There is nothing amusing about this important result, but those who looked
carefully at flask A—very carefully—saw a stick-figure fisherman catching a fish
(use of a magnifying glass helps). The visual prank aside, this diagram is a good
example of a common visual aid in modern scientific documents—the schematic
of an experimental arrangement. By means of this arrangement of components in a

two-dimensional flowchart, the reader gets a quick overview of just how the experiment works. In this case, the figure also represents the architectural plan for an experiment that would eventually help win a Nobel Prize in chemistry for one of the authors, Melvin Calvin (1961).

Scientific Verse

....

J. F. Bunnett and F. J. Kearley, Jr., 1971. "Comparative mobility of halogens in reactions of dihalobenzenes with potassium amide in ammonia." *Journal of Organic Chemistry*, Vol. 36, pp. 184–86.

Speaking of the birth of potassium nitrate in his long scientific poem *The Botanic Garden* (1791), Erasmus Darwin, the poet and grandfather to Charles, wrote:

> Hence orient Nitre owes its sparkling birth,
> And with prismatic crystals gems the earth,
> O'er tottering domes the filmy foliage crawls,
> Or frosts with branching plumes the mouldering walls;
> As woos Azonic Gas the virgin Air,
> And veils in crimson clouds the yielding Fair,
> Indignant Fire the treacherous courtship flies,
> Waves his light wing, and mingles with the skies.

In a modern revival of this practice, the authors of this article composed it in blank verse. We excerpt the resounding conclusion, ending in a rhymed couplet:

> The haloanilines do not react
> Extensively with excess amide ion
> As shown in Table III. In harmony
> Appears the fact that yields of halide ion
> With surplus amide ion slightly exceed
> One ion from each dihalobenzene molecule
> (Table I). However, *ortho*-iodo
> Substrates afford much more halide ion
> Than can be attributed to subsequent
> Attack on haloanilines that form.
> An unexpected pathway of reaction,
> Unclear in its details, is thus revealed.
> This complication, our thanks to him,
> Is under study by Jhong Kook Kim.

In a footnote on the first page of this contemporary scientific poem, the editor of the *Journal of Organic Chemistry* wisely warned potential authors interested in such literary experimentation:

Although we are open to new styles and formats for publication, we must admit to surprise upon receiving this paper. However, we find the paper to be novel in its chemistry, and readable in its verse. Because of the somewhat increased space requirements and possible difficulty for some of our nonpoetically inclined readers, manuscripts in this format face an uncertain future in this office.

In other words, pending the return of Erasmus Darwin, once for this novelty is enough for the *Journal*.

Ode to Fluorescent Dyes

H. M. Shapiro, 1977. "Fluorescent dyes for different counts by flow cytometry: Does histochemistry tell us much more than cell geometry?" *Journal of Histochemistry and Cytochemistry*, Vol. 25, pp. 976–89.

Another chemical research paper included the musical notation used when it was *sung* in Chicago at the symposium on automation and cytochemistry sponsored by the Histochemical Society. Subsequently, the music appeared as part of the published article:

Despite the flouting of convention, Shapiro organized his article in the standard manner: Introduction, Materials and Methods, Results, Discussion, and Acknowledgments (chapter 6). Two stanzas from Materials and Methods, quoted below, demonstrate his clever interplay between the musical lyric and standard scientific prose, two very different genres. As a consequence, we obtain, in addition to the science, a bit of humorous professional autobiography:

> The first photometer we built employed an arc lamp source,
> And while, electrooptically, it was a tour de force,
> Mechanically intricate, conceptually unique,
> It took three physicists to make it run two days a week.

> We've rectified our prior error (so, to us, it seems),
> By building a photometer with triple laser beams;
> Computerized, much like our first, the gadget can be run
> By any histochemist who can swing a grant for one.

FLUORESCENT DYES FOR DIFFERENTIAL COUNTS BY FLOW CYTOMETRY:
DOES HISTOCHEMISTRY TELL US MUCH MORE THAN CELL GEOMETRY?

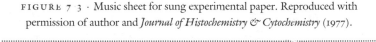

(music © Howard M. Shapiro - used by permission)

FIGURE 7·3 · Music sheet for sung experimental paper. Reproduced with permission of author and *Journal of Histochemistry & Cytochemistry* (1977).

There is one more humorous point to be mentioned. In the Acknowledgments, the author thanks not only colleagues but his mother.

A Thermodynamic Calculation of the Celestial and Infernal States

Anonymous, 1972. "Heaven is hotter than hell." *Applied Optics*, Vol. 11, no. 8, p. A14.

When Galileo defended himself against accusations of heresy for his scientific argument supporting the heliocentric universe, he took the position that the Bible was the authority in matters of faith, but not in matters of science. This is essentially the position of the Church now. But suppose that we took the Bible seriously as a guide to science? This parody of a scientific article gives us one result. In it, the author applies thermodynamics to propositions backed by Biblical authority:

> The temperature of Heaven can be rather accurately computed from available data. Our authority is the Bible, Isaiah 30:26 reads, *Moreover, the light of the Moon shall be as the light of the Sun and the light of the Sun shall be sevenfold, as the light of seven days*. Thus, heaven receives from the moon as much radiation as we do from the sun, and in addition seven times seven (forty nine) times as much as the earth does from the sun, or fifty times in all. The light we receive from the moon is a ten-thousandth of the light we receive from the sun, so we can ignore that. With these data we can compute the temperature of heaven. The radiation falling on heaven will heat it to the point where the heat lost by radiation is just equal to the heat received by radiation. In other words, heaven loses fifty times as much heat as the earth by radiation. Using the Stefan-Boltzmann fourth-power law for radiation
>
> $$(H/E)^4 = 50,$$
>
> where E is the absolute temperature of the earth—$300°C$ ($273 + 27$). This gives H as 798 absolute ($525°C$).
>
> The exact temperature of hell cannot be computed but it must be less than $444.6°C$, the temperature at which brimstone or sulphur changes from a liquid to a gas. Revelations 21:8: *But the fearful, and unbelieving … shall have their part in the lake which burneth with fire and brimstone*. A lake of molten brimstone [sulphur] means that its temperature must be at or below the boiling point, which is $444.6°C$. (Above this point it would be a vapor, not a lake.)
>
> We have then, temperature of heaven, $525°C$ ($977°F$). Temperature of hell, less than $445°C$ ($833°F$). Therefore, heaven is hotter than hell.

We think there is a lesson here on the perils of trying to combine the principles of religion and science.

PERTURBATIONS: BELLIGERENCE

A Cosmologist's Polemic against Cosmology

H. Dingle, 1953. "Science and modern cosmology." *Monthly Notices of the Royal Astronomical Society*, Vol. 113, pp. 393–407.

Not all articles in scientific journals are contributions to scientific knowledge; some are reflections about science, such as this commentary on the state of cosmology in the 1940s and 1950s by the then-president of the Royal Astronomical Society. In it, Herbert Dingle, an astrophysicist turned philosopher of science, speaks out firmly against cosmological theories that, in his opinion, favor theoretical explanations over hard facts:

> The universe we contemplate today is no longer the observable world that for 2000 years was the sole object of astronomical study; it is a hypothetical entity of which what we can observe is an almost negligible part. The assertions we make about it, if susceptible to test at all, require observation over periods of millions of years or in the remote past, and are therefore beyond any sort of practical check. In these circumstances, there is nothing that can control speculation, and preserve legitimate theory from idle fancy, but a strict adherence to the essential principles of science, those principles that in the seventeenth century started the course of ever accelerated progress by which the scientific philosophy is most obviously distinguished from the philosophies that were then its rivals.

Dingle's basic argument is that no scientific claim should be made that cannot at the same time be subjected to rigorous empirical test. In his view, cosmological theorists use mathematics to deduce untestable claims from arbitrary maxims selected by personal taste. Dingle viewed this practice as "pseudo-science" and "unscientific romanticizing"—a step backwards from the empirically based science of earlier centuries.

While critical of modern cosmological models for the universe, Dingle reserves most of his ire for the steady state theory, the main rival to the big bang theory for several decades. In this now discredited theory, the universe has "no vestige of a beginning,—no prospect of an end" (a phrase borrowed from Hutton selection in chapter 4), does not change in appearance at large scales over time, *and* is expanding. At first glance that would seem logically impossible, but the steady state theorists got around that problem by invoking the spontaneous creation of matter out of nothing at a rate "far too low for direct observation." This did not sit at all well with Dingle.

After several pages of castigating the steady state theorists for shoddy reasoning and naïveté about the philosophy of science, he finds himself obligated to justify his deviations from the communal standards of objective prose:

> It is very difficult to describe [steady state theory] objectively without giving the impression of satire, but I have no intention whatever of doing so. I have

the extremest admiration for the single-minded devotion and superlative mathematical skill which are evident throughout the development of this grandiose scheme, but I find it impossible to describe it in the light of the accepted principles of science without making it appear fantastic. And the reason is simply that it is fantastic. Repeated attempts to call attention to what appear as its fallacies have been futile. They have never been answered, but have simply been dismissed as "trivial" or even "frivolous." No criticism has been admitted as valid that is not a criticism of the internal consistency of the scheme. The status of its foundations, and its relevance to the traditional object of scientific enquiry, the world of experience, are held to be idle questions. . . . It has no other basis than the fancy of a few mathematicians who think how nice it would be if the world were made that way.

In his criticism, Dingle does not defend an existing theory or offer a new one. Rather he prescribes a methodological rule: no scientific statement should be made about the natural world "for which there is no evidence."

Dingle's article was reprinted in the leading American journal *Science* and abstracted in the German *Naturwissenschaftliche Rundschau*. We interpret this as a sign that the scientific community at large took Dingle's argument seriously. In retrospect, even through the steady state theory was eventually discarded, Dingle's advice may seem wrongheaded. Given the successes of modern cosmology over the following decades, it is now clear that science is not about to deprive itself of directly undetectable entities or events such as a big bang billions of years ago, spontaneously generated particles, dark matter, quantum gravity, black holes, or whatever, so long as they are useful. Still, we need to remember that the discovery of cosmic microwave background radiation—the deathblow of steady state theory and the apparent confirmation of big bang theory—conforms to Dingle's standard of evidence in favor of a theory.

Extreme Erratum

R. G. Breene, Jr., 1967. "Erratum: Charge exchange between homonuclear diatomic molecules and protons." *Journal of Chemical Physics*, Vol. 47, p. 1882.

The literature of science is as much about wrong turns, aborted lines of inquiry, failure to thrive, and outright failure, as it is about success. Even the greatest of scientists fails now and again, just like the rest of us. Indeed, unless one is selective to the point of unrepresentativeness, it must be admitted that some nontrivial fraction of past science is a collection of dead ends and well-intentioned mistakes, eventually discarded.

Our selection here is a highly unusual erratum. In it, the author acknowledges major flaws in a chemical article he published in the previous year's *Journal of Chemical Physics*. We reproduce the erratum in full. The author is indeed belligerent; his belligerence, however, is self-directed:

> The material contained in this article is nonsense and should be discarded *in toto*. As will be shown elsewhere, one of the most serious defects is the poor transformation given by Eq. (9). It is to be emphasized that the printer set the manuscript precisely as the author had concocted it.

The Rift over Continental Drift

V. V. Beloussov, 1970. "Against the hypothesis of ocean-floor spreading." *Tectonophysics*, Vol. 9, pp. 489–511.

On occasion, scientists combine argument with venom. We have already seen this in the case of Hermann Kolbe and Thomas Huxley (chapter 3). But the passing of the nineteenth century did not mean the passing of scientific polemic entirely. Our selection here is from the conclusion of a twentieth-century article whose sole purpose was to argue against the theory that the movement of the continents was a consequence of the spreading of the ocean floor. This theory was key to the eventual acceptance of Alfred Wegener's much earlier speculation that at one time the earth's surface housed a single supercontinent, which broke apart some 225 million years ago. For decades, Wegener's speculation was dismissed because it lacked a plausible mechanism capable of propelling such large land masses over thousands of miles, or at least a mechanism backed by enough hard evidence to vanquish other contenders (see chapter 5).

That all changed in the 1960s. Now it is believed that convection currents involving the circular movement of molten rock far beneath the earth's surface move the tectonic plates above. As a consequence of this mechanism, new oceanic crust emerges through fissures in ridges on the ocean floor, ancient crust disappears in trenches, and the continents move at the glacial pace of about two to four centimeters per year and are—in the words of J. Tuzo Wilson, a founding father of the ocean-floor spreading theory—"carried about along with the ocean floor like logs frozen in ice."

While for decades a theory believed only by a devoted few, continental drift took root in the 1960s and 1970s when numerous members of the geologic community, both neophytes and veterans, became converts. At the same time, a small group of holdouts led by a distinguished Soviet geophysicist, V. V. Beloussov, and a father-son team from Oklahoma, the Meyerhoffs, fought tenaciously against the rising tide of drift theory and proposed their own alternatives. In the scientific equivalent of Custer's Last Stand, these holdouts at times resorted to an openly confrontational style.

In a fiery counterargument to ocean-floor spreading, for example, Beloussov begins with an abstract composed of a single blunt sentence: "The presently fashionable hypothesis of ocean-floor spreading is examined and conclusion is drawn that this hypothesis is unacceptable." The first paragraph in his conclusion elaborates:

> It is evident that not a single aspect of the ocean-floor spreading hypothesis can stand up to criticism. This hypothesis is based on a hasty generalization of certain data whose significance has been monstrously overestimated. It is replete with distortions of actual phenomenon of nature and with raw statements. It brought into the earth sciences an alien rough schematization permeated by total ignorance of the actual properties of the medium.

It is a common rhetorical tactic to praise one's opponent before, or even in the middle of, damning them: to damn them, as English poet Alexander Pope famously wrote, with "faint praise." We see that at work, for example, in Henry Oldenburg's attack on the French regarding priority for blood transfusion (chapter 1). A more recent example in this chapter is Herbert Dingle, who notes his admiration for the "single-minded devotion and superlative mathematical skill" of his cosmological rivals. So great is Beloussov's contempt and dismay, however, that he cannot summon up even faint praise. Our next chapter treats in more depth the harsh exchanges in two other twentieth-century controversies.

PERTURBATIONS: WRITING WITH STYLE

The Science of Sunrise

William Thomson (Lord Kelvin), 1899. "Blue ray of sunrise over Mont Blanc." *Nature*, Vol. 60, p. 411.

Today, *Nature* is widely considered one of the most prestigious periodicals in all of science, with a weekly print circulation of fifty thousand. Founded in England in 1869, it started out a much different publication than the highly technical and specialized one we know today. Its first stated objective was "to place before the general public the grand results of Scientific Work and Scientific Discovery, and to use the claims of science to a more general recognition in Education and in Daily Life." Only secondly was it mentioned that the journal would "aid

Scientific men themselves, by giving early information of all advances made in the branch of Natural knowledge throughout the world." The journal *Science* in the United States has followed a similar evolutionary path.

An example of an article capturing both original objectives is Lord Kelvin's 1899 meditation on the physical properties of light and its perception by humans. In his mid-seventies, about to retire after fifty-three years at Glasgow University, he took what we assume was a summer vacation in the Alps. But like other highly driven individuals, he could not completely forsake his profession even while on holiday. So he rose before dawn to make solar observations. Fortunately for us he wrote them down in a wonderful piece that is a combination of a prose poem and scientific report:

> Hotel du Mont-Revard, August 27, 1899
>
> LOOKING out at 4 o'clock this morning from a balcony of this hotel, 1545 metres above sea-level, and about 68 kilometres W. 18° S. from Mont Blanc, I had a magnificent view of Alpine ranges of Switzerland, Savoy, and Dauphiné; perfectly clear and sharp on the morning twilight sky. This promised me an opportunity for which I had been waiting five or six years; to see the earliest instantaneous light of sunrise through very clear air, and find whether it was perceptibly blue. I therefore resolved to watch an hour till sunrise, and was amply rewarded by all the splendours I saw. Having only vague knowledge of the orientation of the hotel, I could not at first judge whereabouts the sun would rise; but in the course of half an hour rosy tints on each side of the place of strongest twilight showed me that it would be visible from the balcony; and I was helped to this conclusion by Haidinger's brushes when the illumination of the air at greater altitudes by a brilliant half-moon nearly overhead, was overpowered by sunlight streaming upwards from beyond the mountains. A little later, beams of sunlight and shadows of distant mountains converged clearly to a point deep under the very summit of Mont Blanc. In the course of five or ten minutes I was able to watch the point of convergence travelling obliquely upwards till in an instant I saw a blue light against the sky on the southern profile of Mont Blanc; which, in less than one-twentieth of a second became dazzlingly white, like a brilliant electric arc-light. I had no dark glass at hand, so I could not any longer watch the rising sun.

This paragraph is largely free of the technical terminology and quantifications that would come to dominate scientific articles in the twentieth century. It exhibits the novelist's flair for language: you are there with Lord Kelvin on that balcony on that summer morning. The order of the sentences is exactly coordinated with the order of his perceptions. Of course, a novelist would ordinarily omit the details evident in the first sentence.

You might wonder about the technical term "Haidinger's brushes" (we did anyway). According to a lecture on optics Kelvin delivered to the Academy of Music, Philadelphia, in 1884: "The discoverer is well known in Philadelphia as a mineralogist, and the phenomenon I speak of goes by his name. Look at the sky in a direction of ninety degrees from the sun, and you will see a yellow and blue cross, with the yellow toward the sun, and from the sun, spreading out like two foxes' tails with blue between, and then two red brushes in the space at right angles to the blue. If you do not see it, it is because your eyes are not sensitive enough, but a little training will give them the needed sensitiveness." These brush-like patterns also appear when a viewer looks at a bright surface through a polarizer.

During the twentieth century both *Nature* and *Science* have emerged as the flagship journals for the worldwide community of scientists. To some extent, the wheel has come full circle: we have experienced the return of the highly influential general science periodical. And they still fulfill the two founding objectives stated above: printing both news articles for a general audience, and specialized scientific articles over a wide range of scientific disciplines. But over the last century, their emphasis has turned from the former to the latter.

Imagine That: The Power of Repetition

Hugh M. Smith, 1935. "Synchronous flashing of fireflies." *Science*, Vol. 82, pp. 151–52.

A favorite axiom of writing guides is "Be brief." Another is "Avoid needless repetition." This selection violates both principles, to good effect. It describes the author's nocturnal observation of thousands upon thousands of fireflies sitting in a line of mangrove trees (*Sonneratia acida*) along a river and flashing in perfect unison "hour after hour, night after night for weeks or even months, without regard for air currents, air temperature, moisture or any of the other meteorologic conditions which have been stated to influence firefly flashing." Hugh Smith observed this "outstanding zoological phenomenon in a country [Thailand] that abounds in zoological features of great interest." He invites his readers to

Imagine a tree thirty-five to forty feet high thickly covered with small ovate leaves, apparently with a firefly on every leaf and all the fireflies flashing in perfect unison at the rate of about three times in two seconds, the tree being in complete darkness between the flashes. Imagine a dozen such trees standing close together along the river's edge with synchronously flashing fireflies on every leaf. Imagine a tenth of a mile of river front with an unbroken line of Sonneratia trees with fireflies on every leaf flashing in synchronism, the insects on the trees at the ends of the line acting in perfect unison with those between. Then, if one's imagination is sufficiently vivid, he may form some conception of this amazing spectacle. By going out into the river far enough from shore to lose sight of the individual flashes, a person may obtain from a single tree, a group of trees or a long line of trees a weird pulsating mass effect.

By means of a series of imperative sentences ("Imagine . . . ," "Imagine . . . ," "Imagine . . ."), Smith's prose conveys the literary equivalent of the biological oscillations observed visually. Within this parallel structure, nevertheless, this paragraph builds to a brilliant crescendo. The first sentence summons a single flashing mangrove tree. The second, a dozen trees. The third, the many trees bordering a river for a tenth of a mile. In the fourth sentence, the author halts in his analytical description of the scene, calling it simply an "amazing spectacle." After the descriptive ingenuity of the previous sentences, the usually vapid adjective "amazing" acquires a remarkable precision. It is the reaction anyone *must have* on viewing this phenomenon, even just to imagining it. The final sentence is like a long shot in a film: it moves the viewer outward until the individual firefly flashes can no longer be distinguished. At that distance, the author recapitulates his perceptual journey, taking the reader through the three perspectives of the previous sentences: single, dozen, very long line.

Translated into the flat, concise language of modern science, the author might have written a single sentence in place of the whole paragraph:

The fireflies in Siam congregate in *Sonneratia acida* at a density of 2.5/ft and distance of 0.1 mile and flash in perfect unison at a rate of 1–1.5/sec by a mechanism as yet unexplained.

Brevity is not always a blessing.

Lepidoptery and Literature

V. Nabokov, 1942. "Some new or little known *Nearctic Neonympha* (Lepidoptera: Satyridae)." *Psyche: A Journal of Entomology*, Vol. 49, pp. 61–80.

Born into an aristocratic, scholarly family in Czarist Russia, Vladimir Nabokov lived a life of serial exile. In 1919, after the Bolshevik Revolution, he fled the anarchy of St. Petersburg for peaceful student life at Cambridge University. Following his graduation in 1922, he settled in the large Russian émigré community of Berlin, but fled after Hitler came into power. Settling in Paris, he departed for New York in 1940 just as the Nazi troops were about to enter the city. A year later, traveling by car on the way to Stanford University to lecture on Russian literature, he spent several weeks hunting butterflies. It is no wonder diaspora is a major theme in his novels.

Throughout his life, this Russian master of English prose remained a dedicated lepidopterist. Over the next thirteen years he would publish a series of entomological articles that would make him the only major literary artist of the twentieth century to have also made an original contribution to a specialized field of science.

In the selected passage, Nabokov reports having captured what he believed to be a new species of butterfly (later classified as a subspecies). He named it *Neonympha dorothea* in honor of one of his Russian-language students, Dorothy Leuthold, who drove him and his family from New York to California in search of work at Stanford University, with frequent stops for butterfly hunting in May and June 1941. (Nabokov never learned to drive.) The prize catch was made on the Bright Angel Trail of the Grand Canyon in Arizona.

Nabokov memorialized his having become "godfather to an insect and its first describer" in an early English poem of his ("A Discovery," 1943):

> I found it in a legendary land
> all rocks and lavender and tufted grass,
> where it was settled on some sodden sand
> hard by the torrent of a mountain pass.
>
> The features it combines mark it as new
> to science: shape and shade—the special tinge,
> akin to moonlight, tempering its blue,
> the dingy underside, the chequered fringe.
>
> My needles have teased out its sculptured sex;
> corroded tissues could no longer hide
> that priceless mote now dimpling the convex
> and limpid teardrop on a lighted slide.
>
> Smoothly a screw is turned; out of the mist
> two ambered hooks symmetrically slope
> or scales like battledores of amethyst
> cross the charmed circle of the microscope.
>
> I found it, and I named it, being versed
> in taxonomic Latin; thus became

godfather to an insect and its first
describer—and I want no other fame.

Wide open on its pin (though fast asleep),
and safe from creeping relatives and rust,
in the secluded stronghold where we keep
type specimens it will transcend its dust.

Dark pictures, thrones, the stones that pilgrims kiss,
poems that take a thousand years to die
but ape the immortality of this
red label on a little butterfly.

The well-crafted paragraphs that open his scientific article on just such a discovery are also clearly the product of a considerable writerly talent:

The capture in Arizona in June 1941 of what struck me as an undescribed species of *Neonympha* suggested certain investigations, the results of which are given in this paper. A study of about a hundred specimens labelled "*henshawi* Edw.," which I accumulated from different sources, revealed that two pairs of gemmate species [species that spring as buds from an original species], one pair unnamed, the other neglected, occurred in Arizona. Confusion has been due not so much to some chance obscurity in a great entomologist's description 66 years ago [W. H. Edwards, 1876], as to the indifference and consequent lack of precision in regard to this section of *Neonympha* on the part of those who wrote after him. Somehow lepidopterists have never seemed overeager to obtain these delicately ornamented, quickly fading Satyrids that so quaintly combine a boreal-alpine aspect with a tropical-silvan one, the upperside quiet velvet of "browns" being accompanied by an almost Lycaenid glitter on the upper surface. There exists very little information concerning such things as the number of broods, possible seasonal variation, limits of distribution, allied Mexican and Central American forms, haunts, habits and early stages.

What follows is an attempt to set down the peculiarities of these four insects as a tentative basis for further research that would amplify the comparatively meager facts at my disposal.

We would be remiss not to mention that the elegant prose of this passage is *not* characteristic of the bulk of Nabokov's scientific writings. We would have to concur with the opinion of Kurt Johnson and Steve Coates, authors of a book about Nabokov as a lepidopterist: "his most important [scientific] works are extremely difficult to read, in places nearly impenetrable. Aside from a few telltale flashes, there are few signs of the lucid, graceful, charming author loved by so many." This criticism notwithstanding, Nabokov would have been proud to learn that other entomologists have continued to "amplify the comparatively meager facts" then at his disposal.

Good as Gould

S. J. Gould and R. C. Lewontin, 1979. "The spandrels of San Marco and the Panglossian paradigm: A critique of the adaptationist programme." *Proceedings of the Royal Society of London, Series B*, Vol. 205, pp. 581–98.

Most scientific articles are utilitarian. Not this article. It bristles like a startled porcupine, and it had just that effect when it was presented at a two-day meeting sponsored by the venerable Royal Society of London. At that now-legendary meeting, Gould and his coauthor Richard Lewontin launched an attack on those who believe that natural selection is the sole cause of evolutionary change ("the adaptionist paradigm").

Gould himself referred to this article as "an opinion piece" because it is obviously not a typical measurement-driven article or review of the literature. Gould further notes that the opinion piece "has a long history in the evolutionary literature" because many of the research problems in that specialty are, at the core, as much philosophical as scientific. For that reason, evolutionary biology is one of the niches in which trained philosophers have been able to make a real contribution to the progress of science.

But even as an "opinion piece," the Gould-Lewontin selection is an anomaly in that it is not only openly polemical and provocative, but also brazenly literary. Both its title and introduction, for example, move easily from architecture to creative literature to science. In our selection below, taken from the introduction, Gould and Lewontin weave together examples chosen from these three disciplines into an argument whose point is that the "adaptationist programme" fails to explain all evolutionary changes:

> The great central dome of St. Mark's Cathedral in Venice presents in its mosaic design a detailed iconography expressing the mainstays of Christian faith. Three circles of figures radiate out from a central image of Christ: angels, disciples, and virtues. Each circle is divided into quadrants, even though the dome itself is radically symmetrical in structure. Each quadrant meets one of the four spandrels in the arches below the dome. Spandrels— the tapering triangular spaces formed by the intersection of two rounded

arches at right angles—are necessary architectural by-products of mounting a dome on rounded arches. Each spandrel contains a design admirably fitted into its tapering space. An evangelist sits in the upper part flanked by the heavenly cities. Below, a man representing one of the four Biblical rivers (Tigris, Euphrates, Indus and Nile) pours water from a pitcher into the narrowing space below his feet.

The design is so elaborate, harmonious and purposeful that we are tempted to view it as the starting point of any analysis, as the cause in some sense of the surrounding architecture. But this would invert the proper path of analysis. The system begins with an architectural constraint: the necessary four spandrels and their tapering triangular form. They provide a space in which the mosaicists worked; they set the quadripartite symmetry of the dome above.

Such architectural constraints abound and we find them easy to understand because we do not impose our biological biases upon them. Every fan vaulted ceiling must have a series of open spaces along the mid-line of the vault, where the sides of the fans intersect between the pillars. Since the spaces must exist, they are often used for ingenious ornamental effect. In King's College Chapel in Cambridge, for example, the spaces contain bosses alternately embellished with the Tudor rose and portcullis. In a sense, this design represents an 'adaptation', but the architectural constraint is clearly primary. The spaces arise as a necessary by-product of fan vaulting; their appropriate use is a secondary effect. Anyone who tried to argue that the structure exists because the alteration of rose and portcullis makes so much sense in a Tudor chapel would be inviting the same ridicule that Voltaire heaped on Dr Pangloss: 'Things cannot be other than they are. Everything is made for the best purpose. Our noses were made to carry spectacles, so we have spectacles. Legs were clearly intended for breeches, and we wear them.' Yet evolutionary biologists, in their tendency to focus exclusively on immediate adaptation to local conditions, do tend to ignore architectural constraints and perform just such an inversion of explanation.

In this passage, Gould and Lewontin argue that while the spandrels of St. Mark's Cathedral in Venice are a prominent architectural characteristic, easily discernable and very impressive, it would be a mistake to assume that these are therefore functional, an integral part of the building's architecture. In fact, they maintain that they are only the byproduct of the building of the arches on which the cathedral dome rests. In the opinion of Gould and Lewontin, an analogous sort of reasoning pervades evolutionary biology. It is typical of those who reason in this Panglossian way to assert that, because the chin is a prominent feature of the human face, it must be a product of natural selection, winning out over other, less adaptive features. In fact, the chin is only a byproduct of the formation of the lower jaw. According to Gould and Lewontin, we do not live in the best of all possible Darwinian worlds. (You can view their alternative scientific explanations for evolutionary change at www.aaas.org/spp/dser/evolution/history/spandrel.shtml, a reproduction of the entire article on the web.)

Like all good opinion pieces, this article worked not in the sense of persuading all doubters of the authors' position, but in making a strong enough argument to generate spirited debate on the evolutionary issues at stake. To undermine Gould and Lewontin's thesis, several critics even questioned whether they had used the architectural term "spandrel" correctly. (It would appear they did not.) Gould's own characterization of the critical evolutionary biologists' immediate response was amusingly self-deprecating: "That asshole again! Oh well, at least he can write."

A Feeling for the Genome

Barbara McClintock, 1984. "The significance of responses of the genome to challenge." *Science*, Vol. 226, pp. 792–801; also, http://nobelprize.org/medicine/laureates/1983/mcclintock-lecture.pdf.

On rare occasions, scientists are asked to review their own scientific accomplishments in light of research by others. Winning the Nobel Prize is one such occasion. The selection here is by Barbara McClintock, who won the Nobel Prize for Physiology or Medicine in 1983 for her decades-long experimental research with corn plants. Her work led to many insights into genetic regulatory mechanisms. Here are the eloquent opening paragraphs to the acceptance speech she gave in Sweden, shortly thereafter published in *Science* after some minor editing changes:

> There are "shocks" that a genome must face repeatedly, and for which it is prepared to respond in a programmed manner. Examples are the "heat shock" responses in eukaryotic organisms and the "SOS" responses in bacteria. Each of these initiates a highly programmed sequence of events within the cell that serves to cushion the effects of the shock. Some sensing mechanism must be present in these instances to alert the cell to imminent danger, and to set in motion the orderly sequence of events that will mitigate this danger. But there are also responses of genomes to unanticipated challenges that are not so precisely programmed. The genome is unprepared for these shocks.

Nevertheless, they are sensed, and the genome responds in a discernible but initially unforeseen manner.

An experiment conducted in the mid 1940's prepared me to expect unusual responses of a genome to challenges that the genome is unprepared to meet in an orderly, programmed manner. In most known instances of such challenges, the types of response are not predictable in advance of initial observations of them. Moreover, it is necessary to subject the genome repeatedly to the same challenge in order to observe and appreciate the nature of the changes it induces. Familiar examples are the production of mutation by x-rays and by some mutagenic agents.

It is the purpose of this discussion to consider some observations from my early studies that revealed programmatic responses to threats that are initiated within the genome itself, as well as others similarly initiated, that lead to new and irreversible genomic modifications . . .

Because I became actively involved in the subject of genetics only 21 years after the rediscovery, in 1900, of Mendel's principles of heredity, and at a stage when acceptance of these principles was not general among biologists, I have had the pleasure of witnessing and experiencing the excitement created by revolutionary changes in genetic concepts that have occurred over the past sixty-odd years. I believe we are again experiencing such a revolution. It is altering our concepts of the genome: its component parts, their organizations, mobilities, and their modes of operation. Also, we are now better able to integrate activities of nuclear genomes with those of other components of a cell. Unquestionably, we will emerge from this revolutionary period with modified views of components of cells and how they operate, but only, however, to await the emergence of the next revolutionary phase that again will bring startling changes in concepts.

In the first few paragraphs, McClintock personifies the plant genome; that is, she endows it with typical human qualities. It is shocked, challenged, cushioned from an attack, alerted to danger, taken by surprise, and initiates an unexpected response. These are not the usual drab, neutral, or technical expressions of modern scientific prose.

In the last quoted paragraph, McClintock briefly turns philosophical. She portrays science as a series of revolutionary changes along the lines of Thomas Kuhn's *Structure of Scientific Revolutions*. She also gives a little personal history concerning two such revolutions. Even more revealing is her attitude toward her work as expressed in Evelyn Fox Keller's biography, *A Feeling for the Organism*: "I start with the seedling, and I don't want to leave it. I don't feel I really know the story if I don't watch the plant all the way along. So I know every plant in

the field. I know them intimately, and I find it a great pleasure to know them."
This attitude is well exemplified in another paragraph from her Nobel speech:

> The conclusion seems inescapable that cells are able to sense the presence
> in their nuclei of ruptured ends of chromosomes and then to activate a
> mechanism that will bring together and then unite these ends, one with
> another. And this will occur, regardless of the initial distance in a telophase
> nucleus that separated the ruptured ends. The ability of a cell to sense these
> broken ends, to direct them toward each other, and then to unite them so
> that the union of the two DNA strands is correctly oriented, is a particularly
> revealing example of the sensitivity of cells to all that is going on within
> them. *They make wise decisions and act upon them.*

This is the paragraph as McClintock spoke it in Stockholm. Before the lecture
was published in *Science*, however, someone omitted the closing "pathetic fal-
lacy" we have italicized. Apparently, on second thought at least, biological cells
could be personified, but they ought not be promoted to rational beings, even
metaphorically, at least not in a world-class science periodical.

Handicapping the Origin of Galaxies

P. J. E. Peebles and Joseph Silk, 1990. "A cosmic book of phenomena." *Nature*,
Vol. 346, pp. 233–39.

The best review articles do not simply summarize past literature: in so doing,
they also synthesize and evaluate it. Our selected review does just that: it offers
a statistically based table with which to evaluate competing theories on the formation of the galaxies against the best available observational evidence. But that
table is not our main concern here; rather, it is the article's opening statement,
where Peebles and Silk strive for a grand style absent from the typical scientific
introduction. Here is how they eloquently define what they did, and why:

> There is a tendency at scientific meetings, when a particularly important
> but tentative result is presented, to demand of one's colleagues what odds
> they would give for eventual confirmation. Many bottles of the finest cham-
> pagnes and malt whiskies, and even more esoteric stakes, rest in abeyance
> while observers struggle to count rare photons from remote galaxies or ex-
> perimentalists devote decades to designing new types of detectors. To enrich,
> enlighten and even amuse those of our colleagues who are trying to assess
> the merits of the rival cosmogonies, we have begun a modest programme of

setting up a cosmic book of odds. Our first book focused almost exclusively on the large-scale structure of the Universe. This [present] one is devoted to the observable phenomena that theorists customarily invoke (or ignore) in developing models for the formation of the galaxies. Which observations are reliable? Which observations can be used to discriminate between alternative theories? As issues in extragalactic astronomy rarely are settled in one measurement, an array of measurements . . . may be what finally leads us to a true standard model for the origin of galaxies and the large-scale structure of the Universe.

Peebles and Silk employ a popular strategy for evoking reader interest, one somewhat rare in scientific communications: start with a humorous anecdote. From there they turn to the typical strategy for scientific introductions, as spelled out in our sixth chapter: defining a limited research problem and establishing the gist of the solution. Their solution, a "cosmic book of odds," is a large table of statistics derived from a review of the astronomical literature. The authors conclude from this table that no one theoretical model stands out above the rest. In their quantitatively based judgment, "we would not give very high odds that any of these theories is a useful approximation of how galaxies were actually formed."

In part, this introductory passage works as well as it does because each ascending sentence narrows the audience. The first sentence implicitly addresses all readers of *Nature*; the second, those with a general interest in astronomy; the third and fourth, specialists interested in the theory behind the formation of the universe; and the fifth and beyond, specialists interested in systematically sorting out rival theories on how galaxies formed. This perturbation from the norm makes for a refreshing departure from the usual scientific introduction aimed at specialists alone.

From Theorem to Proof in 365 Years

Andrew Wiles, 1995. "Modular elliptic curves and Fermat's Last Theorem." *Annals of Mathematics*, Vol. 141, pp. 443–551.

Modern novels and scholarly books often begin with an epigraph, a short quotation relevant to the main theme of the work. An epigraph sets the tone for, or comments upon, the text to come. Although rare in scientific or strictly mathematical articles, it is not nonexistent. Indeed, an epigraph heads one of the preeminent mathematical papers of the twentieth century. Below the by-line in the cited article, Princeton mathematician Andrew Wiles quotes the seventeenth-century lawyer and mathematician (Pierre de Fermat) whose theorem, known as Fermat's Last Theorem, or FLT for short, Wiles proved.

This theorem is deceptively simple: there is no integer solution for $x^n + y^n = z^n$ when n is greater than 2. Since $n = 2$ is the Pythagorean theorem taught to every schoolchild, it seemed strange that no larger integer would work. In the seventeenth century, Fermat had tantalizingly written in the margins of a Greek mathematics text that "I have discovered a truly marvelous proof, which this margin is too narrow to contain." He never published his "marvelous proof." The QED to FLT eluded mathematicians, amateurs and professionals alike, for more than 350 years. Their mathematical efforts were not for naught, though. Failed attempts pushed number theory to greater and greater heights, particularly in the twentieth century.

As an epigraph for his long mathematical article, Wiles reproduced Fermat's marginal note in the original Latin. This may well be the easiest statement to decipher in all of Wiles's "truly marvelous proof":

Cubum autem in duos cubos, aut quadratoquadratum in duos quadratoquadratos, et generaliter nullam in infinitum ultra quadratum potestatem in duos ejusdem nominis fas est dividere: cujus rei demonstrationem mirabilem sane detexi. Hanc marginis exiguitas non carperet.

[It is impossible to separate a cube into two cubes, or a biquadrate (fourth degree) into two biquadrates, or in general any power higher than the second into powers of like degree; I have discovered a truly marvelous proof, which this margin is too narrow to contain.]

8

CONTROVERSY
AT WORK:
TWO CASE STUDIES

I N THE PRECEDING CHAPTERS, we concentrated on scientific articles in isolation—that is, separate from any earlier articles that may have motivated them, or later articles that contradicted or built upon them. To better exemplify scientific argument in action, we will briefly depart from our previous narrative approach and examine networks of scientific texts: we will present two cases in which a published scientific report elicited counterarguments in subsequent articles. The first concerns the role of natural selection in biological evolution; the second, the evidential basis of Sigmund Freud's theory of dreams.

To do science is to assert that something is true of the natural world, and to be prepared to defend that claim before a community of peers. In short, to do science is to make arguments and to argue. And nowhere is the argumentative nature of science more apparent than in the give-and-take of a controversy. It is during such episodes that arguments for or against a new knowledge claim are most severely put to the test. One can also find emotional outbursts at odds with the image of the scientist as dispassionate seeker of truth.

Though argument is an integral part of the scientific process, openly confrontational exchanges rarely turn up in scientific articles. Thumb through any volume of a scientific periodical, and you will find nary a disparaging word.

In normal everyday science, authors fashion knowledge claims by constructing a politely worded argument that seeks to establish new facts or explanations. Whether the scientific community at large accepts, rejects, or ignores the new claim, a published exchange between its authors and any critical readers does not normally ensue. And yet, the life of the scientist is not as conflict-free as it would appear from the literature. According to philosopher of science David Hull, "The polemics that make it into print are but the residue of the actual exchanges that go on at meetings, in private correspondence, and in manuscripts before they are sanitized by editors and referees."

On occasion, extended disputes such as the two scrutinized in this chapter do break out in print over what constitutes legitimate scientific facts and plausible explanations for facts. These usually erupt when an individual or segment of the community views an important new knowledge claim as undermining or threatening its own research agenda. In such cases, the rules of the game change in that the main purpose of written communications becomes to defend a claim or to discredit competing claims.

Scientific controversies end, or at least reach a point of temporary stability, in any number of ways. One side may become convinced of the error in its ways because of new evidence or may simply lose interest in continuing the debate. Both sides may agree to disagree and leave it at that, or they may alter their original positions and move toward a common ground. The side with powerful scientist-proponents may prevent lesser-known opponents from receiving a fair hearing. Or a consensus may build in the relevant community as to which side has mounted the better explanation of the available facts. After this, the "losing" side may find itself absent from reference lists, turned down on research grants, and rejected when it sends articles to prestigious scientific journals.

EVOLUTION CONTROVERSY

Moth Wars

R. A. Fisher and E. B. Ford, 1947. "The spread of a gene in natural conditions in a colony of the moth *Panaxia Dominula* L." *Heredity*, Vol. 1, pp. 143–74.

Sewall Wright, 1948. "On the roles of directed and random changes in gene frequency in the genetics of populations." *Evolution*, Vol. 2, pp. 279–94.

R. A. Fisher and E. B. Ford, 1949. "The 'Sewall Wright effect.'" *Heredity*, Vol. 4, pp. 117–19.

Sewall Wright, 1951. "Fisher and Ford on 'The Sewall Wright effect.'" *American Scientist*, Vol. 39, pp. 452–58, 479.

In the late nineteenth century no controversy raged so fiercely as that over Darwin's theory of evolution. Scientists have long since reached a consensus that, in essence, Darwin's theory matches the facts of the fossil and geological record, along with studies of biological organisms in their natural habitats and in controlled environments. Nonetheless, disagreements continue unabated over competing mechanical and mathematical explanations regarding the details of evolution and the role of natural selection.

One of the most prominent of these controversies involved R. A. Fisher and Sewall Wright. A highly regarded statistician from Cambridge University, Fisher developed a mathematical model that grafted Mendel's genes onto Darwinian natural selection. From his model he concluded that very slight selection pressures in large populations could have a major impact on evolution of a species. Fisher's adversary in this controversy, Wright, was a zoologist from the University of Chicago. Extrapolating from laboratory genetics experiments and experience with artificial selection, Wright held that, as a result of chance variations, new mutants could arise in small, partially isolated populations and then spread to the general population. When first proposed in the 1930s, however, neither Fisher's theory emphasizing large populations nor Wright's theory emphasizing small ones possessed much supporting evidence from experimental studies under natural conditions.

In 1939 Fisher teamed up with a colleague from Oxford University, E. B. Ford, to gather the necessary field evidence for testing his evolutionary mechanism against Wright's. Butterfly nets in hand—every summer during World War II, as well as the summers before and after—Fisher and Ford set out to the English countryside in search of *Panaxia dominula*, a large day-flying moth with scarlet hind wings. This moth colony was near at hand in a relatively small and isolated area. The authors chose it for examining evolution in progress because a variant gene had undergone a statistically significant jump in the population for reasons unknown.

Fisher and Ford published their results in a 1947 article in a brand-new biology journal called *Heredity*. They begin with a long narrative that gives enormous presence to the moths being studied—their physical description, habits, and history—along with the authors' methods in capturing, marking, and releasing them. The short segment below will give a flavor of this straight narrative description, more typical of the style in earlier centuries than the twentieth:

> The moth flies actively in the sunshine with a characteristic undulating flight which makes it rather difficult to catch. Often it rises to a considerable height

and may be seen circling the treetops. It also sits on the herbage and is especially attracted by the flowers of Hemp Agrimony, *Eupatorium cannabinum*
L. In warm weather it is, in general, easily disturbed when resting. Towards
evening it usually becomes sedentary, though large numbers may occasionally be seen flying round the tops of certain favourite trees at that time, if the
temperature be high. The insect flies little on dull wet days. It seeks shelter
from rain by hiding under leaves and low-growing herbage, but even then
it is not difficult to find. We have ourselves made large catches in drenching
rain by crawling through the vegetation and boxing the specimens as they
sit. . . .

When the moths fly they show their scarlet hind wings and display a typical
"warning coloration"; but the pattern of the fore wings makes them rather
inconspicuous at rest, particularly on a flower head. We have occasionally
seen imagines [insects at the adult reproductive stage] pursued and caught by
birds, rather more often by dragon-flies, but neither event seems common.
We ourselves do not find them unpleasant to eat, though it could always be
maintained that they are distasteful to predators; certainly they may be less
palatable than other food as easily, or more easily, secured. Some specimens
are unwilling to fly after capture, or when they have spent some time in the
dark. If touched, they will then display the scarlet hind wings and produce
two large drops of clear fluid from the front of the prothorax, just behind
the head. It would be interesting to investigate the glands which give rise to
this substance. To us its flavour is bland, certainly not repellent.

The chief habitats of the moth are river banks and island marshes, though
it is found in one quite exceptional locality on the coast of Kent. In such
situations it usually forms localized colonies, often widely separated from one
another. Within them the insects may be very abundant, though they seldom
stray beyond these confines. In Britain the species is limited to England,
where it is found from Kent to South Devon. Its main area of distribution lies
south of the Thames, though it has been taken as far north as Staffordshire.

One of the authors' figures shows four pairs of moths posed in two columns
and four rows for ease of comparison, much like a typical data table. Taken as
a whole, the eight specimens make apparent the family resemblances necessary
for identification as *P. dominula*. Contrasting one pair with another by the
markings of white and dark spots on the wings reveals important differences.
The top pair represents the typical *P. dominula*; the three pairs below are genetic
deviants. The middle two pairs are the focus of the research project.

The natural world occupies center stage in the article's first half. Then,
an abrupt shift in style occurs. In the text that follows, the moths are meta-

FIGURE 8-1 · Photographs of *Panaxia dominula*.
Reprinted from *Heredity* (1947), with permission
of Nature Publishing Group.

morphosed into tables of data, a few equations, and mathematical matrices. We reproduce one example of the data displays arrayed in support of their argument.

This table is reasonably complex, though the authors do explain its intricacies in their article. How is it to be read? As an example, locate the "5" at the table's center and follow the diagonal upward toward the left to the date, the fourteenth of July, and then downward to the number "46" in the right-hand diagonal row. This means that forty-six specimens were released on the fourteenth. Now follow the diagonal from "5" rightward and upward to the eighteenth of July. This means that of forty-six specimens released on the fourteenth, five were

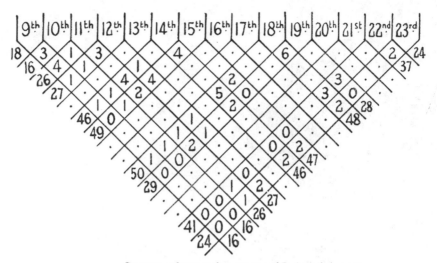

·Captures, releases and recaptures of *P. dominula* in 1943.

FIGURE 8-2 · Schematic of experimental matrix of *Panaxia dominula* study. Reprinted from *Heredity* (1947), with permission of Nature Publishing Group.

recaptured on the eighteenth. This innovative matrix makes visible, at a glance, key aspects of the research method.

Fisher and Ford's main conclusion from their analysis of the accumulated data is that "the observed fluctuations in gene-ratio [in the moth over the eight summers in the study] are much greater than could be ascribed to random survival only. Fluctuations in natural selection (affecting large and small populations equally) must therefore be responsible for them." Furthermore, and more important to our discussion, they wrote, this "fact is fatal to the [Sewall Wright] theory which ascribes particular evolutionary importance to such fluctuations in gene-ratio as may occur by chance in very small isolated populations." And by implication, this "fact" is wholly supportive of Fisher's theory.

A little over a year later Wright's spirited defense appeared in *Evolution: International Journal of Organic Evolution*, also a recently founded journal on biological evolution. Wright opens with his personal philosophy of science, his attempt to define what counts as an argument in evolutionary biology. From there he goes on to stake out his position with respect to Fisher and Ford's frontal assault:

Many writers on evolution have been inclined to ... discuss the subject as if it were a matter of choosing between single factors. My own studies on population genetics have been guided primarily by the belief that a mathematical model must be sought which permits simultaneous consideration of all possible factors. Such a model must be sufficiently simple to permit a rough gasp of the system of interactions as a whole and sufficiently flexible to permit elaboration of aspects of which a more complete account is desired.

On attempting to make such a formulation (Wright, 1931), it was at once apparent that any one of the factors might play the dominating role, at least for a time, under specifiable conditions, but it was concluded that in the long run "evolution as a process of cumulative change depends on a proper balance of the conditions which at each level of organization—gene, chromosome, cell, individual, local race—make for genetic homogeneity or genetic heterogeneity of the species."

The purpose of the present paper is to reiterate this point of view in connection with *certain misapprehensions* which have arisen and in particular to analyze certain data which have been presented recently *by R. A. Fisher and E. B. Ford* (1947) as *invalidating* what *they consider* my point of view (emphasis added).

In the following discussion Wright attempts to undermine the cogency of Fisher and Ford's case by undermining the relevance of their arguments. First, he claims his actual theory has been misrepresented in small but significant ways. This introduces a new issue, a self-reflexive concern, not with the science, but with its representation in texts. By citing, discussing, and quoting his own published writings, Wright makes the case that he never asserted that there was a monocausal explanation for evolution involving random shifts in a small isolated population, but instead viewed this mechanism as one of a number of possible causes of selection. Wright makes three additional points. He claims that Fisher and Ford's findings are of questionable trustworthiness; moreover, even if they are valid, he asserts, his own theory remains viable; finally, he states that Fisher and Ford are not entitled to generalize from a single study of a single species, either in support of their own theory or in opposition to that of Wright.

In contrast to the Fisher-Ford article, only the second half of which is quantitative, Wright's whole article relies heavily upon mathematics: mathematics, not moths, fills the reader's mind. And Wright's quantitatively derived conclusion echoes the philosophy he expounded in the article's first paragraph:

Evolution is a process in which many diverse factors are acting simultaneously.... The fluctuations of some genes are undoubtedly governed largely by violently shifting conditions of selection [Fisher's theory] but for others in the same population, accidents of sampling should be much more important and for still others *both* may play significant roles. It is a question of the *relative values of certain coefficients* (emphasis added).

Wright defines the debate as a contest between his pluralistic explanation for evolution and Fisher and Ford's monocausal one. For Wright, it was not a matter of either his mechanism or Fisher's, but both.

In the next year (1949) Fisher and Ford attempt to undermine the cogency of Wright's counter-arguments. Interestingly, they title their response to Wright's response "The 'Sewall Wright effect,'" with the words *Sewall Wright effect* in quotes. There is more than a hint of sarcasm in the quotation marks, authorizing the reader to infer that this is an effect that exists only in the mind of Sewall Wright. As the title implies, Fisher and Ford yield no ground to Wright. They stand by their eight-year study and do not give an inch: one fact, they say, is

completely fatal to the theory in question, namely that it is not only small isolated populations, but also large populations, that experience fluctuations in gene ratio. . . .

This fact, fatal to "The Sewall Wright Effect," appeared in our own researches from the discovery that the year-to-year changes in the gene ratio in a wild population were considerably greater than could be reasonably ascribed to random sampling, in a population of the size in question. We presumed that random sampling fluctuations must always be present, but that other causes must be acting too, with an intensity, which, even in a population of no more than 1000, seems to be greater than the effects of random sampling . . . This central criticism seems to have escaped Wright's attention, so that in a recent article in *Evolution* he has attributed to us opinions entirely contrary to those which we hold and clearly express in our paper.

In a 1951 *American Scientist* article, Wright has his final say in this episode of the dispute, his last attempt to undermine the arguments of his opponents. Tellingly, he does so in a more public forum: a long-respected journal aimed at the scientific community in general. He begins by mentioning that Fisher and Ford's response completely ignores his contention that the moth study "was highly questionable in its biological premises and in the statistical handling of the data." Wright also complains that his opponents quoted him out of context and willfully misunderstood his intended meaning. An example of such alleged misunderstanding is this skirmish over syntax: "The antecedent of 'this' in the quoted sentence was thus 'intergroup selection' instead of 'nonadaptive radiation' as stated by Fisher and Ford." That's certainly not a sentence one would expect to find in a normal scientific text.

In the Fisher-Wright controversy, we witness an intense battle over what counts as proof in evolutionary biology. Wright's objections to Ford and Fisher's two generalizations—from one gene to all genes, and from one set of experiments to all similar experiments—put into question the cogency of proofs whose experimental basis is so slight. Yet, whatever the weakness of Ford and Fisher's criticism, Wright's own theory still was without a strong experimental foundation. For Fisher and Wright, this stalemate meant that this controversy would very probably not be resolved in their lifetimes, given the extended periods needed to conduct evolutionary studies in nature. As it turns out, while both scientists have long since departed, their theories live on in modified versions. In that sense, both sides "won."

The actual quarrel between Fisher and Wright, the clash of personalities as opposed to ideas, is another matter. Their use of argument from inconsistency and their *ad personam* and *ad hominem* attacks reflect their strong emotional investment. For example, Ford and Fisher's attempt to undermine Wright by accusing him of inconsistency forces both parties to indulge in interpretative arguments more common to literary criticism than science. This tactic turns the debate into a war of words. Wright's texts authorize readers to infer from his indignation the irresponsible ways in which Fisher and Ford have misrepresented his views. The texts of Fisher and Ford, for their part, authorize readers to infer from their open hostility the inadequacy of the case Wright has made in favor of his views.

At the same time, this emotional battle also leads to the intellectual clarification necessary to advance the science. As in Newton's optics controversy in the seventeenth century (chapter 1), the warring parties were forced to think more deeply about their own arguments from the perspective of a critical outsider and shore up any weak links. In particular, Wright appears to have continuously revised his basic argument as the controversy progressed. Moreover, mutual antagonism between the competing parties appears to have actually energized the research field. Although ending the friendship between Fisher and Wright, their antagonism created pressures in favor of more precise and more cogent arguments.

DREAM CONTROVERSY

The Stuff of Dreams: Random Thoughts or Hidden Meanings?

Sigmund Freud, 1922 (1900). *Die Traumdeutung* (Interpretation of Dreams). 3rd ed. F. Deuticke: Leipzig und Wien, pp. 190–91.

Seymour Fisher and Roger P. Greenberg, 1977. *The Scientific Credibility of Freud's Theories and Therapy*. New York: Basic Book, p. 63.

J. Allan Hobson and Robert W. McCarley, 1977. "The brain as a dream state generator: An activation-synthesis hypothesis of the dream process." *American Journal of Psychiatry*, Vol. 134, pp. 1335–48.

Robert W. McCarley and J. Allan Hobson, 1977. "The neurobiological origins of psychoanalytic dream theory." *American Journal of Psychiatry*, Vol. 134, pp. 1211–21.

Anthony L. Labruzza, 1978. "The activation-synthesis hypothesis of dreams: A theoretical note." *American Journal of Psychiatry*, Vol. 135, pp. 1536–38.

Gerald W. Vogel, 1978. "An alternative view of the neurobiology of dreaming." *American Journal of Psychiatry*, Vol. 135, pp. 1531–35.

Gordon G. Globus, 1991. "Dream content: random or meaningful." *Dreaming*, Vol. 1, pp. 27–40.

In 1900, Freud published his masterpiece, *The Interpretation of Dreams*. Before that time, most physicians and scientists viewed dreams as meaningless physiological processes, but no longer. As Freud explained,

> All previous attempts to deal with the problems posed by dreams have been tied directly to what we remember of the manifest content of dreams; they try to extract the meaning of dreams from this content, or if they refrain from attributing any meaning to it, nevertheless ground their judgment with reference to this content. Only we take something else into account: the *latent* content of the dream or the dream-thoughts, revealed only by means of our method. It is from the latent, and not from the manifest content that we discover the meaning of dreams. As a consequence, we are faced with a new task—discovering the relationship between the manifest and latent content of dreams and tracking down the means by which the latter arises from the former.
>
> The manifest and latent content of dreams are really two representations of the same content expressed in two different languages or, to put it more precisely, the content of dreams seems to us like the translation of dream-thoughts into another language, the expressions and laws of combination with which we can become acquainted only through the comparison of the original with the translation. The latent content of dreams can be grasped as soon as we encounter it: the manifest content is presented so to speak in hieroglyphics whose symbols must be translated, one by one, into the language of the latent content. It would clearly be a mistake to read these symbols literally rather than according to their underlying relationships. Suppose I am looking at a picture puzzle or rebus: I see a house with a boat

on top of it, then a single letter, then a running man whose head is missing, and so on. I could now make the mistake of pronouncing this combination and its components nonsense. After all, a boat doesn't belong on top of a house and a man without a head can't run; also, in the rebus the man is bigger than the house and a single letter of the alphabet certainly doesn't belong in a natural landscape. The correct interpretation of the rebus will be revealed to me only if I raise no objection to the whole or to its parts, but try to replace each image with a syllable or word according to whichever relationship it depicts. The combination of words discovered in this way is no longer nonsense, but rather a deeply satisfying and richly ambiguous poetic speech. A dream is like a rebus. Our predecessors in this field of study have made the mistake of judging dreams the way one judges painting. As such, they seemed like nonsense to them, and useless as an indicator of inner turmoil.

Freud advanced a radical new theory founded on the premise that dreams were wish fulfillments. In Freud's vision, the unconscious mind harbors personally disturbing impulses and desires, whose entry into the dreamer's awareness is blocked by a mechanism personified as "the censor." The "dream work" transforms these repressed impulses and wishes into images that will pass by the censor and enter the sleeper's awareness as dreams. Dreams thus operate on two levels: the manifest content—the bizarre distortions resulting from the dream work—and the latent content—what is concealed, the real meaning of dreams. The dream work is functional; it guards sleep by keeping the unsettling wishes and impulses at bay, even in the case of apparent counterinstances such as anxiety dreams and nightmares. On the basis of this theory, Freud developed an interpretive methodology, one that revealed the otherwise concealed inner life of the dreamer and consequently exposed that life to therapeutic treatment.

Freud's landmark theory is specifically psychological. Nevertheless, Freud hoped that the physiology of dreaming, when better understood, would support his psychological theory.

Some hundred years later this particular wish has yet to be fulfilled. In fact, by 1977, when Seymour Fisher and Roger Greenberg assessed the scientific credibility of Freud's theories in detail, his dream theory was found to be largely unsupported by the laboratory and clinical evidence that had accrued over the intervening years. In their view, dreams did not have a latent content distinct from their manifest content, they were by no means all wish fulfillments, they did not have a unique interpretation that could be discovered through psychoanalysis, and their function was not to guard sleep by disguising repressed wishes. Just as important, those portions of Freud's theory that survived this search for corroborating evidence were far from robust. Take, for instance, the function of dreams as a vent for psychological disturbance:

> A central aspect of Freud's dream theory, which asserts that a dream provides outlet for impulses and tensions, seems to be supported by the scientific evidence that can be mustered. To be more exact, the evidence indicates, first of all, that when people are deprived of dream time they show signs of psychological disturbance. Secondly, it indicates that conditions that produce psychological disequilibrium result in increased signs of tension and concern about specific themes in subsequent dreams. One can say these findings are *congruent* with Freud's venting model. But it should be added that they do not specifically document the model.

At the same time as Fisher and Greenberg were demonstrating the seriously flawed character of Freud's theory, two neurobiologists, Allan Hobson and Robert McCarley (1977), were offering an alternative with the same explanatory scope—an alternative that was more compatible with the existing neurological evidence. They introduced their "activation-synthesis" theory in a pair of articles purposely intended to raise a ruckus. As Hobson later confessed in an interview, "I would admit to having created some heat where light might have been more useful, but I can tell you, they [the psychoanalytic community] weren't paying any attention until I turned the heat up a bit."

In the first article (1977), McCarley and Hobson begin with a little history of science, assessing the state of neurobiology in the 1890s, when Freud built his dream theory. In particular, they claim Freud and many of his fellow scientists believed—as McCarley and Hobson do—in a mind-body isomorphism: namely, "that the *same* rules of description and functioning might be applied to both mind and body and that both had the same structures, the same basic operating principles, and the same rules of energy transfer, or dynamics. Whether one talked about physiological or psychological data or concepts seemed to be a matter of convenience and ease, for there were simple rules for mapping the concepts and operations of one domain into the other." Next, by examining a manuscript written by Freud in 1895, "Project for a Scientific Psychology," they link Freud's dream theory with his neurobiological understanding of how the brain works. Finally, they demolish his theory on the grounds that it is based on a grossly inaccurate account of the neurobiological processes involved as they are presently understood:

For the reader whose introduction to Freud has been through the more clin-
ically oriented papers, it is a great shock and revelation to read the "Project"
since the embryos of most of Freud's major theoretical concepts are to be
found here. Twenty years of concepts elaborated in clinical papers are collapsed into a few lines, and to add to the sense of wonder, they appear in a
physiological paper. For the modern neurophysiologist the astonishment is
of a different sort. So many of Freud's ideas about the function of neurons
were simply and fundamentally wrong. We shall show that these erroneous
physiological assumptions had important consequences for the formation of
Freud's psychological theory.

Most important of all, "Freud did not conceive of the possibility of inhibitory
neurons that could *cancel* excitatory signals; in his model energy could only be
diverted or repressed." Not only do McCarley and Hobson claim that Freud's
neurobiological assumptions were faulty, they also claim that dreams—the D
state—occur not during sleep generally, but only during those periods of sleep
characterized by rapid eye movements (better known as REM sleep). While
maintaining a dispassionate veneer, they do not mince words:

> There is also serious difficulty with Freud's theory that the primary motivat-
> ing force for the dream language and dream plot is disguise of a repressed
> wish. The driving force for D sleep is a biologically determined and *moti-
> vationally neutral* activation of cells in the pons [a band of nerve fibers in
> the brain], not a repressed wish. There is no evidence whatsoever that these
> cellular mechanisms of generation are in any way driven by hunger, sex, or
> any other instinct or by repressed wishes for consummation of instinctual
> drives. The *primary motive* for the dream language and dream process cannot
> be disguise if the prime force of dreams is not an instinct or repressed "wish"
> in need of disguise.

In their second article, published a month later in the same journal (1977),
Hobson and McCarley unveil their alternative to Freud, constructed from the
vantage point of nearly eighty years additional research on brain physiology
and dreaming. Their activation-synthesis theory posits the periodic activation
of neuronal dream generators during REM sleep and the simultaneous blocking
of both sensory input and motor output "to account for the maintenance of
sleep in the face of strong central *activation* of the brain" (emphasis added). This
brain stem stimulus is activated "by the perceptual, conceptual, and emotional
structures of the forebrain." This is "primarily a *synthetic* constructive process,

FIGURE 10
Two Models of the Dream Process

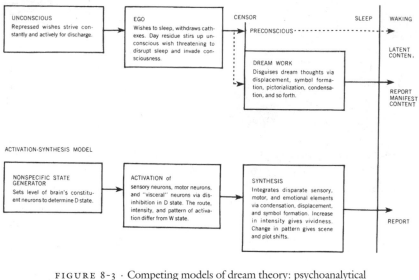

FIGURE 8-3 · Competing models of dream theory: psychoanalytical
and activation synthesis. Reprinted from *American Journal of Psychiatry*.
Copyright (1977) American Psychiatric Association.

rather than a distorting one as Freud presumed" (emphasis added). The partial
coherence of some dreams is merely an artifact of the forebrain doing "the best
of a bad job" in making some sense out of the "relatively noisy" signals transmitted from the brain stem.

The figure above, from the second Hobson-McCarley article, compares
Freud's psychoanalytical model of dreaming with the activation-synthesis
model. The key difference is that the latter requires no censor or mechanism for
disguising the manifest content of dreams. It also calls into doubt the Freudian
schemes for symbolic dream interpretation as part of psychotherapy. According
to Hobson and McCarley's activation-synthesis model, as laid out in this chart,
dreams have no meaning beyond their surface appearance. There can thus be no
hermeneutics of dreams, except in so far as it is a hermeneutics of the perceptual,
conceptual, and emotional activity of the sleeping forebrain. Moreover, dreams
have no psychological role in preserving sleep. (In the diagram, D is the dream
state, W, the waking state.)

Needless to say, the Hobson-McCarley articles did not sit well with the
legions of dream researchers and psychoanalysts who have followed in Freud's

footsteps, and whose very livelihood depended on the credibility of Freud's theory and its accompanying analytical method. Not surprisingly, a flood of indignant articles and letters to the editor soon followed in response.

In our brief analysis of this response, we are not concerned with the status of the activation-synthesis theory within the discipline of neurophysiological psychology. As Gerald Vogel (1978) emphasizes in his defense of the Freudian position, this status concerns the validity of such predictions as the generally bizarre quality of dreams generated by essentially random neural firings during REM sleep. The long-term survival of this theory depends on whether such predictions are borne out and, finally, whether the anomalous findings that dog even the most successful of theories become sufficient to capsize this particular one. Even today, nearly three decades later, that issue remains unresolved.

What we are concerned with is the nature of the Freudian defense: namely, that Freud's theory does not rely upon a mind-body isomorphism as Hobson and McCarley claim, and that they cannot legitimately move, as they routinely do, from psychological explanation to physiological explanation and back as though they were interchangeable.

For Anthony Labruzza (1978), such excursions are "without sound logical justification" and examples of "faulty logic and conceptual unclarity." Whatever is happening in the nervous system during dreaming, he asserts, "cannot be described in the language of dream meaning and motive, nor can the psychological purpose of a dream be explained in terms of the firing of cells of the pons or forebrain. In contrast to Hobson and McCarley, Freud was firm in his resolve to allow 'psychology the right to account for its most common facts . . . by its own means.'" Such confusions lead to "the unwarranted implication that Hobson and McCarley's neurobiological discoveries will logically necessitate major revisions in psychoanalytical dream theory."

In his concurring response, Vogel goes even further, asserting "that purely neurophysical data—such as that used in the activation-synthesis hypothesis— cannot refute a purely psychological theory such as Freud's," since "if [a] direct psychological test reliably supported [a] psychological theory and [a] psychophysiological test did not, we would conclude that the physiological variables were not valid indicators of the psychological variables." Trying to beat Hobson and McCarley at their own game, Vogel also presents published research contradicting their physiological explanations. In particular, he opines that there is no empirical evidence "that dream distortion is caused by a partially random discharge of neuronal units," and that the "kaleidoscopic barrage of unconnected images one would expect" from the activation-synthesis mechanism simply "cannot explain the remarkable coherence of most dreams."

While the Freudians spend much time in trying to poke holes in the activation-synthesis theory, they appear hard pressed to offer corroborating evidence for Freud's theory. Indeed, Vogel freely admits that, after nearly a hundred years, "in the case of Freud's dream theory, a direct psychological test *has yet to be made*" (emphasis added). Barring such a test, the Freudians must fall back upon other argumentative strategies. One form of indirect support offered in Freud's defense is a pragmatic argument of Gordon Globus: "the validity of [a dream] interpretation depends upon its usefulness. Different interpretations of the same dream which are equally fruitful in furthering the patient's under-standing and growth, and in bringing up new material, are equally valid. So there is no canonical interpretation of a dream and different interpretations are differently disclosive."

It is our contention that *both* the Freudian and the activation-synthesis theo-ries *were* good scientific theories—each in its day adapting to the psychological and neurophysiological facts at hand, and doing so with some success. Both also seem to expect to be judged by the same methodological and epistemological standards. In this arena, the Freudian theory appears to have been seriously battered by the neurophysiological facts acquired since its conception. Today, it would appear that Freud's is essentially a discarded theory, except among some psychoanalysts, but "activation-synthesis" has not proved as robust as first thought. Indeed, in an area of interest so abiding, an area of experience so widespread and so mysterious, we are bound to see continuing work and continuing controversy.

9

SELECT MODERN
CLASSICS

THE YEAR 1905 might be considered a turning point in the long history of the scientific article. In that year Albert Einstein—a technical expert in the Swiss patent office freelancing on theoretical physics problems—published four groundbreaking articles in *Annalen der Physik*. They included not only his first two articles on relativity theory, but also one giving Brownian motion a mathematical explanation (a phenomenon vividly described in the Perrin passage, chapter 5), and another making the counterintuitive argument that waves of light also behave as particles. Einstein's stunning quartet opens a century in which nearly all major scientific advances appeared first and foremost in scientific articles, not books: the special and general theories of relativity, quantum mechanics of the micro-universe, the plate tectonic explanation for how the earth's continents formed, the big bang origin of the universe, the double helix of the DNA molecule, and the blueprint for the human genome, to highlight a mere half dozen.

In this chapter, we present and discuss select extracts from twelve classic scientific articles, the oldest being Einstein's 1905 relativity article and the most recent, one of the two 2001 articles mapping the human genome. We make no claim that these are the "top dozen" over the last century, only that they deserve to be called "classics"—that it is impossible to imagine a time in which those interested in the achievements of science will be able to pass them by.

We organize this chapter around related topics: discovering crucial facts, providing theoretical explanations for these facts, extending theory by means of thought experiments, and applying science to technology and technology to science. In the selections for our first section, Thomas Morgan discovers sex-linking in chromosomes, James Watson and Francis Crick, the structure of DNA, and Richard Smalley and his collaborators, a new form of carbon. In our second section, Hermann Muller advocates bold theorizing as the path to understanding the function of the gene in the formation of living organisms, Edwin Hubble calculates that the galaxies are speeding away from us (the more distant, the faster) but shies away from the obvious theoretical explanation, Lise Meitner and Otto Frisch posit nuclear fission to explain anomalous experimental results, and Raymond Davis counts solar neutrinos to explain (or try to explain) a theory on why the sun shines.

In our third section, Albert Einstein and Werner Heisenberg expand theoretical horizons by means of thought experiments: Einstein to establish special relativity and Heisenberg, quantum mechanics. Then in a collaborative thought experiment with Boris Podolsky and Nathan Rosen, Einstein attacks Heisenberg's formulation of quantum mechanics, finding it incoherent, even self-contradictory.

Our fourth section concerns itself with the relationship between science and technology. Because Enrico Fermi turns science into technology, the atomic pile under Stagg Field at the University of Chicago goes critical, and controlled nuclear fission becomes a reality and an atomic bomb, a real possibility. In our last selection, the International Human Genome Sequencing Consortium reverses the process; it applies technology to science, the sequencing of the human genome, a feat impossible to imagine in less technologically sophisticated times.

DISCOVERING CRUCIAL FACTS

Fruitful Fruit Flies

T. H. Morgan, 1910. "Sex limited inheritance in *Drosophila*." *Science*, Vol. 32, pp. 120–22.

Friar Mendel's personal saga is the often-told one of the neglected genius: his published findings on the genetics of peas (chapter 4) received almost no attention from the scientific community at the time. Despite this neglect, his choice of peas turned out to be inspired, revolutionizing the study of inheritance.

Less well known is Mendel's unfortunate misstep: at the urging of the famous botanist Karl Nägeli, Mendel continued his breeding experiments with hawkweed—a plant that, as it turned out, has an unusual reproductive behavior not well suited for such experiments. After several years of laborious research, Mendel found that his hawkweed results did not conform to his earlier pea hybridizations. Thereafter, he turned his considerable talents from the scientific to the religious world. Sad to say, he never knew the wide applicability of his initial scientific work. Neither did the rest of the world, until the rediscovery of his research in the early twentieth century.

Working on genetics in a Columbia University laboratory in 1910, Thomas Morgan made a similar inspired choice of a model organism to Mendel's pea, the common fruit fly (*Drosophila*). Until then primarily known as a nuisance, the lowly fly made for the ideal experimental subject because it had a short life span (about two weeks), was easily cared for and fed, reproduced rapidly, required minimal laboratory space, and possessed only eight chromosomes (by comparison, the human genome has 46; butterflies, 380). The fly was a model organism: results obtained from flies could be generalized to other creatures, including humans. As a consequence, an argument about flies is also an argument about genetics.

Normally, Morgan writes, fruit flies have "brilliant red eyes," but one day a white-eyed male mutant turned up in his "fly room" at Columbia University. By breeding this single white-eyed male with red-eyed females, Morgan demonstrated the genetic inheritance of a single physical trait, eye color in fruit flies. Here is how he summarizes his experiments through three generations of inbreeding—giving prominence to his key results by presenting them in columns separated by white space:

In a pedigree culture of *Drosophila* which had been running for nearly a year through a considerable number of generations, a male appeared with white eyes. The normal flies have brilliant red eyes.

The white-eyed male, bred to his red-eyed sisters, produced 1,237 red-eyed offspring (F_1), and 3 white-eyed males. The occurrence of these three white-eyed males (F_1) (due evidently to further sporting [mutation]) will, in the present communication, be ignored.

The F_1 hybrids, inbred, produced:
2,459 red-eyed females,
1,011 red-eyed males,
782 white-eyed males.

No white-eyed females appeared. The new character showed itself therefore to be sex limited in the sense that it was transmitted only to the grandsons.

But that the character is not incompatible with femaleness is shown by the following experiment.

The white-eyed male (mutant) was later crossed with some of his daughters (F$_1$), and produced:

129 red-eyed females,
132 red-eyed males,
88 white-eyed females,
86 white-eyed males.

The results show that the new character, white eyes, can be carried over to the female by a suitable cross, and is in consequence in this sense not limited to one sex. It will be noted that the four classes of individuals occur in approximately equal numbers (25 per cent).

In the remainder of the article, Morgan presents evidence and arguments linking the gene for eye color to the X chromosome. His year-long research established that genes really do exist and are not just convenient theoretical constructs to explain inheritance, that they pass on physical characteristics to subsequent generations through chromosomes, and that the X chromosome controls eye color in flies.

Dawn of Molecular Biology

J. D. Watson and F. H. C. Crick, 1953. "The structure of DNA." *Cold Spring Harbor Symposia on Quantitative Biology*, Vol. 18, pp. 123–31.

A common misconception is that James Watson and Francis Crick discovered DNA, less well known as deoxyribonucleic acid. They did not. In 1869 Johann Friedrich Miescher first isolated this molecule in white blood cells from human pus. But he never understood its important biological function. In 1944, the team of Oswald Avery, Colin MacLeod, and Macyln McCarty ran an ingenious series of experiments indicating that DNA is the molecule responsible for the transfer of genetic information (see chapter 6). But they did not know anything much about its chemical structure or mechanism of replication. In 1949, Erwin Chargaff reported that the four bases of DNA follow a simple compositional rule: the amount of adenine equals that of thymine, while that of guanine equals that of cytosine (or in shorthand, A = T and G = C). But he also did not know its structure or mechanism of replication. Finally, in 1953, Watson and Crick proposed the double-helical structure for the DNA molecule, along with a simple replication mechanism.

Watson and Crick published four papers on DNA in the same year. At fewer than a thousand words, the first and best known is often praised for its brevity, mistakenly in our view. The implication is that brevity somehow equates with quality. In fact, Watson and Crick wanted to get the word out as quickly as possible on their discovery, so they chose to do so by a "letter" in *Nature*, its relative shortness dictated by the expectations of the genre. It is a beautifully written technical letter to be sure (as are their longer publications)—the authors manage eloquence without resorting to openly literary language, and follow the basic arrangement of parts we outlined in chapter 6.

The initial letter establishes their theoretical structure for this important biological molecule; nothing more. A few months later, they followed this initial publication with another letter in *Nature* describing DNA's mechanism of self-duplication. In the same year they wrote two full-length articles on DNA: one for the *Proceedings of the Royal Society*, another for the *Cold Spring Harbor Symposia on Quantitative Biology*. In this selection from the latter publication, the authors deal with their double-helix DNA structure, but only insofar as it suggests an explanation of self-duplication: the two helical strands unwind, each becoming a stamping machine or template that latches on only to the basic units (monomers) that fit physically (sterically) into the open chain, turning them into new DNA (a polymer):

A Mechanism for DNA Replication

The complementary nature of our structure suggests how it duplicates itself. It is difficult to imagine how like attracts like, and it has been suggested (see Pauling and Delbrück, 1940; Friedrich-Freksa, 1940; and Muller 1947) that self duplication may involve the union of each part with an opposite or complementary part. In these discussions it has generally been suggested that protein and nucleic acid are complementary to each other and that self replication involves the alternate syntheses of these two components. We should like to propose instead that the specificity of DNA self replication is accomplished without recourse to specific protein synthesis and that each of our complementary DNA chains serves as a template or mould for the formation onto itself of a new companion chain.

For this to occur the hydrogen bonds linking the complementary chains must break and the two chains unwind and separate. It seems likely that the single chain (or the relevant part of it) might itself assume the helical form and serve as a mould onto which free nucleotides (strictly polynucleotide precursors) can attach themselves by forming hydrogen bonds. We propose that polymerization of the precursors to form a new chain only occurs if the resulting chain forms the proposed structure. This is plausible because steric reasons would not allow monomers "crystallized" onto the first chain to approach one another in such a way that they could be joined together in a new chain, unless they were those monomers which could fit into our structure. It is not obvious to us whether a special enzyme would be required to carry out the polymerization or whether the existing single helical chain could act effectively as an enzyme.

In these paragraphs, Watson and Crick incorporate by means of citation the views of specific members of a community of researchers, views that are no longer, in their opinion, the best explanation of the facts. Their preferred explanatory candidate is grounded in analogical argument: what looks like a template acts like one. Since there is no direct evidence of the phenomenon in question, the account that follows is carefully nuanced to reflect relative uncertainty: "it seems likely"; "this is plausible". Nuance is carried to perfection when the metaphorical predicate "crystallized" is surrounded by quotes. The tentative manner in which Watson and Crick present their claim is meant throughout to imply their care as scientists. Indeed, the section that follows is entitled "Difficulties in the Replication Scheme."

In this passage, Watson and Crick make sure that their choice of language reflects the degree of certainty they can legitimately claim. In contrast to some of the passages we quoted from the nineteenth century (chapter 3), they avoid direct confrontation by muting their disagreement with their predecessors, and by contextualizing their claim as an alternative to those of others.

What's in a Name?

H. W. Kroto, J. R. Heath, S. C. O'Brien, R. F. Curl, and R. E. Smalley, 1985. "C_{60}: Buckminsterfullerene." *Nature*, Vol. 318, pp. 162–63.

Chemists discover new chemical compounds and structures with at least the same frequency as biologists and geologists find new species of animals, plants, and minerals. Most such discoveries receive little attention from the targeted scientific community, much less the general public. But the finding of a new form of a pure element is a cause for a champagne celebration, particularly if that element is carbon, a constituent of protein, petroleum, paper, coal, the tissue of plants and animals, and even champagne. Before 1985, chemistry textbooks reported only two crystalline materials made up of carbon atoms alone: diamond and graphite. After 1985, the textbooks had to be rewritten. Kroto and his colleagues added a third all-carbon crystalline compound, "buckminsterfullerene," with an "unusually beautiful (and perhaps unique)" structure.

The article opens with a tale of discovery by happy accident. To investigate experimentally the mechanisms by which long chains of carbon form in outer space, the research team decided to vaporize graphite with a laser, carry the condensed products in a helium stream to a mass spectrometer for analysis, and see what turned up. Among various large clusters, they detected a super-stable one made up of sixty carbon atoms. It had the same basic structure as a soccer ball—a hollow sphere composed of interlocking hexagons and pentagons. This same structure appears in the geodesic dome designed by R. Buckminster Fuller for the 1967 Montreal World Exhibition. Hence, the name "buckminsterfullerene" seemed natural.

The final paragraphs of this article address the possible uses of this new molecule, possible fruitful areas of future research, and the long-windedness of its baptismal name:

> If a large-scale synthetic route to this C_{60} species can be found, the chemical and practical value of the substance may prove extremely high. One can readily conceive of C_{60} derivatives of many kinds—such as C_{60} transition metal compounds, for example, $C_{60}Fe$ or halogenated species like $C_{60}F_{60}$ which might be a super-lubricant. We also have evidence that an atom (such as lanthanum and oxygen) can be placed in the interior, producing molecules which may exhibit unusual properties. For example, the chemical shift in the NMR [nuclear atomic resonance, the absorption of electromagnetic radiation] of the central atom should be remarkable because of the ring currents [of the earth's magnetosphere]. If stable in macroscopic, condensed phases, this C_{60} species would provide a topologically novel aromatic nucleus for new branches of organic and inorganic chemistry. Finally, this especially stable and symmetrical carbon structure provides a possible catalyst and/or intermediate to be considered in modeling prebiotic chemistry [the transition as a result of which carbon becomes central element in living creatures].
>
> We are disturbed at the number of letters and syllables in the rather fanciful but highly appropriate name we have chosen in the title to refer to this C_{60} species. For such a unique and centrally important molecular structure, a more concise name would be useful. A number of alternatives come to mind (for example, ballene, spherene, soccerene, carbosoccer), but we prefer to let this issue of nomenclature be settled by consensus.

Why did the authors not prefer a shorter name to start with? The attention-getting quality of "buckminsterfullerene" must have proved as irresistible to them as that of "quark" had been to Murray Gell-Mann (see chapter 6). Today, C_{60} is called "buckyball" and "fullerene" for short. Why the concern over a mere name? Because the chance to name a discovery is part of the spoils from making an original discovery, though there is no guarantee the name will stick, as the authors acknowledge.

In his Nobel Lecture, with the hindsight of nearly a decade of further research and development, Richard Smalley put the significance of this chemical discovery into better perspective. First, it gives us an insight into origins: "Carbon has wired within it, as part of its birthright ever since the beginning of this universe, the genius for spontaneously assembling into fullerenes." Second, it opens up new technological vistas: "Essentially, every technology you have ever heard of where electrons move from here to there, has the potential to be revolutionized by the availability of molecular wires made up of carbon [as fullerene or some related form]."

PROVIDING THEORETICAL EXPLANATIONS

Theory as a Guide to Research

H. J. Muller, 1922. "Variation due to change in the individual gene." *American Naturalist*, Vol. 56, pp. 32–50.

Theory can serve many purposes. In the case of Einstein, the purpose of special relativity is to resolve an anomaly in Maxwellian electrodynamics and to offer a consistent explanation that discarded the aether along with Newtonian notions of absolute space and time. In the quoted article Hermann Muller is interested in theory too, but less as an Einsteinian-type explanation than as a spur to formulating new research problems and approaches for their solution. In the passage we have selected, the general problem he identifies is the role of genetics in the formation of living organisms:

> The present paper will be concerned rather with problems, and the possible means of attacking them, than with details of cases and data. The opening up of these new problems is due to the fundamental contribution which genetics has made to cell physiology within the last decade. This contribution, which has so far been scarcely assimilated by the general physiologists themselves, consists in the demonstration that, besides the ordinary proteins, carbohydrates, lipoids, and extractives, of their several types, there are present within the cell *thousands* of distinct substances—the "genes"; these genes exist as ultramicroscopic particles; their influences nevertheless permeate the entire cell, and they play a fundamental rôle in determining the nature of all cell substances, cell structures, and cell activities.

On the basis of theoretical considerations, Muller recognizes that central to that role is the structure of the molecule that constitutes the gene and its mechanism for replication, eventually unraveled by James Watson and Francis Crick three decades later:

> What sort of structure must the gene possess to permit it to mutate in this way? Since through change after change in the gene, this same phenomenon persists, it is evident that it must depend upon some general feature of gene construction—common to all genes—which gives each one a *general* autocatalytic power—a "carte blanche"—to build material of whatever specific sort it itself happens to be composed of. This general principle of gene structure might, on the one hand, mean nothing more than the possession by each gene of some very simple character, such as a particular radical or "side-chain"—alike in them all—which enables each gene to enter into combination with certain highly organized materials in the outer protoplasm, in such a way as to result in the formation, "by" the protoplasm, of more material like this gene which is in combination with it. In that case the gene itself would only initiate and guide the direction of the reaction. On the other hand, the extreme alternative to such a conception has been generally assumed, perhaps gratuitously, in nearly all previous theories concerning hereditary units; this postulates that the chief feature of the autocatalytic mechanism [genes directing their own synthesis] resides in the structure of the genes themselves, and that the outer protoplasm does little more than provide the building material. In either case, the question as to what the general principle of gene construction is, that permits this phenomenon of mutable autocatalysis, is the most fundamental question of genetics.

If this problem is to be solved, Muller urges, it will be through designing experiments based upon theoretical speculation of the kind this article exemplifies. Despite the long-standing conventional wisdom that theory should follow careful experimentation and observation, Muller was unapologetic concerning the use of theory as a guiding force in science. In fact, Muller is assertive about this use; by several decades, he anticipates the modern philosophical insight that all scientific observation is theory-driven. For Muller, the difference between scientists who acknowledge this truth and those who deny it is the difference between the masters of a theory and its slaves:

> Where results are thus meager, all thinking becomes almost equivalent to speculation. But we can not give up thinking on that account, and thereby give up the intellectual incentive to our work. In fact, a wide, unhampered treatment of all possibilities is, in such cases, all the more imperative, in

order that we may direct these labors of ours where they have most chance to count. We must provide eyes for action.

The real trouble comes when speculation masquerades as empirical fact. For those who cry out most loudly against "theories" and "hypotheses"— whether these latter be the chromosome theory, the factorial "hypothesis," the theory of crossing over, or any other—are often the very ones most guilty of stating their results in terms that make illegitimate *implicit* assumptions, which they themselves are scarcely aware of simply because they are opposed to dragging "speculation" into the open. Thus they may finally be led into the worst blunders of all. Let us, then, frankly admit the uncertainty of many of the possibilities we have dealt with, using them as a spur to the real work.

Muller's paper differs from the typical scientific paper in that its purpose is not to present new science, but a strategy for scientific discovery at a particular time in the history of a line of inquiry.

"Proof" That the Universe is Expanding

Edwin Hubble, 1929. "A relation between distance and radial velocity among extra-galactic nebulae." *Proceedings of the National Academy of Sciences*, Vol. 15, pp. 168–73.

The first key observational evidence for an expanding universe and the big bang came from an astronomer, Edwin Hubble, who never fully embraced either interpretation of his data. To see why this was the case tells us something about the relationship between scientific theories and the empirical evidence supporting them.

In the cited article, Hubble analyzed the available observational data from forty-six nebulae and came to a surprising conclusion: these nebulae do not remain stationary or move around randomly, but the farther away a particular nebula, the faster it appears to be moving away from planet earth. In technical terms, the results establish a "linear correlation between distances and velocities." Hubble's main evidence for this relationship is communicated in a line graph with three sets of data from which two lines of central tendency are plotted, a line graph that, as it was finally understood, embodied an argument for the expanding universe.

The solid circles in Hubble's famous graph show "reasonable estimates" for twenty-four nebulae with distances and velocities measured earlier. As another take on the same data, the open circles represent the twenty-four nebulae lumped into nine groups "according to proximity in direction and in distance."

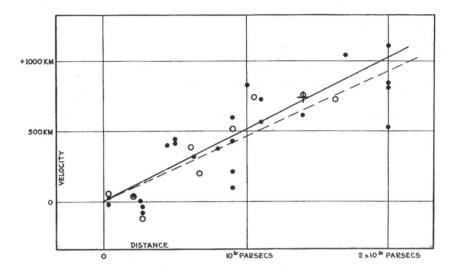

Velocity-Distance Relation among Extra-Galactic Nebulae.

FIGURE 9-1 · Relationship between velocity and distance among
distant galaxies. University of Minnesota Library.

Finally, the single cross represents twenty-two additional nebulae concerning which velocities were available, but not reliable distances. Hubble plots the mean for their velocities versus the mean distance estimated from "apparent" luminosities. The solid line reflects the best fit of the solid circles; the dashed line, the open circles. Along the vertical axis, *KM* is velocity in kilometers per second; along the horizontal axis, *parsecs* is a unit of measurement equivalent to approximately 200,000 times the distance between the earth and the sun.

Clearly, the agreement between the circles and the lines is anything but close. This messy scatter concerned Hubble. As further evidence for the velocity-distance relationship, Hubble then calculated individual distances for twenty-one of the twenty-two nebulae from their known radial velocities and the linear correlation. Agreement with the data from the twenty-four measured nebulae turned out to be excellent. Nonetheless, in his conclusion, Hubble shows extreme caution, understandably. He characterizes the linear velocity-distance relationship as "a first approximation representing a restricted range in distance." What's more, he refuses to speculate on the most obvious interpretation of its significance: "New data to be expected in the near future may modify the significance of the present investigation or, if confirmatory, will lead to a solution having many times the weight. For this reason it is thought premature to discuss in detail the obvious consequences of the present results." Even when the data did later pour in supporting his "first approximation" Hubble resisted the explanation of an expanding universe, instead viewing it as only one of several plausible explanations consistent with theory. In other words, what was retrospectively regarded as proof of an expanding universe was never so regarded by Hubble. Even when his results were confirmed, he persisted in regarding the relationship he discovered as stubbornly empirical. Despite this restraint, the name of Hubble will forever be linked with the expanding universe.

The Frission behind Fission

Lise Meitner and O. R. Frisch, 1939. "Disintegration of uranium by neutrons: A new type of nuclear reaction." *Nature*, Vol. 143, p. 239.

"Our Marie Curie" is how Einstein characterized the brilliant experimentalist and theoretician Lise Meitner, then working in Berlin. Despite the formidable barriers to women pursuing a life in science, Meitner persevered to become one of the premier physicists in a Germany that fielded a group of male physicists, the equivalent of the New York Yankees from the same era.

Working at the Kaiser Wilhelm Institute with radiochemist Otto Hahn and others in the 1930s, Meitner had been experimenting with heavy elements like uranium, bombarding them with uncharged subatomic particles called "neutrons," then examining the byproducts. In the midst of this research (1938), Meitner, having been born Jewish and fearing for her continued well-being, fled Nazi Germany for the safe harbor of Stockholm. Suddenly, in her late fifties, having devoted her life to science with an almost religious fervor, Meitner found herself a refugee—cut off from her friends, family, and research group, not to mention her worldly possessions.

But the practice of theoretical science requires neither passport nor laboratory nor earthly possessions. Soon after her arrival in Stockholm, Meitner received a letter from Hahn reporting a perplexing experimental result: the neutron bombardment of uranium had apparently yielded isotopes of radium, an element with an atomic number significantly less than uranium. The alchemist's dream of the transmutation of one element into another had apparently become a reality. But in terms of nuclear physics, this transmutation did not yet make sense. At a secret meeting in Copenhagen, Meitner urged Hahn to go back to the lab and investigate further. He did so, using different analysis methods.

This time the findings were even more perplexing: the radium was not radium at all, but barium, an element with an atomic number about half that of uranium. This time Hahn had confidence in his experimental results but was more puzzled than ever. Meitner wrote to Hahn his results were "very startling," but not to worry: "in nuclear physics we have experienced so many surprises, that one cannot unconditionally say: it is impossible."

Now collaborating with her nephew Otto Frisch, also in exile, Meitner arrived at a plausible theoretical explanation: the neutrons had caused the uranium to "fission," that is, break into two approximately equal parts, a split accompanied by a release of energy. As an analogue for this fission, Meitner and Frisch applied the liquid-drop model of the nucleus proposed earlier by Niels Bohr: they pictured the uranium nucleus as a liquid drop, which becomes unstable when hit by a neutron, oscillates and vibrates, elongates into a dumbbell shape, then divides into two smaller drops, which fly apart at great speed. The two "drops" are barium and krypton. Borrowing a biological term for cell division, the authors named this process "fission." Here is the passage presenting their theoretical model:

On the basis, however, of present ideas about the behaviour of heavy nuclei, an entirely different and essentially classical picture of these new disintegration processes suggests itself. On account of their close packing and strong energy exchange, the particles in a heavy nucleus would be expected to move in a collective way which has some resemblance to the movement of a liquid drop. If the movement is made sufficiently violent by adding energy, such a drop may divide itself into two smaller drops.

In the discussion of the energies involved in the deformation of nuclei, the concept of surface tension of nuclear matter has been used and its value has been estimated from simple considerations regarding nuclear forces. It must be remembered, however, that the surface tension of a charged droplet is diminished by its charge, and a rough estimate shows that the surface tension of nuclei, decreasing with increasing nuclear charge, may become zero for atomic numbers of the order of 100.

It seems therefore possible that the uranium nucleus has only small stability of form, and may, after neutron capture, divide itself into two nuclei of roughly equal size (the precise ratio of sizes depending on finer structural features and perhaps partly on chance). These two nuclei will repel each other and should gain a total kinetic energy of c. 200 Mev [200 million electron volts], as calculated from nuclear radius and charge. This amount of energy may actually be expected to be available from the difference in packing fraction [the algebraic difference between the relative atomic mass and the total number of protons and electrons in the nucleus] between uranium and the elements in the middle of the periodic system. The whole 'fission' process can thus be described in an essentially classical way, without having to consider quantum-mechanical 'tunnel effects' [the ability of electrons to show up in unexpected places], which would actually be extremely small, on account of the large masses involved.

After division, the high neutron/proton ratio of uranium will tend to readjust itself by beta decay to the lower value suitable for lighter elements. Probably each part will thus give rise to a chain of disintegrations. If one of the parts is an isotope of barium, the other will be krypton (Z [atomic number] = 92 - 56), which might decay through rubidium, strontium and yttrium to zirconium. Perhaps one or two of the supposed barium-lanthanum-cerium chains are then actually strontium-yttrium-zirconium chains.

Bohr's liquid-drop model is a metaphor, linking two otherwise alien worlds: the liquid and the atomic. Meitner and Frisch argue for its scientific fruitfulness, its ability to account for their otherwise anomalous experimental results. What was startling is not less startling, but it is no longer a mystery defying explanation.

Why the Sun Shines

Raymond Davis, Jr., 1964. "Solar neutrinos: II. Experimental." *Physical Review Letters*, Vol. 12, pp. 303–5.

Contemporary science textbooks tell us that the sun shines because of nuclear fusion within its core. Behind that simple assertion lies seventy-five years of painstaking theoretical and experimental research by hundreds of men and women in several disciplines: physics, astronomy, chemistry, engineering, and (in the last several decades anyway) computer science. At the center of this solar research is a subatomic particle called the "neutrino," or "little neutron," a byproduct of the nuclear fusion that has kept the sun shining brightly for billions of years. The neutrino is metaphorically referred to as a "ghost particle" because it has essentially no mass and no charge. So ghostly is the neutrino that it can pass through the earth unimpeded, penetrating from one side to the other at close to the speed of light without interacting with any other matter. And the sun rains neutrinos on the earth at the astonishing rate of 400 billion per second per square inch! If someone were to write a scientific edition of "Ripley's Believe It or Not," neutrinos would deserve a prominent space.

Our selection concerns the first solar neutrino experiment, spearheaded by Raymond Davis at Brookhaven National Laboratory. In retrospect, this experiment was a very successful failure. To understand this paradox, we need some theory. That theory comes from a companion article written by John Bahcall and published in the pages of *Physical Review Letters* immediately preceding Davis's short article. Bahcall succinctly explained the experimental problem facing those looking to test the neutrino theory and hinted at a possible solution based on theoretical calculation. Bahcall's calculated prediction of neutrino capture rate came to $(4 \pm 2) \times 10^{-35}$ per second per ^{37}Cl atom, a number whose significance we will clarify shortly.

This theoretical calculation inspired Davis's preliminary experiment involving 1,000 gallons of cleaning fluid (technically speaking, perchloroethylene, C_2Cl_4) buried in a mine 2,300 feet underground. In this experiment, a solar neutrino passes through the earth's surface, strikes the tank of fluid, reacts with the ^{37}Cl isotope in it, and forms $^{37}Ar[gon]$ isotope. Special devices record and count every such ^{37}Ar atom, and thus allow the experimenters to count solar neutrinos. The experiment is conceptually simple, yet fiendishly difficult to execute. Davis sketches out his experimental procedure and findings in a single long paragraph:

The apparatus consists of two 500-gallon tanks of perchloroethylene, C_2Cl_4, equipped with agitators and an auxiliary system for purging with helium. It is located in a limestone mine 2300 feet below the surface (1800 meters of water equivalent shielding, [or] m.w.e.). Initially, the tanks were swept completely free of air argon by purging the tanks with a stream of helium gas. ^{36}Ar carrier (0.10 cm^3) was introduced and the tanks exposed for periods of four months or more to allow the 35-d[ay] ^{37}Ar activity to reach nearly the saturation value. Carrier argon along with any ^{37}Ar produced were removed from the tanks by sweeping them in series with 5000 liters of helium. Argon was extracted from the helium gas stream with activated charcoal at 78 °K [-319° F]. Finally the argon was desorbed from the charcoal, purified and counted. The overall efficiency of the processing was determined by ^{36}Ar isotopic analysis of the recovered argon. The recovery of carrier argon was always greater than 95%. The entire argon sample placed in a small proportional counter 1.2 cm long and 0.3 cm in diameter to measure the ^{37}Ar activity. Pulse-height analysis was used, and counters were recorded in anticoincidence with a ring of proportional counters, and an enveloping NaI crystal. The counter was provided with an end window to permit exposure of the counting volume to ^{55}Fe x rays for energy calibration and determination of the resolution of the counter. The resolution, full width at half-height for the 2.8-keV Auger electrons from the ^{37}Ar decay, was 26%. The over-all counter efficiency for ^{37}Ar in the full peak was 46%. The counting rate with the sample was 3 counts for 18 days and this is the same as the background rate for the counter filled with air argon. Therefore the observed counting rate of 3 counts for 18 days is probably entirely due to the background activity. However, if one assumes that this rate corresponds to real events and uses the efficiencies mentioned, the upper limit of the neutrino capture rate in 1000 gallons of C_2Cl_4 is ≤ 0.5 per day or $\phi\sigma \leq 3 \times 10^{-34}$ sec^{-1} (^{37}Cl atom)$^{-1}$. From this value, Bachall has set an upper limit on the central temperature of the sun and other relevant information.

In other words, his experiment detected no neutrinos above the natural background, and this null result was a *positive* finding because it did not negate Bahcall's earlier theoretical prediction. In this case, nothing was actually better than something; the possibility still existed that more work along these lines would be fruitful. While Bahcall's and Davis's argument in favor of solar neutrino theory was weak, it was strong enough to make a case for continued experiments on a much grander scale.

Bahcall and Davis successfully made their case; they and others raised the sizable funding needed to perform a more elaborate experiment. This time, in 1968, the experiment did detect some neutrinos above the background. But the authors were disappointed to find that the measured number of neutrinos per day was only about one-third that predicted by Bahcall's theory. Many large-scale experiments and theoretical calculations followed over the next three decades. Yet, the discrepancy persisted. In 2001, however, there was a breakthrough: the Sudbury Neutrino Observatory, which traps solar neutrinos in 1,000 metric tons of heavy water, buried deep underground, announced a possible solution to the puzzle. Art McDonald, the SNO Project Director, announced that "we now have high confidence that the discrepancy is not caused by problems with the models of the Sun but by changes in the neutrinos themselves as they travel from the core of the Sun to Earth."

PERFORMING THOUGHT EXPERIMENTS

Relatively Speaking

....

A. Einstein, 1905. "Zur Elektrodynamik bewegter Körper" (On the electrodynamics of moving bodies). *Annalen der Physik*, Vol. 17, pp. 891–921. Translated by John Stachel in *Einstein's Miraculous Year: Five Papers That Changed the Face of Physics*. Princeton: Princeton University Press (1998), pp. 123–60.

When portraying a theoretical scientist like Einstein, the Madison Avenue and Hollywood image makers tend to show him or her posed deep in thought, chalk in hand, standing before a blackboard full of scribbled equations, as impenetrable to mere mortals as hieroglyphics or atonal music. And there is much truth behind that picture. Indeed, the hundreds of scientific articles written by Einstein attest to it. However, the first quarter of his most famous article—the June 1905 article on special relativity in *Annalen der Physik*—does not. Therein we find only five simple algebraic equations within the first four pages, minimal technical language for a physics paper, and essentially no discussion of the past theoretical or experimental research out of which Einstein's great work sprung.

In his introduction, Einstein sets forth two postulates from which his theory flows. The first is relativity: all physical processes depend upon the frame of reference chosen. A brief example of what this means should suffice. A woman jogging on a train might measure her forward progress as, say, five miles an hour. But a man standing on the ground observing the train pass would clock the jogger's forward motion at sixty-five miles per hour, five for the jogger plus sixty for the train. Galileo had first formulated this basic principle several centuries ago for mechanics. In an extraordinary leap of imagination, Einstein extended relativity to electric, magnetic, and optical phenomena. Einstein's second postulate is more radical. The speed of light in a vacuum is always the same, regardless of the speed of the observer: the speed of light in a spaceship moving at a quarter the speed of light is exactly the same as that for a stationary observer.

These two postulates appear logically inconsistent. Why should the speed of light always be the same while the speed of other material bodies depends on the frame of reference? Einstein brilliantly resolves that paradox in a thought experiment involving the definition of *simultaneity*. As we shall see, a thought experiment is not an experiment at all; it is really an argument.

At first glance, it would appear obvious that simultaneity is simply two events happening at the same time. So, one might ask, what's the problem? In response, Einstein asks two questions of childlike simplicity rooted in the physical world: What is time? And more important, how might one, in principle, go about determining that distant events happened at the same time? Here is Einstein's original explanation:

> If we want to describe the *motion* of a particle, we give the values of its coordinates as functions of time. However, we must keep in mind that a mathematical description of this kind only has physical meaning if we are already clear as to what we understand here by "time." We have to bear in mind that all our judgments involving time are always judgments about

simultaneous events. If, for example, I say that "the train arrives here at 7 o'clock," that means, more or less, "the pointing of the small hand of my watch to 7 and the arrival of the train are simultaneous events."

It might seem that all difficulties involved in the definition of "time" could be overcome by my substituting "position of the small hand of my watch" for "time." Such a definition is indeed sufficient if a time is to be defined exclusively for the place at which the watch is located; but the definition is no longer satisfactory when series of events occurring at different locations have to be linked temporally, or—what amounts to the same thing—when events occurring at places remote from the clock have to be evaluated temporally.

To be sure, we could content ourselves with evaluating the time of events by stationing an observer with a clock at the origin of the coordinates who assigns to an event to be evaluated the corresponding position of the hands of the clock when a light signal from that event reaches him through empty space. However, we know from experience that such a coordination has the drawback of not being independent of the position of the observer with the clock. We reach a far more practical arrangement by the following argument.

If there is a clock at point A in space, then an observer located at A can evaluate the time of events in the immediate vicinity of A by finding the positions of the hands of the clock that are simultaneous with these events. If there is another clock at point B that in all respects resembles the one at A, then the time of events in the immediate vicinity of B can be evaluated by an observer at B. But it is not possible to compare the time of an event at A with one at B without a further stipulation. So far we have only an "A-time" and a "B-time," but not a common "time" for A and B. The latter can now be determined by establishing *by definition* that the "time" required for light to travel from A to B is equal to the "time" it requires to travel from B to A. For, suppose a ray of light leaves from A for B at "A-time" t_A, is reflected from B toward A at "B-time" t_b, and arrives back at A at "A-time" t_A'. The two clocks are synchronous by definition if

$$t_B - t_A = t_A' - t_B$$

We assume that it is possible for this definition of synchronism to be free of contradictions, and to be so for arbitrarily many points, and therefore that the following relations are generally valid:

1. If the clock at B runs synchronously with the clock at A, the clock at A runs synchronously with the clock at B.

2. If the clock at A runs synchronously with the clock at B as well as with the clock at C, then the clocks at B and C also run synchronously relative to each other.

Thus if light leaves at noon (t_A) and arrives at its destination at 12:02 (t_B) and returns at 12:04 (t'_A) precisely, then times 12:00 at A and 12:02 at B are simultaneous by Einstein's definition (12:02 - 12:00 = 12:04 - 12:02). Note that, in this example, points A and B would also be about 36 million kilometers apart because the speed of light is roughly 300,000 kilometers per second.

Armed with that operational definition of simultaneity for clocks at rest, Einstein goes on in the next section to treat the meaning of simultaneity for one clock moving uniformly with respect to another. There he concludes that "we cannot ascribe *absolute* meaning to the concept of simultaneity; instead, two events that are simultaneous when observed from some particular coordinate system can no longer be considered simultaneous when observed from a system that is moving relative to that system." The implication here is that time (and related to that, length too) is not absolute, but relative to the velocity of the observer. The faster one observer moves away from another, the slower the time, and the shorter the length of objects until the observer reaches the speed of light, where time and length vanish. Because of this Alice-in-Wonderland effect, the speed of light is the same for all observers, no matter what their speed with respect to one another. From that foundation, Einstein goes on to rewrite the laws of mechanics, electromagnetism, and optics in the remainder of this landmark article.

Uncertainty in Copenhagen and the Micro-universe

Werner Heisenberg, 1927. "Über den anschaulichen Inhalt der quantentheoretischen Kinematik und Mechanik" (The physical content of quantum kinematics and mechanics). *Zeitschrift für Physik*, Vol. 43, pp. 172–98. Translated by J. A. Wheeler and W. H. Zurek, in *Quantum Theory and Measurement*. Princeton: Princeton University Press (1983), pp. 62–84.

In September 1941, two world-famous physicists met in Nazi-occupied Copenhagen. It was there and then that the great Danish scientist Niels Bohr met with his friend and former protégé Werner Heisenberg, visiting from Nazi Germany. Heisenberg's intention behind visiting Bohr is the subject of a prize-winning and controversial play, *Copenhagen*, by Michael Frayn. Was his purpose to pry from Bohr information on how far advanced the Allies were in developing a fission-based weapon? Was it to coax him into giving into the inevitable and riding the victorious wave of the German juggernaut? Was it to discuss the moral implications of developing such an unimaginably destructive weapon? Or was it simply to visit an old friend to catch up on physics and gossip? Their recollections differ. While Heisenberg recalled that he brushed aside any possibility of practical atomic weapons during the war, Bohr recalled that he was shocked at Heisenberg's report that "under [his] leadership, everything was being done in Germany to develop atomic weapons."

Another topic of the play concerns not politics but physics: the uncertainty principle of the quantum world, first developed by Heisenberg in 1927, when he worked as Bohr's assistant in Copenhagen. In the passage we excerpt, Heisenberg attempts to come to grips with "how microscopic processes can be understood by way of quantum mechanics." It is in connection with this that he developed his uncertainty principle: the more precisely the position of a subatomic particle like an electron is measured, the greater the uncertainty of its momentum, and vice versa. From this simple-sounding principle, Heisenberg concluded that the future of events in the micro-world is fundamentally indeterminate because "even in principle we cannot know the present in all detail. For that reason everything observed is a selection from a plenitude of possibilities and a limitation on what is possible in the future." In the quantum world, uncertainty rules.

The selection presents Heisenberg's demonstration of the uncertainty principle by means of a thought experiment. Heisenberg asked: how can one actually determine the position of an electron with a powerful microscope? To do so, he imagines, requires reflecting a beam of light or radiation off the electron, an action that will disrupt its momentum: the more powerful the microscope, the greater the uncertainty of momentum. Measuring momentum has an analogous effect on position. Similarly, an electron's path within an atom is not precisely calculable: the behavior of individual subatomic particles is unpredictable by definition.

Here is the thought experiment in which Heisenberg formulated his famous principle. It is really, as in the case of Einstein, an argument for the principle:

> When one wants to be clear about what is to be understood by the words "position of the object," for example of the electron (relative to a given frame of reference), then one must specify definite experiments with whose help one plans to measure the "position of the electron"; otherwise this word has no meaning. There is no shortage of such experiments, which in principle even allow one to determine the "position of the electron" with arbitrary accuracy. For example, let one illuminate the electron and observe it under a microscope. Then the highest attainable accuracy in the measurement of position is governed by the wavelength of the light. However, in principle one can build, say, a λ [gamma]-ray microscope and with it carry out the determination of position with as much accuracy as one wants. In this measurement

there is an important feature, the Compton effect. Every observation of scattered light coming from the electron presupposes a photoelectric effect (in the eye, on the photographic plate, in the photocell) and can therefore also be so interpreted that a light quantum hits the electron, is reflected or scattered, and then, once again bent by the lens of the microscope, produces the photoeffect. At the instant when position is determined—therefore, at the moment when the photon is scattered by the electron—the electron undergoes a discontinuous change in momentum. This change is the greater the smaller the wavelength of the light employed—that is, the more exact the determination of the position. At the instant at which the position of the electron is known, its momentum therefore can be known up to magnitudes which correspond to that discontinuous change. Thus, the more precisely the position is determined, the less precisely the momentum is known, and conversely. . . .

We turn now to the concept of "path of the electron." By path we understand a series of points in space (in a given reference system) which the electron takes as "positions" one after the other. As we already know what is to be understood by "position at a definite time," no new difficulties occur here. Nevertheless, it is easy to recognize that, for example, the often used expression, the "1s orbit of the electron in the hydrogen atom [the first and simplest orbital in the Bohr model of the atom with one electron]," from our point of view has no sense. In order to measure this 1s "path" we have to illuminate the atom with light whose wavelength is considerably shorter than 10^{-8} cm. However, a single photon of such light is enough to eject the electron completely from its "path" (so that only a single point of such a path can be defined). Therefore here the word "path" has no definable meaning. This conclusion can already be deduced, without knowledge of the recent theories, simply from the experimental possibilities.

While impressed, Einstein himself firmly believed that another explanation would eventually do away with the "uncertainties" within quantum mechanics. In a letter to Max Born, another architect of quantum mechanics, Einstein famously insisted: "The theory says a lot, but does not really bring us any closer to the secret of the 'old one.' I, at any rate, am convinced that *He* is not playing at dice." In the ensuing decades, an isolated Einstein followed his dream of a unified field theory, to no avail, while a large segment of the theoretical physics community productively tilled the field of quantum mechanics. They did so, however, in the shadow of another thought experiment of Einstein's, one that purported to display an incoherence at the heart of quantum mechanics. For a discussion of this article, read on.

Experimenting in Your Head

A. Einstein, B. Podolsky, and N. Rosen, 1935. "Can quantum-mechanical description really be considered complete?" *Physical Review*, Vol. 47, pp. 777–80.

Quantum mechanics is the physics of the microworld: the atom and its constituents. Among physicists there is general agreement that this microworld is subject to Heisenberg's uncertainty principle: if we measure the momentum of an elementary particle like the electron, we cannot determine without uncertainty the position it held at the time that measurement was made. This limit is built into the microworld: to measure the parameters of this tiny universe is so to disturb this state that our ordinary assumption of an objectivity independent of our measurements, the standing assumption in our macro-world, must be suspended.

EPR—named after the paper's three authors—is a thought experiment, one that takes place entirely in the minds of the authors and their readers. It asks us to imagine a world in which the test of a theory is its completeness, its ability to match its content with the whole content of the material world:

> Any serious consideration of a physical theory must take into account the distinction between the objective reality, which is independent of any theory, and the physical concepts with which the theory operates. These concepts are intended to correspond with the objective reality, and by means of these concepts we picture this reality to ourselves.
>
> In attempting to judge the success of a physical theory, we may ask ourselves two questions: (1) "Is the theory correct?" and (2) "Is the description given by the theory complete?" It is only in the case in which positive answers may be given to both of these questions, that the concepts of the theory may be said to be satisfactory. The correctness of the theory is judged by the degree of agreement between the conclusions of the theory and human experience. This experience, which alone enables us to make inferences about reality, in physics takes the form of experiment and measurement. It is the second question that we wish to consider here, as applied to quantum mechanics.
>
> Whatever the meaning assigned to the term *complete,* the following requirement for a complete theory seems to be a necessary one: *every element of the physical reality must have a counterpart in the physical theory.* We shall call this the condition of completeness. The second question is thus easily answered, as soon as we are able to decide what are the elements of the physical reality.

The elements of the physical reality cannot be determined by *a priori* philosophical considerations, but must be found by an appeal to results of experiments and measurements. A comprehensive definition of reality is, however, unnecessary for our purpose. We shall be satisfied with the following criterion, which we regard as reasonable. *If, without in any way disturbing a system, we can predict with certainty (i.e., with probability equal to unity) the value of a physical quantity, then there exists an element of physical reality corresponding to this physical quantity.* It seems to us that this criterion, while far from exhausting all possible ways of recognizing a physical reality, at least provides us with one such way, whenever the conditions set down in it occur. Regarded not as a necessary, but merely as a sufficient, condition of reality, this criterion is in agreement with classical as well as quantum-mechanical ideas of reality.

In the EPR thought experiment, we are to imagine two particles in the microworld, each with a position and a momentum. They interact then go their separate ways. Then the thinking begins. The measurement of the momentum of particle 1, in accord with Heisenberg's principle, makes it impossible for us to simultaneously determine its position without uncertainty. But EPR show that at any time *after* interaction between the two particles has ceased, if we measure the momentum of particle 1, it is an easy matter to calculate the momentum for particle 2 from that knowledge alone. The same holds for the particle position. This finding is strange indeed. It means that we can know the momentum and position of particle 2 without disturbing its state by measuring those properties for particle 1, followed by a little mathematics. But how can the behavior of particle 1 continue to be linked with that of particle 2 after all interactions between them have stopped? This ESP-like kinship between the particles exists no matter how distant or how long they are apart. In the language of EPR, the momentum and position of particle 2 so determined are "elements of physical reality." But the principles of quantum mechanics do not allow them both to be so. As a consequence, EPR conclude, quantum theory must be incomplete; it cannot be the final word in physics.

Not surprisingly, Niels Bohr—one of the founding fathers of quantum theory—disagreed with the EPR findings. He insisted that EPR were guilty of a serious equivocation in their definition of thinking physically. Quantum mechanics demonstrates that we must think physically in a new way. It teaches us that the observer is part of the system; to think physically about the microworld, therefore, we must include the influence of the observer in our measurements. Whatever the Heisenberg principle entails, whatever observer dependence entails, we must learn to live with it. It is not quantum mechanics that is an incomplete, but EPR's view of what it is to think physically.

In the end, the philosophical differences between Einstein and Bohr were settled, not by more philosophy, but by more physics; they were decided nearly half a century later by a theoretical formulation, Bell's inequality, and a series of experiments demonstrating that the elementary particles of the microworld consistently violated Bell's inequality. In this reconceptualization, the Einstein-Bohr philosophical dilemma proved amenable to experimental resolution. Einstein appears to have been misguided in his criticisms of quantum mechanics. But his thought experiment, his argument in favor of the determinateness of the world beneath the atom, survived for half a century to trouble and enrich physics.

TURNING TO TECHNOLOGY

Dawn of the Nuclear Age

Enrico Fermi, 1946. "The development of the first chain reacting pile." *Proceedings of the American Philosophical Society*, Vol. 90, pp. 20–24.

More than a half century ago Enrico Fermi was standing in the squash courts under the stands at Stagg Field, a football arena abandoned at the end of the 1939 season, the last inglorious year of Division I football at the University of Chicago. It was the morning of December 2, 1942, and the first atomic pile was about to go critical on the south side of Chicago. Fermi led this historic engineering feat. Here is his description of the events of that day and their wider significance:

> On the morning of December 2, 1942, the indications were that the critical dimensions had been slightly exceeded and that the system did not chain react only because of the [neutron] absorption of the cadmium strips. During the morning all the cadmium strips but one were carefully removed; then this last strip was gradually extracted, close watch being kept on the intensity. From the measurements it was expected that the system would become critical by removing a length of about eight feet of this last strip. Actually when about seven feet were removed the intensity rose to a very high value but still stabilized after a few minutes at a finite level. It was with some trepidation that the order was given to remove one more foot and a half of the strip. This operation would bring us over the top. When the foot and a half was

pulled out, the intensity started rising slowly, but at an increasing rate, and kept on increasing until it was evident that it would actually diverge. Then the cadmium strips were again inserted into the structure and the intensity dropped to an insignificant level.

This prototype of a chain reacting unit proved to be exceedingly easy to control. Intensity of its operation could be adjusted with extreme accuracy to any desired level. All the operator has to do is to watch an instrument that indicates the intensity of the reaction and move the cadmium strips in if the intensity shows a tendency to rise, and out if the intensity shows a tendency to drop. To operate a pile is just as easy as to keep a car running on a straight road by adjusting the steering wheel when the car tends to shift right or left. After a few hours of practice an operator can keep easily the intensity of the reaction constant to a very small fraction of 1 percent.

The first pile had no device built in to remove the heat produced by the reaction and it was not provided with any shield to absorb the radiations produced by the fission process. For these reasons it could be operated only at a nominal power which never exceeded two hundred watts. It proved, however, two points: that the chain reaction with graphite and natural uranium was possible, and that it was very easily controllable.

A huge scientific and engineering development was still needed to reduce to industrial practice the new art. Through the collaboration of all the men [and women] of the Metallurgical project and of the Du Pont Company, only about two years after the experimental operation of the first pile large plants based essentially on the same principle were put in operation by the Du Pont Company at Hanford, producing huge amounts of energy and relatively large amounts of the new element, plutonium.

These are the final paragraphs from the first publication about the first controlled chain reaction. Fermi read his paper in Philadelphia on November 17, 1945, before a large audience drawn from the American Philosophical Society and the National Academy of Sciences.

Fermi paints his picture in the typically neutral tone of scientific writing, as if describing an everyday experiment. But while his style understates his content, for anyone with a smattering of knowledge about the historical context, the underlying drama of the situation shines through: "Everybody was conscious early in 1939 of the imminence of a war of annihilation. There was well founded fear that the tremendous military potentialities that were latent in the new scientific developments might be reduced to practice first by the Nazis." Perhaps it was too obvious to also mention that the United States had dropped two atomic weapons on Japan a mere three months before Fermi's talk.

"We Shall Not Cease from Exploration . . ."

International Human Genome Sequencing Consortium, 2001. "Initial sequencing and analysis of the human genome." *Nature*, Vol. 409, pp. 860–921.

For most articles, only time will tell whether or not they will become "classics" in the history of science. Nevertheless, a few reach that status immediately upon publication. We suspect this instant classic also holds the record for what must be the longest article ever published by *Nature* magazine (over sixty pages), where articles seldom exceed four or five pages. Its length is certainly justified by the content. This decoding of the 3.2 billion letters in the DNA of the human genome is the work of twenty laboratories and more than 250 scientists and computer experts around the world.

Why is this achievement important? Unlocking the secrets of the human genome cannot tell us why we are able to write novels and create scientific theories, or to ponder our own mortality, while other sentient organisms cannot, but this information does show us how we are genetically similar to, and different from, other organisms. In the near future this knowledge will be explored and exploited the better to understand the causes of death and disease. (Of course, whether society uses genetic information for good only, or good and ill, remains to be seen.)

Our selection is introductory background information that, first, contextualizes the importance of the article within the history of biological science over the previous century and, second, gives us a history of the international effort to sequence the three billion nucleotides of the human genome:

> The rediscovery of Mendel's laws of heredity in the opening weeks of the 20th century sparked a scientific quest to understand the nature and content of genetic information that has propelled biology for the last hundred years. The scientific progress made falls naturally into four main phases, corresponding roughly to the four quarters of the century. The first established the cellular basis of heredity: the chromosomes. The second defined the molecular basis of heredity: the DNA double helix. The third unlocked the informational basis of heredity, with the discovery of the biological mechanism by which cells read the information contained in genes and with the invention of the recombinant DNA technologies of cloning and sequencing by which scientists can do the same.
>
> The last quarter of a century has been marked by a relentless drive to decipher first genes and then entire genomes, spawning the field of genomics. The fruits of this work already include the genome sequences of 599 viruses

and viroids, 205 naturally occurring plasmids, 185 organelles, 31 eubacteria, seven archaea, one fungus, two animals and one plant.

Here we report the results of a collaboration involving 20 groups from the United States, the United Kingdom, Japan, France, Germany and China to produce a draft sequence of the human genome. The draft genome sequence was generated from a physical map covering more than 96% of the euchromatic part of the human genome and, together with additional sequence in public databases, it covers about 94% of the human genome. The sequence was produced over a relatively short period, with coverage rising from about 10% to more than 90% over roughly fifteen months. The sequence data have been made available without restriction and updated daily [via the World Wide Web] throughout the project. The task ahead is to produce a finished sequence, by closing all gaps and resolving all ambiguities. Already about one billion bases are in final form and the task of bringing the vast majority of the sequence to this standard is now straightforward and should proceed rapidly . . .

The idea of sequencing the entire human genome was first proposed in discussions at scientific meetings organized by the US Department of Energy and others from 1984 to 1986. A committee appointed by the US National Research Council endorsed the concept in its 1988 report, but recommended a broader programme, to include: the creation of genetic, physical and sequence maps of the human genome; parallel efforts in key model organisms such as bacteria, yeast, worms, flies and mice; the development of technology in support of these objectives; and research into the ethical, legal and social issues raised by human genome research. The programme was launched in the US as a joint effort of the Department of Energy and the National Institutes of Health. In other countries, the UK Medical Research Council and the Wellcome Trust supported genomic research in Britain; the Centre d'Etude du Polymorphisme Humain and the French Muscular Dystrophy Association launched mapping efforts in France; government agencies, including the Science and Technology Agency and the Ministry of Education, Science, Sports and Culture supported genomic research efforts in Japan; and the European Community helped to launch several international efforts, notably the programme to sequence the yeast genome. By late 1990, the Human Genome Project had been launched, with the creation of genome centres in these countries. Additional participants subsequently joined the effort, notably in Germany and China. In addition, the Human Genome Organization (HUGO) was founded to provide a forum for international coordination of genomic research. Several books provide a more comprehensive discussion of the genesis of the Human Genome Project.

Through 1995, work progressed rapidly on two fronts... The first was construction of genetic and physical maps of the human and mouse genomes, providing key tools for identification of disease genes and anchoring points for genomic sequence. The second was sequencing of the yeast and worm genomes, as well as targeted regions of mammalian genomes. These projects showed that large-scale sequencing was feasible and developed the two-phase paradigm for genome sequencing. In the first, "shotgun," phase, the genome is divided into appropriately sized segments and each segment is covered to a high degree of redundancy (typically, eight- to tenfold) through the sequencing of randomly selected subfragments. The second is a "finishing" phase, in which sequence gaps are closed and remaining ambiguities are resolved through directed analysis. The results also showed that complete genomic sequence provided information about genes, regulatory regions and chromosome structure that was not readily obtainable from c[complementary]DNA studies alone.

In 1995, genome scientists considered a proposal that would have involved producing a draft genome sequence of the human genome in a first phase and then returning to finish the sequence in a second phase. After vigorous debate, it was decided that such a plan was premature for several reasons. These included the need first to prove that high-quality, long-range finished sequence could be produced from most parts of the complex, repeat-rich human genome; the sense that many aspects of the sequencing process were still rapidly evolving; and the desirability of further decreasing costs.

Instead, pilot projects were launched to demonstrate the feasibility of cost-effective, large-scale sequencing, with a target completion date of March 1999. The projects successfully produced finished sequence with 99.99% accuracy and no gaps. They also introduced bacterial artificial chromosomes (BACs), a new large-insert cloning system that proved to be more stable than the cosmids and yeast artificial chromosomes (YACs) that had been used previously. The pilot projects drove the maturation and convergence of sequencing strategies, while producing 15% of the human genome sequence. With successful completion of this phase, the human genome sequencing effort moved into full-scale production in March 1999.

The idea of first producing a draft genome sequence was revived at this time, both because the ability to finish such a sequence was no longer in doubt and because there was great hunger in the scientific community for human sequence data. In addition, some scientists favoured prioritizing the production of a draft genome sequence over regional finished sequence because of concerns about commercial plans to generate proprietary databases of human sequence that might be subject to undesirable restrictions on use.

The consortium focused on an initial goal of producing, in a first production phase lasting until June 2000, a draft genome sequence covering most of the genome. Such a draft genome sequence, although not completely finished, would rapidly allow investigators to begin to extract most of the information in the human sequence. Experiments showed that sequencing clones covering about 90% of the human genome to a redundancy of about four- to fivefold ("half-shotgun" coverage)... accomplish this. The draft genome sequence goal has been achieved, as described below.

The second sequence production phase is now under way. Its aims are to achieve full-shotgun coverage of the existing clones during 2001, to obtain clones to fill the remaining gaps in the physical map, and to produce a finished sequence (apart from regions that cannot be cloned or sequenced with currently available techniques) no later than 2003.

The consortium's most surprising finding is deflating: the human genome has 30,000–40,000 protein-encoding genes (recently downgraded to about 25,000)—a far cry from the more than 100,000 previously estimated. Indeed, this number is not that much higher than less complex forms of life like fruit flies (13,000), microscopic round worms (19,000), and mustard weeds (26,000).

This sixty-two-page-long marathon of an article ends on a decidedly literary note:

Finally, *it has not escaped our notice that* the more we learn about the human genome, the more there is to explore.

Here, the authors employ a form of understatement called litotes that echoes a provocative closing sentence in Watson and Crick's famous, and much shorter, DNA article from the 1953 issue of *Nature* (see www.nature.com/genomics/human/watson-crick). Our added italics mark the repetition of wording between the two articles. (We eliminated a typographical error in the original.) The word "explore" at the very end of the above quotation provides the impetus for a closing quotation from the final stanza in T. S. Eliot's "Four Quartets":

We shall not cease from exploration.
And the end of all our exploring

Will be to arrive where we started,
And know the place for the first time.

At the beginning of the article, we learn the story behind the international consortium's project. It is presented as the dramatic culmination of an effort that began with the rediscovery of Mendel's laws of heredity. At the end of the article, the allusion to Watson and Crick's DNA article pays an oblique homage to that landmark discovery (Watson also being the genome project's first director), and hints that the present accomplishment may possibly belong in the same general category. The quotation from Eliot reminds us that this is not the end of the story.

Bibliography

....

Fifty Books We Recommend in Science Studies

1. Aldersey-Williams, Hugh. 2005. *Findings: Hidden Stories in First-Hand Accounts of Scientific Discovery*. Norwich, UK: Lulox Books.

2. Anderson, Wilda C. 1984. *Between the Library and the Laboratory: The Language of Chemistry in Eighteenth-Century France*. Baltimore: Johns Hopkins University Press.

3. Bazerman, Charles. 1988. *Shaping Written Knowledge: The Genre and Activity of the Experimental Article in Science*. Madison: University of Wisconsin Press.

4. Berkenkotter, Carol, and Thomas N. Huckin. 1995. *Genre Knowledge in Disciplinary Communication: Cognition, Culture, Power*. Northvale, NJ: L. Erlbaum Associates.

5. Bowler, Peter J. 1984. *Evolution, the History of an Idea*. Berkeley: University of California Press.

6. Brannigan, Augustine. 1981. *The Social Basis of Scientific Discoveries*. Cambridge: Cambridge University Press.

7. Brown, Richard Harvey. 1998. *Toward a Democratic Science: Scientific Narration and Civic Communication*. New Haven: Yale University Press.

8. Ceccarelli, Leah. 2001. *Shaping Science with Rhetoric: The Cases of Dobzhansky, Schrödinger, and Wilson*. Chicago: University of Chicago Press.

9. Condit, Celeste Michelle. 1999. *The Meanings of the Gene: Public Debates about Human Heredity*. Madison: University of Wisconsin Press.

10. Collins, Harry. 2004. *Gravity's Shadow: The Search for Gravitational Waves*. Chicago: University of Chicago Press.

11. Darian, Steven. 2003. *Understanding the Language of Science*. Austin: University of Texas Press.

12. Dear, Peter, ed. 1991. *The Literary Structure of Scientific Argument: Historical Studies*. Philadelphia: University of Pennsylvania Press.

13. Fahnestock, Jeanne. 1999. *Rhetorical Figures in Science*. New York: Oxford University Press.

14. Feyerabend, Paul K. 1975. *Against Method: Outline of an Anarchistic Theory of Knowledge*. London: NLB.

15. Fine, Arthur. 1986. *The Shaky Game: Einstein, Realism, and the Quantum Theory*. Chicago: University of Chicago Press.

16. Fleck, Ludwik. 1979. *Genesis and Development of a Scientific Fact*, translated by Fred Bradley and Thaddeus J. Trenn. Chicago: University of Chicago Press.

17. Galison, Peter Louis. 1987. *How Experiments End*. Chicago: University of Chicago Press.

18. Galison, Peter. 1997. *Image & Logic: A Material Culture of Microphysics*. Chicago: University of Chicago Press.

19. Gilbert, G. Nigel, and Michael Mulkay. 1984. *Opening Pandora's Box: A Sociological Analysis of Scientists' Discourse*. Cambridge: Cambridge University Press.

20. Gross, Alan G. 1996. *The Rhetoric of Science*, 2nd edition. Cambridge: Harvard University Press.

21. Gross, Alan G., Joseph E. Harmon, and Michael Reidy. 2002. *Communicating Science: The Scientific Article from the 17th Century to the Present*. Oxford: Oxford University Press.

22. Gruber, Howard E. 1981. *Darwin on Man: A Psychological Study of Scientific Creativity*, 2nd edition. Chicago: University of Chicago Press.

23. Halliday, M. A. K., and J. R. Martin, eds. 1993. *Writing Science: Literacy and Discursive Power*. Pittsburgh: University of Pittsburgh Press.

24. Harré, Rom. 1981. *Great Scientific Experiments: Twenty Experiments That Changed Our View of the World*. Oxford: Oxford University Press.

25. Heidegger, Martin. 1977. *The Question Concerning Technology, and Other Essays*, translated by William Lovitt. New York: Harper & Row.

26. Holton, Gerald James. 1973. *Thematic Origins of Scientific Thought: Kepler to Einstein*. Cambridge: Harvard University Press.

27. Hull, David L. 1988. *Science as a Process: An Evolutionary Account of the Social and Conceptual Development of Science*. Chicago: University of Chicago Press.

28. Keller, Evelyn Fox. 1985. *Reflections on Gender and Science*. New Haven: Yale University Press.

29. Knorr-Cetina, Karin D. 1981. *The Manufacture of Knowledge: An Essay on the Constructivist and Contextual Nature of Science*. Oxford: Oxford University Press.

30. Kuhn, Thomas S. 1977. *The Essential Tension: Selected Studies in Scientific Tradition and Change*. Chicago: University of Chicago Press.

31. Kuhn, Thomas S. 1970. *The Structure of Scientific Revolutions*, 2nd edition. Chicago: University of Chicago Press.

32. Lakatos, Imre. 1976. *Proofs and Refutations: The Logic of Mathematical Discovery*. Cambridge: Cambridge University Press.

33. Latour, Bruno. 1987. *Science in Action: How to Follow Scientists and Engineers through Society*. Cambridge: Harvard University Press.

34. Latour, Bruno, and Steve Woolgar. 1979. *Laboratory Life: The Social Construction of Scientific Facts*. Beverly Hills, CA: Sage Publications.

35. Locke, David. 1992. *Science as Writing*. New Haven: Yale University Press.

36. Longino, Helen E. 1990. *Science as Social Knowledge: Values and Objectivity in Scientific Inquiry*. Princeton: Princeton University Press.
37. Medawar, P. B. 1982. *Pluto's Republic*. Oxford: Oxford University Press.
38. Merton, Robert King. 1973. *The Sociology of Science: Theoretical and Empirical Investigations*. Chicago: University of Chicago Press.
39. Montgomery, Scott L. 1996. *The Scientific Voice*. New York: Guilford Press.
40. Montgomery, Scott L. 2000. *Science in Translation: Movements of Knowledge through Cultures and Time*. Chicago: University of Chicago Press.
41. Moss, Jean Dietz. 1993. *Novelties in the Heavens: Rhetoric and Science in the Copernican Controversy*. Chicago: University of Chicago Press.
42. Myers, Greg. 1990. *Writing Biology: Texts in the Social Construction of Scientific Knowledge*. Madison: University of Wisconsin Press.
43. Pera, Marcello. 1994. *The Discourses of Science*. Chicago: University of Chicago Press.
44. Pera, Marcello, and William R. Shea, eds. 1991. *Persuading Science: The Art of Scientific Rhetoric*. Canton, MA: Science History Publications.
45. Pickering, Andrew. 1984. *Constructing Quarks: A Sociological History of Particle Physics*. Chicago: University of Chicago Press.
46. Rudwick, M. J. S. 1985. *The Great Devonian Controversy: The Shaping of Scientific Knowledge among Gentlemanly Specialists*. Chicago: University of Chicago Press.
47. Rudwick, M. J. S. 1972. *The Meaning of Fossils: Episodes in the History of Palaeontology*. New York: Elsevier.
48. Shapin, Steven, and Simon Schaffer. 1985. *Leviathan and the Air-Pump: Hobbes, Boyle, and the Experimental Life*. Princeton: Princeton University Press.
49. Simons, Herbert W., ed. 1990. *The Rhetorical Turn: Invention and Persuasion in the Conduct of Inquiry*. Chicago: University of Chicago Press.
50. Traweek, Sharon. 1988. *Beamtimes and Lifetimes: The World of High Energy Physicists*. Cambridge: Harvard University Press.

Secondary Literature Sources by Chapter
Key secondary sources plus direct quotations in commentaries

CHAPTER ONE. First English Periodical

Anonymous. 1665. Review of *"Philosophical Transactions*, à Londres, chez Jean Martin & James Allistry, Imprimeurs de la Societé Royale, & se trouve à Paris chez Jean Cusson, ruë S. Jacque." *Journal des Sçavans* 1: 156. Quotation on p. 156.
Atkinson, Dwight. 1999. "The Royal Society and its *Philosophical Transactions*: A brief institutional history," in *Scientific Discourse in Sociohistorical Context: The Philosophical Transactions of the Royal Society of London, 1675–1975*. Mahwah, NJ: Lawrence Erlbaum Associates, pp. 15–55.
Bazerman, Charles. 1988. "Between books and articles: Newton faces controversy," in *Shaping Written Knowledge: The Genre and Activity of the Experimental Article in Science*. Madison: University of Wisconsin Press, pp. 80–127.

Bluhm, R. K. 1960. "Henry Oldenburg, F. R. S." *Notes and Records of the Royal Society of London* 15: 183–97.

Dear, Peter. 1985. " *Totius in verba:* Rhetoric and authority in the early Royal Society." *Isis* 76: 145–61.

Gascoigne, Robert Mortimer. 1985. *A Historical Catalogue of Scientific Periodicals, 1665–1900, with a Survey of their Development*. New York: Garland Publishing.

Heilbron, J. L. 1983. *Physics at the Royal Society during Newton's Presidency*. Los Angeles: William Andrews Clark Memorial Library, University of California.

Hill, John. 1750. *Lucina sine Concubitu*. Berkshire: Golden Cockerel Press, 1930. Quotations on pp. i and 16.

Jardine, Lisa. 2003. *The Curious Life of Robert Hooke: The Man Who Measured London*. New York: HarperCollins, pp. 183–87.

Johns, Adrian. 1998. *The Nature of the Book: Print and Knowledge in the Making*. Chicago: University of Chicago Press.

Kronick, David A. 1976. *A History of Scientific & Technical Periodicals: The Origins and Development of the Scientific and Technical Press, 1665–1790*. Metuchen, NJ: Scarecrow Press.

Ornstein, Martha. 1928. *The Role of Scientific Societies in the Seventeenth Century*. Chicago: University of Chicago Press.

Pardies, Ignace. 1672. "A second letter of P. Pardies, written to the editor from Paris, May 21, 1672, to Mr. Newton's answer made to his first letter, printed in No 84." *Philosophical Transactions* 7 (84): 5012–13. Quotation on p. 5012.

Shapin, Steven, and Simon Schaffer. 1985. "Seeing and believing: The experimental production of pneumatic facts," in *Leviathan and the Air-Pump: Hobbes, Boyle, and the Experimental Life*. Princeton: Princeton University Press, pp. 22–79.

Shapiro, Alan. 1996. "The gradual acceptance of Newton's theory of light and color, 1672–1727." *Perspectives on Science* 4: 59–140.

Stigler, Stephen M. 1986. "Arbuthnot and the sex ratio at birth," in *The History of Statistics: The Measurement of Uncertainty before 1900*. Cambridge: Harvard University Press, pp. 225–26.

Stimson, Dorothy. 1968. *Scientists and Amateurs: A History of the Royal Society*. New York: Greenwood Press.

Tebeaux, Elizabeth. 1999. "Technical writing in seventeenth-century England: The flowering of a tradition." *Journal of Technical Writing and Communication* 29: 209–53.

Valle, Ellen. 1997. "A scientific community and its texts: A historical discourse study," in *The Construction of Professional Discourse*, Britt-Louise Gunnarsson et al., eds. London: Longman, pp. 76–97.

CHAPTER TWO. First French Periodicals

Birn, Raymond. 1965. "Le *Journal des Savants* sous l'Ancien Regime." *Journal des Savants*, January-March, pp. 15–35.

Crosland, Maurice. 1992. *Science under Control: The French Academy of Sciences 1795–1914*. Cambridge: Cambridge University Press.

Daston, Lorraine, and Katharine Park. 1998. "Strange facts," in *Wonders and the Order of Nature: 1150–1750*. New York: Zone Books, pp. 215–53.

Davy, H. 1821. "On the magnetic phenomena produced by electricity." *Philosophical Transactions* III: 7–19. Quotation on p. 17.

Fox, Robert. 1992. *The Culture of Science in France, 1700–1900*. Aldershot, Great Britain: Variorum.

Fox, Robert, and George Weisz, eds. 1988. *The Organization of Science and Technology in France, 1808–1914*. Cambridge: Cambridge University Press.

Gould, Stephen Jay. 1998. "Capturing the center: Antoine-Laurent Lavoisier's scientific contributions." *Natural History* 107 (December): 14–25. www.findarticles.com/ p/articles/mi_m1134/is_10_107/ai_53378966/print.

———. 1998. "The man who invented natural history." *The New York Review of Books* 45 (16): 83–90. Quotation on p. 83.

Hahn, Roger. 1971. *The Anatomy of a Scientific Institution: The Paris Academy of Sciences, 1666–1803*. Berkeley: University of California Press. Quotation on closing of French academies on p. 238.

Holmes, Frederic Lawrence. 1985. "Respiration and a general theory of combustion," in *Lavoisier and the Chemistry of Life: An Exploration of Scientific Creativity*. Madison: Wisconsin University Press, pp. 91–128.

———. 1989. "Argument and Narrative in Scientific Writing," in *The Literary Structure of Scientific Argument*, Peter Dear, ed. Philadelphia: University of Pennsylvania Press, pp. 164–81.

Licoppe, Christian. 1997. "Théâtres de la preuve expérimentale en France aux XVIIIe siècle: De la pertinence d'un lien entre sciences et sociabilités." *Bulletin de la Société d'Histoire Moderne et Contemporaine* 3–4: 29–35.

McCutcheon, Roger Philip. 1924. "The *Journal des Sçavans* and the *Philosophical Transaction* of the Royal Society." *Studies in Philology* 21: 626–28.

Marchant, Nicholas. 1707. "Dissertation sur une rose monstrueuse." *Mémoires de l'Academie Royale des Sciences*, Paris, pp. 488–89. Quotation on p. 488.

Perrault, Claude. 1733. "Description anatomique de trois cameleons," in *Mémoires de l'Académie Royale des Sciences depuis 1666 jusqu'à 1699*, Paris. Quotation from Vol. 3, Part 1, p. 35.

Roberts, Lissa. 1989. "Setting the table: The disciplinary development of eighteenth-century chemistry as read through the changing structure of its tables," in *The Literary Structure of Scientific Argument*, Peter Dear, ed. Philadelphia: University of Pennsylvania Press, pp. 99–132.

Roger, Jacques. 1997. *Buffon: A Life in Natural History*, Sarah Lucille Bonnefoi, trans. Ithaca: Cornell University Press.

Sallo, Denis de. 1665. "L'imprimeur au lecteur." *Journal des Sçavans* 1 (5 January): i–ii. Quotation on p. ii.

Stroup, Alice. 1990. *A Company of Scientists: Botany, Patronage, and Community at the Seventeenth-Century Parisian Royal Academy of Sciences*. Berkeley: University of California Press. Quotation on p. 34.

Terrall, Mary. 2002. *The Man Who Flattened the Earth: Maupertius and the Sciences in the Enlightenment*. Chicago: University of Chicago Press.

CHAPTER THREE. Internationalization and Specialization

Beer, John J. 1958. "Coal tar dye manufacture and the origins of the modern industrial research laboratory." *Isis* 49: 123–31.

Bell, Thomas. 1860. "Presidential address to the Linnean Society on the anniversary of Linnaeus's birth, May 24, 1859." *Journal of the Proceedings of the Linnean Society* 4: viii–xx. Quotation on p. viii.

Ben-David, Joseph. 1971. *The Scientist's Role in Society: A Comparative Study*. Englewood Cliffs: Prentice Hall. Quotation on p. 108.

Berry, Eliot M. 1981. "The evolution of scientific and medical journals." *New England Journal of Medicine* 305: 400–402. Quotation on p. 400.

Blum, Ann Shelby. 1993. *Picturing Nature: American Nineteenth-Century Zoological Illustrations*. Princeton: Princeton University Press.

Boig, Fletcher S., and Paul W. E. Howerton. 1952. "History and development of chemical periodicals in the field of organic chemistry: 1877–1949." *Science* 115: 25–31.

———. 1952. "History and development of chemical periodicals in the field of analytical chemistry: 1877–1949." *Science* 115: 555–60.

Cajori, Florian. 1929. *A History of Mathematical Notations*, Vol. 2, Chicago: Open Court Press. Leibniz quoted on p. 184.

Darwin, Charles. 1858. "Abstract of a Letter from C. Darwin, Esq., to Prof. Asa Gray, Boston, U.S., Dated Down, September 5th, 1857." *Journal of the Proceedings of the Linnean Society* 3: 50–53. Quotation on p. 53.

Davis, Natalie Zemon. 1995. "Metamorphosis," in *Women on the Margins: Three Seventeenth Century Lives*. Cambridge: Harvard University Press, pp. 140–202.

Gascoigne, John. 1995. "The eighteenth-century scientific community: A prosopographical study." *Social Studies of Science* 25: 575–81.

Godman, John D. 1829. *Addresses Delivered on Various Public Occasions, with an Appendix, Containing a Brief Exploration of the Effects of Tight Lacing*. Philadelphia: Carey, Lea & Carey. Quotation on pp. 128–29.

Greene, John C. 1984. *American Science in the Age of Jefferson*. Ames: Iowa State University Press.

Herschel, J. F. W. 1833. *Memoirs of the Royal Astronomical Society*. 5: 171–222. Quotation on p. 178.

Hufbauer, Karl. 1982. *The Formation of the German Chemical Community (1720–1795)*. Berkeley: University of California Press.

Huxley, T. H. 1872. "Yeast." *Contemporary Review* 19: 23–36. Also in *Collected Essays*, Vol. VIII, *Discourses: Biological & Geological*. Quotation on p. 125.

Kelves, Daniel J. 1978. *The Physicists: The History of a Scientific Community in Modern America*. New York: Alfred A. Knopf.

Laeven, Hub. 1990. *The "Acta Eruditorum" under the Editorship of Otto Mencke (1644–1707): The History of an International Learned Journal between 1662–1707*. Amsterdam: Holland University Press.

Lynch, John M. 2000. "Introduction," in *Vestiges and the Debate before Darwin*. Bristol: Thoemmes Press.

McClellan, James E. 1979. "The scientific press in transition: Rozier's journal and the scientific societies in the 1770s." *Annals of Science* 36: 425–49.

———. 1985. *Science Reorganized: Scientific Societies in the Eighteenth Century*. New York: Columbia University Press. Abbé Rozier quoted on p. 191.

Porter, Roy. 1978. "Gentlemen and geology: The emergence of a scientific career, 1660–1920." *The Historical Journal* 21: 809–36.

Rocke, Alan J. 1993. *The Quiet Revolution: Hermann Kolbe and the Science of Organic Chemistry*. Berkeley: University of California Press, pp. 325–39. Van't Hoff quoted on p. 330.

Schiebinger, Londa. 1992. "Women in science: Historical perspectives." In *Proceedings of the Women in Astronomy Workshop*, Meg Urry, ed. Baltimore: Space Telescope Science Institute, pp. 11–19.

Secord, James A. 2001. *Victorian Sensation: The Extraordinary Publication, Reception, and Secret Authorship of* Vestiges of the Natural History of Creation. Chicago: University of Chicago Press.

Tilling, Laura. 1975. "Early experimental graphs." *The British Journal for the History of Science* 8: 193–213.

CHAPTER FOUR. Select Pre-Modern Classics

Anonymous. 1896. "Medical Applications of Röntgen's Discovery." *Nature*, 53: 324. Quotation on p. 324.

Bacon, Francis. 1603(?). *Valerius Terminus: Of the Interpretation of Nature*. University of Adelaide Library, Electronic Texts Collection, Quotation in www.etext.library .adelaide.edu.au/b/bacon/francis.

Berg, Paul, and Maxine Singer. 1998. "Inspired choices." *Science* 282: 873–74.

Brock, William H. 1993. "On the constitution and metamorphoses of chemical compounds," in *The Norton History of Chemistry*. New York: W. W. Norton, pp. 241–69.

Brooks, John Langdon. 1984. *Just Before the Origin: Alfred Russel Wallace's Theory of Evolution*. New York: Columbia University Press. Alfred Russel Wallace letter quoted on p. 181.

Carter, K. Codell. 1987. *The Essays of Robert Koch*. New York: Greenwood Press. Paul Ehrlich quoted on p. xvi.

Collins, Randall. 1998. "Cross-breeding networks and rapid-discovery science," in *The Sociology of Philosophies: A Global Theory of Intellectual Change*. Cambridge: Harvard University Press, pp. 523–69. Quotation on pp. 870–71.

Curie, Marie Sklodowski. 1904. "Radium and radioactivity." *Century Magazine*, January, pp. 461–66. Quotation on p. 461.

Dalton, John. 1964. *A New System of Chemical Philosophy*. New York: Philosophical Library. Illustration on p. 164.

Darwin, Charles. 1958. *The Autobiography of Charles Darwin*. New York: Dover. Quotation on p. 43.

————. 1987. *Notebooks: 1836–1844*. Paul H. Barrett et al., eds. Ithaca: Cornell University Press. Quotation on pp. 374–75.

Einstein, Albert. 1951. *Albert Einstein: Philosopher-Scientist*, Paul Arthur Schilpp, ed. New York: Harper and Row. Quotation in Vol. 1 on p. 280 (response to Hadamard).

————. 1954. *Ideas and Opinions*. New York: Bonanza Books. Quotation on p. 272.

Fahnestock, Jeanne. 1998. "Antimetabole," in *Rhetorical Figures in Science*. New York: Oxford University Press, pp. 122–55.

Garber, Elizabeth. 1999. *The Language of Physics: The Calculus and the Development of Theoretical Physics in Europe, 1750–1914*. Boston: Birkhäuser.

Gross, Samuel D. 1857 (1839). *Elements of Pathological Anatomy*, 3rd ed. Philadelphia: Blanchard and Lea. Quotation on p. 36.

Knight, David, and Helge Kraghe, eds. 1998. *The Making of the Chemist: The Social History of Chemistry in Europe: 1789–1914*. Cambridge: Cambridge University Press.

Laudan, Rachel. 1987. *From Mineralogy to Geology: The Foundations of a Science, 1650–1830*. Chicago: University of Chicago Press.

Mazurs, Edward G. 1974. *Graphic Representations of the Periodic System during One Hundred Years*, 2nd edition. University, AL: University of Alabama Press.

Mendeleev, Dimitri Ivanovich. 1891. *Principles of Chemistry*, 5th ed., translated by George Kaminisky. London: Longmans, Green and Co. Quotation from Vol. 2 on p. 16.

Moseley, Henry Gwyn Jeffreys. 1914. "The high-frequency spectra of the elements, Part II." *Philosophical Magazine and Journal of Science*, 6th series, 27: 703–12.

Nye, Mary Jo. 1996. *Before Big Science: The Pursuit of Modern Chemistry and Physics 1800–1940*. New York: Twayne Publishers.

Nyhart, Lynn K. 1995. *Biology Takes Form: Animal Morphology and the German Universities, 1800–1900*. Chicago: University of Chicago Press.

Oldroyd, David R. 1990. *The Highlands Controversy: Constructing Geological Knowledge through Fieldwork in Nineteenth-Century Britain*. Chicago: University of Chicago Press.

Pancaldi, Giuliano. 2003. *Volta: Science and Culture in the Age of Enlightenment*. Princeton: Princeton University Press. Price quoted on p. 203.

Turner, Edward. 1835. *Elements of Chemistry, Including the Recent Discoveries and Doctrines of the Science*, 5th edition. Philadelphia: Desilver, Thomas & Co.

CHAPTER FIVE. Equations, Tables, and Pictures

Bais, Sander. 2005. *The Equations: Icons of Knowledge*. Cambridge: Harvard University Press.

Cleveland, William S. 1994. *The Elements of Graphing Data*. Summit, NJ: Hobart Press.

Crick, Francis. 1988. "The alpha helix," in *What Mad Pursuit: A Personal View of Scientific Discovery*. New York: Basic Books, pp. 53–61.

Eames, Charles and Ray. 1989. "The powers of ten," in *The Films of Charles & Ray Eames*. Chatsworth, CA: Lucia Eames DBA Eames Office, DVD video, Vol. 1.

Einstein, Albert. 1954. *Ideas and Opinions.* New York: Crown Publishers. Quotation on p. 276.

Fahnestock, Jeanne. 2003. "Verbal and visual parallelism." *Written Communication* 20: 123–52.

Farmelo, Graham, ed. 2002. *It Must Be Beautiful: Great Equations of Modern Science.* London: Granta Books.

Frankel, Henry. 1987. "The continental drift debate," in *Scientific Controversies: Case Studies in the Resolution and Closure of Disputes in Science and Technology.* H. Tristram Engelhardt and Arthur L. Caplan, eds. Cambridge: Cambridge University Press, pp. 203–48.

Fritscher, Bernhard. 2002. "Alfred Wegener's 'The Origin of continents,' 1912." *Episodes* 25: 100–106.

Galison, Peter. 1998. "Judgment against objectivity," in *Picturing Science, Producing Art.* New York: Routledge, pp. 327–59.

Gott III, J. Richard, Mario Jurić, David Schlegel, Fiona Hoyle, Michael Vogeley, Max Tegmark, Neta Bahcall, and Jon Brinkmann. 2005. "Map of the universe." *Astrophysical Journal* 624: 463–84.

Gould, Stephen Jay. 1989. "Reconstruction of the Burgess Shale: Toward a new view of life," in *Wonderful Life: The Burgess Shale and the Nature of History.* New York: W. W. Norton, pp. 79–239. Quotation on pp. 2–3.

Hankins, Thomas L. 1999. "Blood, dirt, and nomograms: A particular history of graphs," *Isis* 90: 50–80.

Kelves, Daniel J. 1998. *The Baltimore Case: A Trial of Politics, Science, and Character.* New York: W. W. Norton.

Kemp, Martin. 2000. *Visualizations: The Nature Book of Art and Science.* Berkeley: University of California Press.

Lynch, Michael, and Steve Woolgar, eds. 1990. *Representation in Scientific Practice*, Cambridge: MIT Press.

Pais, Abraham. 1982. "The new kinematics," in *'Subtle is the Lord...': The Science and the Life of Albert Einstein.* Oxford: Oxford University Press, pp. 138–62.

Rédei, George P. 2002. "Vignettes in the history of genetics," in *Quantitative Genetics, Genomics and Plant Biology*, M. S. Kang, ed. Wallingford, UK: CAB International. Punnet quoted on p. 4.

Rudwick, Martin J. S. 1976. "The emergence of a visual language for geological science, 1760–1840." *History of Science* 14: 149–95. Quotation on p. 168.

Tufte, Edward R. 1983. *The Visual Display of Quantitative Information.* Cheshire, CN: Graphics Press.

CHAPTER SIX. Organizing Scientific Arguments

Burns, Gerald. 1992. *High-Temperature Superconductivity: An Introduction.* San Diego: Academic Press, pp. 1–6.

Day, Robert A. 1988. *How to Write and Publish a Scientific Paper*, 3rd edition. Phoenix: Onyx Press.

Franck, Georg. 2002. "The scientific economy of attention: A novel approach to collective rationality of science." *Scientometrics* 55: 3–26.

Gopnik, Myrna. 1972. *Linguistic Structures in Scientific Texts*. The Hague: Mouton.

Gross, Alan G. 1996. "The arrangement of the scientific paper," in *The Rhetoric of Science*, 2nd edition. Cambridge: Harvard University Press, pp. 85–96.

Hau, Lene Vestergaard. 2001. "Frozen light." *Scientific American* 285 (1): 66–73.

Hoffmann, Roald. 1988. "Under the surface of the chemical article." *Angewandte Chemie* 27: 1593–1602.

Kimura, Motoo. 1979. "The neural theory of molecular evolution." *Scientific American* 241 (5): 98–126.

Mayer, M. G. 1949. "On closed shells in nuclei, II." *Physical Review* 75: 1969–70. Quotation on p. 1970.

Siegel, Warren. 1986. "The super G-string," in *Workshop on Unified String Theories*. Singapore: World Scientific Publishing, pp. 729–37. Quotation on p. 729.

Suppe, Frederick. 1998. "The structure of the scientific paper." *Philosophy of Science* 65: 381–405.

Swales, John. 1990. "Research articles in English," in *Genre Analysis: English in Academic and Research Settings*. Cambridge: Cambridge University Press, pp. 110–76.

CHAPTER SEVEN. Scientific Writing Style: Norms and Perturbations

Cipra, Barry. 1995. "Princeton mathematician looks back on Fermat proof." *Science* 268: 1133–34.

Darwin, Erasmus. 1799. *The Botanic Garden: A Poem, in Two Parts*. London: J. Johnson. Quotation in Part I on pp. 85–86.

Fox Keller, Evelyn. 1983. *A Feeling for the Organism*. New York: W. H. Freeman and Company. Quotation on p. 198.

Gamow, George. 1952. *The Creation of the Universe*. New York: Viking Press. Quotation on p. 65.

Gopen, George D., and Judith A. Swan. 1990. "The science of scientific writing." *American Scientist* 78: 550–58.

Gould, Stephen J. 1993. "Fulfilling the spandrels of world and mind," in *Understanding Scientific Prose*, Jack Selzer, ed. Madison: University of Wisconsin Press, pp. 310–36. Quotation on p. 321.

Gross, Alan G., Joseph E. Harmon, and Michael Reidy. 2002. "Style and presentation in the 20th century," in *Communicating Science: The Scientific Article from the 17th Century to the Present*. Oxford: Oxford University Press, pp. 161–86.

Halliday, M. A. K. 1993. "On the language of physical science," in *Writing Science: Literacy and Discursive Power*, M. A. K. Halliday and J. R. Martin, eds. Pittsburgh: University of Pittsburgh Press, pp. 54–68. Quotation on p. 67.

———. 1993. "The analysis of scientific texts in English and Chinese," in *Writing Science: Literacy and Discursive Power*, M. A. K. Halliday and J. R. Martin, eds. Pittsburgh: University of Pittsburgh Press, pp. 124–32.

Johnson, Kurt, and Steve Coates. 1999. *Nabokov's Blues: The Scientific Odyssey of a Literary Genius*. Cambridge, MA: Zone Books. Quotation on p. 104.

Leather, Simon R. 1996. "The case for the passive voice." *Nature* 381: 467.

Leatherdale, W. H. 1974. *The Role of Analogy, Model and Metaphor in Science*. Amsterdam: North-Holland.

Locke, David. 1998. "Voices of science." *The American Scholar*, summer, pp. 103–14.

Martin, J. R., and Robert Veel, eds. 1998. *Reading Science: Critical and Functional Perspectives on Discourses of Science*. New York: Routledge.

Montgomery, Scott L. 2000. "Issues and examples for the study of scientific translation today," in *Science in Translation: Movements of Knowledge through Cultures and Time*. Chicago: University of Chicago Press, pp. 253–70.

———. 2004. "Of towers, walls, and fields: Perspectives on language in science." *Science* 303: 1333–35.

Stewart, John A. 1990. "Opposition by a minority," in *Drifting Continents & Colliding Paradigms*. Bloomington: Indiana University Press, pp. 116–21.

Thomson, William (Lord Kelvin). 1889. "The wave theory of light," in *Popular Lectures and Addresses*. London: Macmillan. Quotation in Vol. 1 on p. 334.

Weinberg, Steven. 1992. "Conceptual foundation of the unified theory of weak and electromagnetic interactions," in *Nobel Lectures, Physics: 1971–1980*, Stig Lundqvist, ed. Singapore: World Scientific Publishing, pp. 543–59. Quotation on p. 545.

Wilson, J. Tuzo. 1968. "A revolution in earth science." *Geotimes* 13: 10–17. Quotation on p. 11.

CHAPTER EIGHT. Controversy at Work: Two Case Studies

Crews, Frederick. 1986. *Skeptical Engagements*. New York: Oxford University Press.

Engelhardt, H. Tristram, and Arthur L. Caplan, eds. 1987. *Scientific Controversies: Case Studies in the Resolution and Closure of Disputes in Science and Technology*. Cambridge: Cambridge University Press.

Grünbaum, Adolf. 1984. *The Foundations of Psychoanalysis: A Philosophical Critique*. Berkeley: University of California Press.

Hull, David L. 1988. *Science as a Process: An Evolutionary Account of the Social and Conceptual Development of Science*. Chicago: University of Chicago Press. Quotation on p. 222.

Leonard, Jonathan. 1998. "Dream-catchers." *Harvard Magazine*, May-June, pp. 58–68. Hobson quoted on p. 59.

Mayr, Ernst. 1982. "Diversity and synthesis of evolutionary thought," in *The Growth of Biological Thought: Diversity, Evolution, and Inheritance*. Cambridge: Harvard University Press, pp. 535–70

Montgomery, Scott L. 1996. "A case of (mis)taken identity: The question of Freudian discourse," in *The Scientific Voice*. New York: Guilford Press, pp. 360–429.

Pera, Marcello. 1994. "The dialectical model of science," in *The Discourses of Science*. Chicago: University of Chicago Press, pp. 129–52.

Provine, William B. 1992. "The R. A. Fisher–Sewall Wright controversy," in *The Founders of Evolutionary Genetics*, Sahotra Sarkar, ed. Netherlands: Kluwer Press, pp. 201–29.

CHAPTER NINE. Select Modern Classics

Bahcall, John N., and Raymond Davis, Jr. 1982. "An account of the development of the solar neutrino problem," in *Essays in Nuclear Astrophysics*, Charles A. Barnes et al., eds. Cambridge: Cambridge University Press, pp. 243–85.

Bell, John S. 2001. *On the Foundations of Quantum Mechanics*, M. Bell et al., eds. Singapore: World Scientific Publishing.

Beller, Mara. 1999. "The polyphony of Heisenberg's uncertainty paper," in *Quantum Dialogue: The Making of a Revolution*. Chicago: University of Chicago Press, pp. 103–16.

Bohr, Niels. 1935. "Can quantum-mechanical description of physical reality be considered complete?" *Physical Review* 48: 696–702.

———. 1957? Unsent letter to Werner Heisenberg. Quotation from www.nbi.dk/NBA/papers/docs/d01tra.htm.

Boyce, Peter B., and Heather Dalterio. 1996. "Electronic publishing of scientific journals." *Physics Today* 49 (January): 42–47.

Einstein, Albert. 1950. *The Meaning of Relativity*, 3rd edition. Princeton: Princeton University Press.

Einstein, Albert, and Max Born. 2005. *The Born-Einstein Letters 1916–1955: Friendship, Politics, and Physics in Uncertain Times*. New York: Walker and Company, 1971. Quotation on p. 91.

Eliot, T. S. 1963. "Four Quartets," in *T. S. Eliot: Collected Poems 1909–1962*. New York: Harcourt, Brace & World. Quotation on p. 208.

Fritscher, Bernhard. 2002. "Alfred Wegener's 'The origin of continents,' 1912." *Episodes* 25 (2): 100–106.

Frayn, Michael. 2000. *Copenhagen*. New York: Anchor Books.

Galison, Peter. 1997. "Cloud chambers: The peculiar genius of British physics," in *Image & Logic: A Material Culture of Microphysics*. Chicago: University of Chicago Press, pp. 65–141.

———. 2003. "Synchrony," in *Einstein's Clocks, Poincaré's Maps: Empires of Time*. New York: W. W. Norton, pp. 13–47.

Genzer, Peter. 2001. "SNO's results indicate solution to 30-year solar-neutrino problem." *The Bulletin: Brookhaven National Laboratory* 55 (21): 1–2. Quotation on p. 1.

Harnad, Stevan. 1991. "Post-Gutenberg galaxy: The fourth revolution in the means of production of knowledge." *Public-Access Computer Systems Review* 2: 39–53.

Kragh, Helge. 1996. "Background: From Einstein to Hubble," in *Cosmology and Controversy: The Historical Development of Two Theories of the Universe*. Princeton: Princeton University Press, pp. 3–21.

Libby, Leona Marshall. 1979. "In Chicago," in *The Uranium People*. New York: Crane Russak & Company, pp. 118–39.

Little, Peter. 2003. "DNA sequencing: The silent revolution," in *A Century of Nature: Twenty-one Discoveries That Changed Science and the World*, Laura Garwin and Tim Lincoln, eds. Chicago: University of Chicago Press, pp. 231–58.

Mermin, N. D. 1981. "Bringing home the atomic world: Quantum mysteries for anybody." *American Journal of Physics* 49: 940–43.

Sime, Ruth Lewin. 1996. "The discovery of nuclear fission," in *Lise Meitner: A Life in Physics*. Berkeley: University of California Press, pp. 231–58. Meitner quoted on p. 235.

Smalley, Richard E. 1997. "Discovering the fullerenes" (Nobel lecture). *Reviews of Modern Physics* 69: 723–30. Quotations on pp. 723 and 725.

Stachel, John. 1998. "Einstein on the theory of relativity," in *Einstein's Miraculous Year: Five Papers That Changed the Face of Physics*. Princeton: Princeton University Press, pp. 101–21.

Watson, J. D., and F. H. C. Crick. 1953. "A structure for deoxyribose nucleic acid." *Nature* 171: 737–38. Quotation on p. 737.

....

World Wide Web Resources

Listing of online collections of historically significant scientific articles and journals along with scholarly articles about them. Note: the selected articles in our book from the late twentieth and the twenty-first centuries are available through the cited journal's Web site, which is normally restricted to subscribers or those willing to pay a fee for one-time online access. However, many university libraries provide free access for students and faculty.

Classics of the Scientific Literature, National Academy of Sciences of the United States of America, http://www.pnas.org/misc/classics.shtml. [Several landmark papers from the twentieth century published by the *Proceedings of the National Academy of Sciences*.]

Digitalisierung der Akademieschriften und Schriften zur Geschichte der Königlich Preussischen Akademie der wissenschaften (1700–1900), Akademiebibliothek, http://www3.bbaw.de/bibliothek/digital/liste1.htm. [Runs of German scientific journals published in the eighteenth and nineteenth centuries, including *Miscellanea Berolinensia* and *Histoire de l'Académie Royale des Sciences et des Belles-Lettres de Berlin*.]

Echo: Exploring and Collecting Online—Science, Technology and Industry, Center for History and New Media, George Mason University, http://echo.gmu.edu. [Extensive annotated listing of scholarly web sites in the history of science, technology, and medicine.]

Evolution: Classic Texts, Blackwell Publishing, http://www.blackwellpublishing.com/ridley/classictexts. [Twenty classic scientific texts from the history of evolutionary biology.]

Foundations of Classical Genetics, Electronic Scholarly Publishing, http://www.esp.org/foundations/genetics/classical. [Extensive database of classic scientific texts from the history of genetics and evolutionary biology.]

Gallica, Bibliothéque nationale de France, http://gallica.bnf/PeriosListe.htm. [Runs of many historic scholarly journals, including *Philosophical Transactions, Acta Eruditorum, Journal des Sçavans*, and *Mémoires de l'Académie Royale des Sciences* from the seventeenth and eighteenth centuries.]

Genome Gateway, Nature Publishing Group, http://www.nature.com/genomics/papers. [Library of genome-related articles appearing in *Nature* and *Nature Genetics*.]

The Newton Project: Newton's Optical Disputes, Imperial College London, http://www .newtonproject.ic.ac.uk/prism.php?id=111. [Seventeenth-century documents related to Isaac Newton and controversy over his new theory of light and colors.]

Nobelprize.org, Official Web Site of Nobel Foundation, http://nobelprize.org. [Complete collection of Nobel Prize speeches in physics, chemistry, and medicine plus biographies of laureates.]

Profiles in Science, National Library of Medicine, http://profiles.nlm.nih.gov. [Classic articles by several important biomedical researchers in the twentieth century.]

Scholarly Journal Archive, JSTOR, http://www.jstor.org. [Runs of several general scientific journals, including *Philosophical Transactions* from 1665 through the late twentieth century. Also, the twentieth-century journals *Science* and *Proceedings of the National Academy of Sciences*. Subscription based.]

Selected Classic Papers from the History of Chemistry, Carmen J. Giunta, LeMoyne University, http://web.lemoyne.edu/~giunta/paper.html. [Numerous classic scientific texts in the physical sciences.]

Permissions

....

Permissions to reproduce illustrations appear at the end of figure captions. Excerpts from the following articles quoted with permission:

Gell-Mann, M.—*Model of baryons and mesons*. Reprinted from *Physics Letters*. Copyright (1964), with permission from Elsevier.

Kimura, M.—*Evolutionary rate at molecular level*. Reprinted by permission from *Nature*. Copyright (1968) Macmillan Publishers Ltd.

Liu, C., et al.—*Halted light pulses*. Reprinted by permission from *Nature*. Copyright (2001) Macmillan Publishers Ltd.

Lowry, O., et al.—*Protein measurement*. Reprinted from *Journal of Biological Chemistry*. Copyright (1951), with permission of the American Society for Biochemistry and Molecular Biology via Copyright Clearance Center.

Stojanovic, M., Stefanovic, D.—*Deoxyribozyme-based molecular automata*. Reprinted with permission of *Nature Biotechnology*. Copyright (2003) Macmillan Publishers Ltd.

Wu, M., et al.—*Superconductivity at 93 K*. Reprinted from *Physical Review Letters*. Copyright (1987), with permission from American Physical Society.

CHAPTER SEVEN

Alpher, R., et al.—*Origin of chemical elements*. Reprinted from *Physical Review*. Copyright (1948), with permission from the American Physical Society.

Baade, W., Zwicky, F.—*Supernovae & cosmic rays*. Reprinted from *Physical Review*. Copyright (1934), with permission from the American Physical Society.

Beloussov, V.—*Against ocean-floor spreading*. Reprinted from *Tectonophysics*. Copyright (1970), with permission from Elsevier.

Bligh, E., Dyer, W.—*Lipid extraction and purification*. Reprinted from *Canadian Journal of Biochemistry and Physiology*. Copyright (1959), with permission from NRC Research Press.

Breene, R.—*Erratum*. Reprinted from *Journal of Chemical Physics* (1967), with permission of the American Institute of Physics.

Bunnett, J., Kearley, F.—*Mobility of halogens*. Reprinted with permission of *Journal of Organic Chemistry*. Copyright (1971) American Chemical Society.

Dingle, H.—*Science and modern cosmology*. Reprinted from *Monthly Notices of the Royal Astronomical Society* (1953), with permission of Blackwell Publishing.

Gould, S., Lewontin, R.—*Spandrels of San Marco* (1979). Quoted with permission from The Royal Society and Art Science Research Laboratory, Inc.

McClintock, B.—*Responses of the genome*. Excerpted from *Nobel Lectures*. Copyright (1984) The Nobel Foundation.

Nabokov, V.—*A discovery*. Reprinted by arrangement with the estate of Vladimir Nabokov. All rights reserved.

Nabokov, V.—*New or little known Nearctic neonympha*. Reprinted with permission from *Psyche* (1942).

Paez, J., et al.—*EGFR mutations in lung cancer*. Excerpted with permission from *Science*. Copyright (2004) AAAS.

Peebles, P., Silk, J.—*Cosmic book of phenomena*. Reprinted by permission from *Nature*. Copyright (1990) Macmillan Publishers Ltd.

Sanger, F., et al.—*Nucleotide sequence of bacteriophage.* Reprinted by permission from *Nature.* Copyright (1977) Macmillan Publishers Ltd.

Smith, H.—*Synchronous flashing of fireflies.* Reprinted with permission of *Science.* Copyright (1935) AAAS.

Wilson, A., Calvin, M.—*Photosynthetic cycle.* Reprinted from *Journal of American Chemical Society.* Copyright (1955) American Chemical Society.

CHAPTER EIGHT

Fisher, R., Ford, E.—*Spread of a gene.* Reprinted from *Heredity* (1947), with permission of Nature Publishing Group.

Fisher, R., Ford, E.—*"Sewall Wright effect."* Reprinted from *Heredity* (1949), with permission of Nature Publishing Group.

Hobson, J., McCarley, R.—*Brain as dream state generator.* Reprinted from *American Journal of Psychiatry.* Copyright (1977) American Psychiatric Association.

McCarley, R., Hobson, J.—*Psychoanalytic dream theory.* Reprinted from *American Journal of Psychiatry.* Copyright (1977) American Psychiatric Association.

Wright, S.—*Genetics of populations.* Reprinted from *Evolution* (1948), with permission of the Society for the Study of Evolution.

CHAPTER NINE

Davis, R.—*Solar neutrinos.* Reprinted from *Physical Review.* Copyright (1964), with permission from the American Physical Society.

Einstein, A.—*Electrodynamics of moving bodies* (1905). Translated by John Stachel, *Einstein's Miraculous Year.* © 1998 Princeton University Press. Reprinted by permission of Princeton University Press.

Einstein, A., et al.—*Quantum-mechanical description.* Reprinted from *Physical Review Letters.* Copyright (1935), with permission from the American Physical Society.

Fermi, E.—*First chain reacting pile.* Reprinted from *Proceedings of the American Philosophical Society* (1946), with permission from the American Philosophical Society.

Heisenberg, W.—*Quantum kinematics and mechanics* (1927). Translated by John A Wheeler, *Quantum Theory and Measurement.* © 1983 Princeton University Press. Reprinted by permission of Princeton University Press and the Werner Heisenberg estate.

Lander, E., et al. (International Human Genome Sequencing Consortium)—*Human genome.* Reprinted by permission from *Nature.* Copyright (2001) Macmillan Publishers Ltd.

Kroto, H., et al.—*Buckminsterfullerene.* Reprinted by permission from *Nature.* Copyright (1985) Macmillan Publishers Ltd.

Meitner, L., Frisch, O.—*Disintegration of uranium by neutrons.* Reprinted by permission from *Nature.* Copyright (1939) Macmillan Publishers Ltd.

Watson, J., Crick, F.—*Structure of DNA.* Reprinted from *Cold Spring Harbor Symposia on Quantitative Biology.* Copyright (1953) Cold Spring Harbor Laboratory Press.

Acknowledgments

....

IN CHAPTER 6, we dealt with the acknowledgments section of the typical scientific paper. It is now time for ours. While acknowledgments in both books and scientific papers pay tribute to others for assistance received in the course of research and writing and bring to the surface underlying social networks, they differ in emphasis. Scientific-paper acknowledgments tend to emphasize those individuals and institutions who helped in conducing the experimental research or fieldwork; in typical books, the emphasis is on those who contributed in some way to the composition of the final text.

While only a name or two may appear imprinted on the cover and spine of any given book, none would ever be shepherded into the world without the help of many other hands. For books of a scholarly bent, that is particularly so in the sense that critical feedback by colleagues is an essential corrective to the many outright errors and expository problems that inevitably creep into the early drafts. In that regard, we were the beneficiaries of astute critical readings by Jeanne Fahenstock, Michael Reidy, and the anonymous readers enlisted by the University of Chicago Press. During the long journey from concept to published book, we also received helpful comments from Shiu-Wing Tam, Marion Wood Covey, Lee Eiden, and Gregory Harmon, as well as expert copyediting from Rosina Busse. It goes without saying that the responsibility for any remaining inconsistencies, factual errors, translation slips, or other problems rests upon our shoulders.

Further, this book would not have been possible to write without the outstanding past work in science studies, especially those with a rhetorical and literary basis, which have grown from a molehill to a mountain over the last several decades. Our bibliographies only list a fraction of that work.

Even though more and more historical materials are continually being added to the World Wide Web as part of a vast virtual library, writing a book such as ours still requires access to a first-class research library with dank stacks warehousing scholarly books and journals in print form. We were lucky enough to have three such places at our disposal: the Argonne National Laboratory, University of Chicago, and University of Minnesota libraries. In particular, we were able to consult the extraordinary collection of rare seventeenth- and eighteenth-century scientific materials in the Special Collections Research Center of the University of Chicago Library and obtain needed photographic images of engravings with the aid of its able staff led by Alice Schreyer and Daniel Meyer.

Family members must bear the brunt of living with significant others preoccupied with the heavy demands of researching and writing a book. So acknowledgments must also be given to Emma and Elizabeth Harmon and Suzanne Gross. And last but not least, we thank our editor at the University of Chicago Press, Christie Henry, for her advice, encouragement, and rapid responses to our queries. We could not imagine having a better editor.

Index